PRACTICAL
SELF-SUFFICIENCY

THE COMPLETE GUIDE TO SUSTAINABLE LIVING TODAY

DK UK

Editor Amy Slack
Editorial team Alice Horne, Kate Berens, Jo Whittingham
Designer Hannah Moore
Senior designer Glenda Fisher
Producer, pre-production Heather Blagden
Producer Stephanie McConnell
Jacket designer Amy Cox
Jacket co-ordiantor Lucy Philpott
Managing editor Stephanie Farrow
Managing art editor Christine Keilty
Art director Maxine Pedliham
Publisher Mary-Clare Jerram

DK INDIA

Senior editor Arani Sinha
Assistant editor Udit Verma
Managing editor Soma B. Chowdhury
DTP designer Manish Upreti
Pre-production manager Sunil Sharma

This edition published in 2020
First published in Great Britain in 2010 by
Dorling Kindersley Limited
80 Strand, London, WC2R 0RL

A CIP catalogue record for this book
is available from the British Library.
ISBN: 978-0-2414-0084-5

Manufactured in Hong Kong

A WORLD OF IDEAS:
SEE ALL THERE IS TO KNOW

www.dk.com

MIX
Paper from
responsible sources
FSC™ C018179

PRACTICAL SELF-SUFFICIENCY

THE COMPLETE GUIDE TO SUSTAINABLE LIVING TODAY

DICK & JAMES STRAWBRIDGE

CONTENTS

AUTHORS' INTRODUCTION

We are lucky enough to be living full lives that have given us enough time to appreciate nature, where our food comes from, and the technology supporting a 21st century lifestyle. We have always had an interest in self-sufficiency. Some of Dick's earliest memories are of helping his father in the garden, or collecting mushrooms with his mother; quality time that produced the tasty vegetables he remembers from his childhood. While James has always been involved with nature, going for walks, collecting berries, or digging over the gardens of the many houses we've lived in. Now we both write, work in television, and run businesses to pay the bills.

For years we lived on a smallholding in a beautiful part of Cornwall, nestled in our own valley and half a mile from the sea. Our move to Newhouse Farm was documented in three BBC television series of *It's Not Easy Being Green*, but it was not inspired by television; it was a family decision that was the culmination of years of experimenting and dreaming about the good life.

Life moves on and you cannot write the script, instead you make decisions that take you on the journey. We must be doing something right as James is by the sea in Cornwall, loving nature and its bounty, and Dick is in a 'castle' in France, where, as well as acres of beautiful parkland, he has orchards, nut groves, and a walled garden which is now abundant all year round.

This book is aimed at everyone who has a desire to be a bit more self-sufficient or to live life more sustainably. There is absolutely no reason not to start today. It doesn't matter where you are living you can reduce your energy bills, make cheese in any kitchen, grow herbs and tomatoes on any windowsill, and learn more about sustainable living and apply it to your current situation. This book is a collection of the information that we have been amassing for years. We don't expect you to jump in and do it all at once, but no matter what stage in your life you are at, we are sure there will be something in here for you. Don't forget to have fun – if you are not smiling you are doing it wrong!

SAFETY WARNINGS

While we want you to feel empowered to try many of the things included in this book, we don't want you to take risks, cut corners, or get out of your depth. We use a wide range of sharp implements and power tools in our projects. Make sure you familiarize yourself with them at the outset and feel competent to use them before you start work. Also read through the advice and cautions included here.

Renewable energy options

We've focused on what is possible on a range of sites and showed you the nuts and bolts of our own solutions, but before you consider putting any of this into practice on your own plot, check what permissions are needed from your local authority. Seek advice from and work with a qualified electrician and your energy supplier at all times.

Biofuels and anaerobic digestion

We've shown how we made our reactor and a digester from an assortment of oddments to prove that you don't need expensive custom-built kit. However, we're able to fall back on years of engineering experience and if you have any doubts about your own competence, get some expert help. When dealing with chemicals, heat, and gas, you can't be too careful.

Animal husbandry

Some advice about hygiene: if you have a young family insist on scrupulous hand-washing after children come into contact with livestock to reduce the risk of e-coli infection. Lambing-time is especially risky for pregnant women as contact with newborn lambs or any of the by-products of the birth carries a risk of infection that can cause miscarriages. Anyone attending lambing ewes should wear protective clothing and gloves and wash thoroughly to remove potential contamination.

Dairy produce

Keep everything you use and your kitchen environment scrupulously clean while you make dairy produce. And if you are pregnant avoid using unpasteurized milk, as products may harbour listeria or campylobacter bacteria.

Making natural remedies

It's always best to seek proper medical advice before you self-treat with home remedies, especially if you are suffering from recurring symptoms or have a serious or long-standing illness. When using tinctures try small amounts first; also test our salves and creams on a small area of skin before using to check for allergic reactions. Be especially cautious if you are pregnant or breastfeeding, as some herbs are contraindicated during these times.

NEW WAY TO SELF-SUFFICIENCY

WHAT ARE THE BASICS FOR LIFE?

We're lucky that life in the 21st century is easier than it has ever been at any other time in history, but convenience can come at a price, and many people now want to get back to basics. So, what are the basics for life? Food and shelter is the simplest answer, but we are looking for a good quality of life, too, and for that we need comfortable shelter, quality food, and a good amount of pleasure.

TURNING DREAMS INTO REALITY

Most people have a dream where their life is very different from how they live now. There are so many things that we would like to do, but time and/or money tends to be the main stumbling block. So work is partly the problem and partly the solution. Should we live to work or work to live? A few of us are able to say that our work is fun, fulfilling, and we would choose to do it even if we weren't being paid. However, the vast majority of us have yet to find a way of living our dream. So what's stopping us?

We've always been struck by John Seymour's idea that he was writing for "dreamers and realists", and we're keen to help dreamers turn their dreams into reality – hopefully sooner rather than later.

LIVING THE DREAM

Rushing around in our modern world, it's tempting to put convenience before quality. But we have very firm views on what we call "quality" food, and while living on our smallholding at Newhouse Farm, where we grew our own fruit and vegetables and reared our own meat, we put this ahead of convenience every time.

We both cook, and we're lucky that we're quite good at it. Time spent in the kitchen preparing meals is very sociable; we look in the fridge, see what needs harvesting outside, and then create a menu. There's nothing quite like fresh, quality (there's that word again) produce, enjoyed just after it's been picked and prepared. It may be old-fashioned, but we also still sit down around the table for lunch and supper, which allows everyone to catch up and talk.

Even without the produce of a smallholding at your fingertips, quality food is achievable. As you will see in this book, everyone can grow something that they can eat.

Rediscovering local food

Modern transport systems keep the supermarkets stocked with an abundance of foods, and it's easy to forget what is seasonal and local. Consumers have demanded variety at a reasonable price and the supermarkets have responded; or have supermarkets led and we have followed? Either way,

it now seems acceptable for food to travel halfway round the world and then to be stored and distributed in the supermarket chain before we get a chance to eat it.

When we buy food, our first choice is local, seasonal, organic produce. If what we need doesn't fit all these criteria we go for local and seasonal, or seasonal and organic. We have been known to deviate from our principles, but let's not forget Douglas Bader's words: "Rules are for the obedience of fools and the guidance of wise men."

Gimme shelter

Our homes come in many shapes and sizes, and offer comfortable shelter, but we question the definition of "comfort". We've been "ecovating" our homes for years, aiming for an environment that's warm and cosy, and generates very few bills, so that as we grow old we won't have to worry about the escalating cost of water and energy. We believe that to be truly comfortable you need to know you can live within your means, even in an uncertain future.

There are many ways to achieve this. The simplest is to concentrate on reducing the amount of fuel and water you use, and then to find ways of providing the utilities you need. This approach is valid for any home; in fact, we would go further and say it's essential for every home.

The pleasure principle

The final requirement for a good quality of life is enjoyment. We're all different and we find happiness, peace, harmony – call it what you will – in our individual ways. We've tried many aspects of sustainable living; some we've really enjoyed, others we haven't. Our philosophy is that you should try as many different things as you can and then continue to do those that suit you.

It's important to remember, too, that you don't have to wait until you've bought your ideal property to embark on a more sustainable lifestyle. Have a go now! The experiences you gain and the fun you have may be the catalyst you need to fulfil your dream of a better life.

1. Using an aqueduct to divert part of our stream allowed us to harness the energy in the water. **2. Our old farmhouse** in winter was warm and well-insulated, and the energy was sourced in a variety of sustainable ways. **3. Seasonal crops,** such as chilli peppers, harvested when they are ripe, taste a hundred times better than early forced crops. **4. We have fun** mixing clay for an earth oven. **5. Local produce** has a low carbon footprint and helps generate income for smallholders like us. **6. Wind turbines** are a dependable source of energy if you live in an exposed area. **7. Home baking** is one job we really enjoy on cold winter days, when the smell of fresh bread lifts the spirits. **8. Learning a new skill,** like basketry, is fun and creative.

SELF-SUFFICIENCY VS SUSTAINABILITY

A sustainable lifestyle takes planning, preparation, and practice, but to be truly self-sufficient and live off the land, means an even bigger commitment. There may be many reasons why you can't fully adopt the good life immediately, but there's nothing to prevent you from pursuing at least part of that dream right now, regardless of where you live or the size of your plot.

ESSENTIAL CONSIDERATIONS

How much time do you have to spend working?
Time spent sitting behind a desk to pay the bills is time that you're not on your plot enjoying your chosen lifestyle. However, you have to go into things with your eyes open and accept that to provide sufficient funding for, say, a smallholding or larger plot, it may be necessary to postpone your full commitment to the paradise you have planned.

Where should you live?
Many of us need to be near a workplace, perhaps in an urban setting, for part of our life. If you have no choice where to live, it's important to make the best use of your environment to try out all those things you want to do later. If you're taking the plunge and moving to the country, it's easier and cheaper to find a smallholding in a remote place away from the modern world, but that also limits your opportunities for paid employment, and usually means you have a very small choice of facilities close by. We like the idea of being part of a community and feel we need some of the trappings of 21st-century living, such as good rail and road links, broadband, and mobile phone coverage.

How much support do you need to fulfil your dream?
On a smallholding there's always more to do than you have time for. Nature is always keen to reclaim any land, so there's an ongoing struggle, which depends on the size of your chosen battlefield. In a rural community you will find lots of local expertise, but that can cost money if you want more than a bit of advice. It's important to assess how fit you are and what you are capable of. One thing's for sure – you'll get fitter.

WHAT DO WE MEAN BY SELF-SUFFICIENCY?

For us, self-sufficiency is a way of life where you endeavour to produce all you need from the resources that are available to you. In the past, such a lifestyle was essential for subsistence farmers; the communication infrastructures we take for granted didn't exist and people had to live on what they could obtain within a few miles of their homes. Communities formed where natural resources were concentrated and people traded their skills and produce to ensure everyone had all they needed to live. It was often a matter of trying to survive rather than having a good quality of life. In many countries this is still the case, but in the developed world today we don't need to do everything ourselves or within the local community. That has some positive benefits, but reaching outside the community, especially across the continents, can have a serious impact on the planet.

WHAT DO WE MEAN BY A SUSTAINABLE LIFESTYLE?

Living a sustainable lifestyle means using no more than our fair share of the planet's resources to meet our needs. In the future we can expect energy and food to increase in price as resources become scarcer, and consequently, a degree of self-sufficiency will become essential.

Self-sufficiency automatically leads to a sustainable lifestyle, as you try to produce everything you need, but sustainability is achievable without being entirely self-sufficient, if you make sure you are acting as a conscientious consumer.

A BETTER WAY OF DOING THINGS

Eco-living is no longer an unusual lifestyle choice that needs justifying. Nowadays, few would deny that we're living beyond the planet's means. The eco-warrior's mantra of "Reduce, Reuse, Recycle" is the best way we can all contribute. Put simply, we have to cut down on the resources that we're using, and stop throwing so much away.

No matter whether you live on a farm or in an urban flat, you can make your lifestyle more sustainable. Reducing what you use in all aspects of your life also makes sense financially – from saving energy in the home to minimizing unused food that's thrown away, and mending old clothes instead of discarding them.

With this book, we want to demystify the information around sustainable living and help people make changes that save them money and benefit the environment.

1. Friends and family help to make light work of a construction job. **2. Smoke from our woodburner** floats above the cold air in the valley on an autumn morning. **3. Working outdoors** is part of our lifestyle and keeps us fit – plus it feels better than going to the gym. **4. The good life** is not all hard work: elderflower champagne is one of the pleasures. **5. Onions** and our intrepid mouse hunter enjoy the warmth of a polytunnel.

HOW MUCH LAND DO YOU NEED?

When we first started out, we were in rented army accommodation with small plots, so we know what it's like to have very little land. We've since lived in terraced and semi-detached houses, a Cornish smallholding and even a château with several acres. In each plot we tried everything from keeping bees to building smokers, and we continue to do the same with the next generation of Strawbridges.

MAKE THE MOST OF WHAT YOU HAVE

We all wish we had more land, whatever our plot size. We'd love to have space for livestock, woodlands, and even for some natural meadows, but we know that our first priority is to make full use of what we've got. You can have a really productive plot, even in a tiny urban garden, and if you're out at work all day, this may be as much as you can handle in your spare time.

Living in a "normal" house can mean space is an issue if you want to be self-sufficient; but it should not be a major problem if you are aspiring to a sustainable lifestyle. Suburban living doesn't have to be about immaculate lawns and washing the car every weekend – you can be very productive on an average urban or suburban plot: your veggie patch, cloches, greenhouse, polytunnel, beehive, and chickens can go a long way to help you achieve a degree of self-sufficiency. In fact, your garden may even make you the envy of your friends and neighbours.

If you live in the city or suburbs the good public transport links mean you don't need a car, so you can reduce your demand on finite fuel reserves. There are plenty of shops, too, where you can buy local, environmentally sound produce and products. Taking all this into account, you may end up with a lower-impact lifestyle than those with a similar-sized rural plot.

WHERE TO ACQUIRE MORE LAND

If you are using your available space to the full, and you have time and energy to take on more land, the next problem is how to get hold of some. The odds are that your local land is expensive, or you may find your ideal plot has been sold, but persevere. Most landowners are not keen to sell, so see if you can rent a fallow or unproductive area from a farmer, and after years – yes, years – of softening him up with boxes of produce, he may give in and decide to sell.

ALLOTMENTS

Britain has an allotment scheme where local councils have a duty to provide land for residents to grow their own fruit and vegetables. The origins of

allotments can be traced back over centuries but the system as we know it came into practice in the late 19th century, when philanthropic Victorians were keen to provide poor people with the chance to grow their own produce, and to stop them from getting drunk in their leisure time.

Allotments are no longer the place where men retreat to their sheds or grow prize onions. Many are worked by women and whole families. In our village some allotments have three generations sharing the labour and the bounty. In Britain, allotments are governed by local byelaws. On some plots it may be possible to keep hens, bees, pigeons, rabbits, and even pigs. But you can expect restrictions on planting trees, erecting fencing, subletting, digging a pond, or using the plot for business purposes. If you are lucky enough to acquire an allotment, ask to see a copy of the rules before you get carried away. And don't let a long waiting list deter you from signing up – it's worth it!

COMMUNITY GARDENS

Land for cultivation also comes in the form of community gardens and city farms, which are on the increase. They tend to be in urban areas, often on reclaimed wasteland, where they provide a valuable space for local people to grow their own food. Generally run by volunteers, they are also a great way of bringing people together.

Councils and local authorities are increasingly making unused land available for community gardens. These schemes take a lot of effort to get off the ground and organize, but if there is one near you they are well worth joining.

LANDSHARE SCHEMES

Landsharing may seem altruistic, but it is practical and many people are happy to allow others to cultivate some of their land. A national UK landshare database shows you what is on offer in your area. Alternatively, there is a very good chance that someone will know who could do with a hand; for example, an elderly person who can no longer cope with a large garden and would appreciate a box of veg in return for use of a plot, or a farmer with some fallow land.

With any luck you'll end up with the extra growing space you need, while landowners benefit from a productively cultivated plot and some produce to sample into the bargain.

1. A productive urban garden, no matter now small, can be as beautiful as it is bountiful with some flowering fruit and vegetables.
2. Allotments are much sought after in cities and suburbs, but it is still worth joining the waiting lists at a few sites.
3. Laying hens can be kept in urban gardens with specially designed chicken coops.
4. If you want cattle you will need a few acres of good pasture, and possibly space for a shed for them in winter.
5. A greenhouse provides that chance to grow all year round and is perfect for salads and herbs. **6. Allotments have rules** but some allow you to keep animals, such as pigs, hens, and bees. Find out what is allowed before buying any animals.

5

6

THE CYCLE OF THE SEASONS

We moved to rural Cornwall and France to live closer to nature, and every day we learn more about the seasons and the nuances of living in the countryside. No matter where you live, there is a real opportunity to make the most of each season, to appreciate the variety of local produce on offer, observe different wildlife, and to enjoy the changeable weather, come rain or shine.

ENJOYING SEASONAL FOOD

With fresh strawberries in summer, warming squash soups over winter, and a turkey at Christmas, we really enjoy the fruits of our labour. Each one offers a real treat and marks the passing of the seasons.

It's equally frustrating, though, to realise that it's May and oysters are out of season until September (the wild ones, not the farmed variety). Or that it's September and yet again we have failed to get our fly rods out and missed the wild trout.

Nothing compares to the taste of local produce in season. Although we're not fully self-sufficient – some things, such as flour, rice, and sugar cane, we can't or won't produce ourselves – we eat extremely well. We grow our food, rear it, harvest it, kill it, forage for it, hunt it, prepare it, cook it – and enjoy every mouthful.

GETTING BACK TO NATURE

We get in touch with our roots in many different ways – we know that sounds a bit "new age", but it is a fair description. Most involve being up close and personal with nature. In autumn, for example, there's nothing more impressive than watching a murmuration of starlings gathering before dusk on a clear evening. Seeing wildlife at different times of the year outside your front door is very special, so don't miss out. Go for walks or enjoy some feathered company while you work on your plot.

ADJUSTING TO THE CLIMATE

We live in a changing world and the seasons appear to be changing, too.

They still follow one another, but global warming seems to be playing havoc. We get hail in June, floods in July, and long periods of sunshine in January. Nature is responding too – birds nest before their food sources are available, and plants flower early and are then killed by frost.

We've tried to choose the best time of year for certain projects in this book based on our experience, but year on year we are having to adjust to fit in with changes in our microclimate, as well as more global changes.

Our experience and advice is definitely not "one size fits all". It takes years to learn about the conditions specific to a given geographical location, and you will need to adjust your dates and activities to suit the microclimate in the area you call home.

LIVING SEASONALLY

No matter what time of year it is or whatever the weather, we always find something that we can enjoy and that can keep us busy. In addition to our suggestions below, don't forget that foraging offers a bountiful harvest all year round, if you know where to look (see pages 186–187). We try to make the most of our wild larder with regular free shopping walks in every season.

During winter we tend to rely on our stored harvest in winter and supplement outdoor veggies with crops from the greenhouse, geodesic dome, and polytunnel. We put the chickens in the polytunnel to feed on the pests and leftover soft fruit. We plant trees in unfrozen soil and leave the potatoes to chit in the potting shed. We maintain our tools, practise indoor crafts, plan projects or study, experiment and research.

At the spring equinox, when the sun is higher, we prepare the vegetable beds and harden off plants we sowed earlier under cover. They can be transplanted when the risk of frost has passed. We also incubate eggs, and hatch and raise chicks, ducklings, and goslings.

By summer, the vegetable beds are in full production, giving us time off "to smell the roses" and go to the nearby seaside. We buy in five-week-old turkeys to fatten for Christmas and spend time preserving and storing – late summer is filled with making jams, chutneys, and cider.

Autumn is when we harvest crops and clear the land. As turkeys and geese fatten up, we kill and process our pigs to ensure the freezer is full and that new salamis and air-dried hams are hanging for next year.

1. **In early summer** the runner beans are in flower and squashes are scrambling over the raised beds. 2. **Late summer** is a bountiful time when we can sell some of our produce. 3. **Preserve your harvest** by fermenting or pickling for a taste of the season to be enjoyed all year round. 4. **Fattening up the turkeys** in summer and autumn reminds us of the pleasures yet to come. 5. **In late winter** it's too cold to plant seed outside, but we can chit potatoes indoors for an early crop. 6. **Spring** is a busy time, as we hatch eggs and rear goslings, ducks, and chicks.

URBAN OPPORTUNITIES

In the UK more people live in urban areas than in the countryside, and whether you see living in the city as a necessary stage before moving to your rural idyll or you're a confirmed city dweller, you'll be amazed how easy it is to reduce your carbon footprint. A few small steps – from growing your own salads to reducing the amount of waste you put in your dustbin – can make a big difference.

MAKE THE MOST OF CITY LIFE

It may come as a surprise but a sustainable lifestyle can be easier to achieve in an urban setting than in the country. For instance, a car uses masses of energy, from the embodied energy used to manufacture it to the fuel to run it, so opting to use a city's mass public transport system or cycling routes is one of the greenest choices you can make.

Densely packed urban houses can also be very energy efficient. Terraced houses tend to have lower heating bills, as neighbouring homes serve as a direct form of insulation, while smaller spaces require less energy to heat them in the first place.

ENERGY AND WASTE IN THE CITY

Add to your home's energy advantage by insulating your loft and hanging thick curtains to keep heat in. Reduce electricity bills by choosing energy-efficient appliances when it's time to replace a fridge or freezer, and replace light bulbs with energy-saving types. You can also do some research and switch to a green energy supplier.

Trying renewables

If you are aspiring to the rural idyll, an urban house or apartment is a long way from your destination, but that doesn't mean that it can't be "en route" to your final dream home. It may be easier to be sustainable in an urban setting where the infrastructure, if selectively used, can reduce the amount of energy needed to support a person or family. That said, it's near impossible to be self-sufficient in an urban setting, although there are many things you can do and learn while living in the city.

Saving energy should be your priority, but there may also be scope to try out a couple of renewable options. If you live in a house, check out your roof. If it has a south-facing slope, consider installing a solar thermal system to heat water. Of all the renewables this would be our top priority; it can pay for itself very quickly and even in winter it will capture enough solar energy to pre-heat your water.

Household waste

Composting prevents waste going to landfill sites and produces the perfect soil improver for raised beds and containers. In a small garden or on a patio the most compact option is a wormery; it produces great compost, plus a potent liquid fertilizer (see page

139). A bokashi tub (see page 116) will compost cooked food and dairy waste.

For other items, check out recycling facilities offered by your local council so that you put only the bare minimum into landfill each week.

Of course, the best change you can make to reduce your household waste is to simply bring less unrecyclable rubbish into your home. Avoid plastic-wrapped food where possible, don't buy single-use plastic items such as straws or plastic bottles that are destined for landfill, and try shopping at zero-waste stores or your local farmers' market.

Save water
Reducing the amount of water you use makes environmental sense and can save you money. Aerated taps, dual-flush loos, water-saving showers, dishwashers, and washing machines, all use less water. If you have a patio or garden, connect a water butt or two to the gutter on the house to catch rainwater and use it to water the garden and indoor plants.

GROWING YOUR OWN
In a city, outdoor space is always at a premium. You may only have a balcony or windowsill, but however small the area, use it to capacity. Fill window boxes, both outside and inside, with herbs, salad leaves, cherry tomatoes, or chillies.

If you don't have anywhere to grow your own produce, put your name down for an allotment or look out for community gardens where your neighbours get together to grow food on spare local authority land (see pages 16–17).

If you can't grow your own at all, you can still shop seasonally at farmers' markets and buy produce with few associated food miles.

City foraging
Gathering wild foods is easier than you think. Look out for hedgerow berries along canal banks, public footpaths, and parks. Nettles are common on wasteland; for an introduction to foraging, see page 186.

IN THE KITCHEN
Try being a bit more self-sufficient in the kitchen, too. How about baking fresh bread (see pages 216–217) or making your own chutney (see pages 174–175)? Churning your own butter or making cheese is fun (see pages 258–261) and gives you a real sense of achievement.

1. Making preserves to stock up the pantry is something we can all do, wherever we live. **2. Start off plants** in trays before planting out in pots or raised beds. **3. Hanging baskets** are ideal for growing strawberries. Suspend them with copper wire to deter slugs and snails; the wire gives the pests a mild electric shock. **4. A wormery** transforms kitchen waste into compost and takes up minimal space. **5. Grow your herbs** in pots on a patio or on a windowsill. **6. Use a garden irrigation system** to water plants in containers. **7. Small solar PV panels** may provide useful energy around the garden. **8. Tomato plants** can be grown in large pots and baskets.

The urban plot

There are lots of different ways to maximize your use of space, however tiny your plot. Think vertically: grow climbing beans up walls, tie in a grapevine along a trellis, and train fruit trees as space-saving espaliers. Tuck a wormery into a corner and catch rainwater in a butt. Don't forget wildlife, even in the city. Grow native nectar-rich flowers in among vegetables to attract pollinating insects and keep a bird table well stocked to encourage birds into your plot that will munch their way through insect pests. If you want to keep livestock, you don't need much space for a couple of chickens in a moveable ark, and you'll have the satisfaction of eating your own eggs for breakfast.

KEY TO PLOT

1 **Tool and potting shed** (pp.142–143)
2 **Espalier fruit trees** (pp.158–59)
3 **Small lean-to greenhouse** (pp.128–129)
4 **Solar panel for irrigation** (pp.122–123)
5 **Wall pots for edibles** (pp.136–137)
6 **Bee B&B** (p.133)
7 **Herb and vegetable bed** (pp.150–157)
8 **Cold frame for protection** (pp.126–127)
9 **Raised bed for vegetables** (p.138)
10 **Climbing plants, such as beans**
11 **Potato bag**
12 **Loft insulation** (pp.42–43)
13 **Solar thermal panels** (pp.72–73)
14 **Edibles in window boxes** (pp.136–137)
15 **Bokashi bin** (p.116)
16 **Recycling bins** (pp.102–103)
17 **Solar dryer** (pp.170–71)
18 **Water butt** (p.88)
19 **Seating area**
20 **Herbs in pots** (pp.180–181)
21 **Flowers to attract pollinating insects**
22 **Trellis for climbing plants**
23 **Hanging baskets for edibles** (pp.136–137)
24 **Wormery** (p.139)
25 **Compost bins** (pp.116–118)
26 **Bed for salads** (p.153)
27 **Fig tree for fruit and shade**
28 **Salad and vegetable bed** (pp.150–157)
29 **Small wildlife pond** (p.133)
30 **Grapevine** (p.161)
31 **Chicken run for fresh eggs** (pp.236–243)
32 **Bird table close to water supply** (p.132)

Vertical chicken coops (31) take up little space and you can enjoy free-range eggs from your hens.

A solar dryer (17) uses the sun's energy to preserve foods, such as tomatoes and apple slices, by drying out all the moisture.

A bee B&B (6) provides shelter for solitary bees, which are vital for pollinating plants and ensuring a crop.

A bokashi bin (15) converts cooked food and dairy scraps into compost – with no unpleasant smells.

Potato bags (11) allow you to grow your own spuds anywhere – even on a balcony.

A cold frame (8) is like a mini greenhouse, while **raised beds (9)** are ideal for city gardens where there is no soil. Put up some trellis at the back for **climbing plants (10)** such as runner beans.

Solar thermal panels (13) warm up water efficiently. Use them to supply a shower in an extension with a south-facing roof.

Plant window boxes (14) with herbs, salad leaves, and tomatoes. Cut-and-come-again salads will keep producing all summer long.

SUBURBAN EXPANSION

Gardens in the suburbs tend to be larger than city plots, offering greater opportunities to become self-sufficient. In fact, we believe that a well-designed suburban garden can be more productive than a larger plot that's difficult to manage efficiently. A house and garden with hens, fruit trees, a polytunnel, and some renewable technologies can provide you with almost everything you need.

BECOMING ENERGY EFFICIENT

Wherever you live, your top priority should be to reduce the amount of energy you use. In the suburbs you may be able to take this a step further with "microgeneration technologies", which generate power from the sun, wind, and other renewable sources on a domestic scale. Combine these with energy-saving ideas and there is absolutely no reason why you shouldn't become completely self-sufficient in electricity.

Assessing the possibilities

To find out if your site is suitable for the microgeneration of energy, start by assessing the possibilities offered by your roof (see renewable energy options, pages 68-69).

One practical option could be solar photovoltaic panels (see pages 70-71), which make electricity from sunshine. However, the panels must be positioned on a south-facing roof to work efficiently.

Heating your home

Using woodburning stoves to heat part or all of your house is feasible in a suburban area where smoke control regulations are currently not too restrictive (see pages 92-93). Regulations do change, however, so you'll need to check with your local authority first. Solar thermal panels use the sun's energy to heat water (see pages 72-73). Like photovoltaic panels, they need to be installed on a south-facing roof.

SAVING AND COLLECTING WATER

Dual-flush toilets, aerated showers, and low-water-usage washing machines are all essential if you want to reduce water consumption, and a water butt is a must for any urban or suburban garden.

You can also expand the ways in which you collect rainwater and how you use it. For example, you could install a rainwater harvesting system, which stores the run-off from the roof in an underground tank (see pages 86-89). This is often known as "grey water", but it's excellent for everything other than drinking or cooking. For instance, it can supply showers, toilets, and other household appliances.

CONVERTING YOUR GARAGE

We have never actually kept our car in the garage: the space is far too valuable. Like many people, we use it as a workshop and a store. For more information about equipping a workshop, see pages 52-53.

It is also possible to turn a corner of your garage or any outbuilding into a biodiesel plant to produce fuel from waste vegetable oil – we find making fuel for our car very empowering (see pages 94-97).

The shed can become more than storage space for tools. Add a lean-to and turn it into a smokehouse for curing foods, or a store room for air-dried hams, pumpkins, and home-made beer and wine.

MAKING THE MOST OF YOUR GARDEN

If your garden has enough space for either a greenhouse or a small polytunnel, you can greatly extend your growing season. They not only protect young plants from frost, enabling you to start crops earlier in the year, but they also allow you to grow salads and herbs over winter. Improve the productivity of your greenhouse or polytunnel by installing a heat sink (see pages 130-131) to keep it frost-free in winter.

In larger gardens you'll have room to plant fruit and nut trees (see pages 168-171). Trees take little tending and, when established, can yield large crops.

Attracting friendly wildlife

Extra space in the average suburban plot means more room for wildlife too. Dig a shallow pond to encourage frogs – they eat slugs. Attract bees, which pollinate fruit and veg, with a patch of wildflowers and bee-friendly plants like lavender (see pages 132-135). A beehive, sited away from busy paths and seating areas, is also possible (see pages 270-271).

EXPANDING YOUR FLOCK

Hens are easy to keep in the suburbs. You will need space for a dedicated area in spring and summer, and can then allow them to roam freely in autumn and winter when the garden is at its least productive (see pages 236-243). Unless you have a large pond, ducks are not an easy option, and you'll be unpopular with the neighbours if you keep geese – they're very noisy.

1. Biodegradable seed pots are easy to make from recycled cardboard tubes. **2. Chickens** are simple to keep if you have a large garden. **3. Start your crops early** by sowing seed in a greenhouse or under a cloche in late winter for an early summer harvest. **4. Raised beds** provide productive spaces for growing rows of your chosen herbs and other plants.

1 2
3 4

The suburban plot

The garden of a suburban semi-detached or detached house can be turned into a very productive plot, with a good-sized vegetable patch, greenhouse and fruit trees, and plenty of room to house a flock of chickens and to try your hand at keeping bees. There's also space to experiment with making your own biodiesel to fuel your car, and to install a combination of renewable energy systems to reduce your electricity bills.

KEY TO PLOT

1 **Beds for crop rotation** (pp.120–121)
2 **Compost bins** (pp.116–117)
3 **Willow trees** (pp.230–231)
4 **Compact fruit trees** (pp.158–159, pp.162–165)
5 **Beehives** (pp.270–271)
6 **Polytunnel for early crops** (pp.128–129)
7 **Trained fruit trees** (pp.158–159, pp.162–165)
8 **Shed and smokehouse** (p.142, p.224)
9 **Smoker for smoked food**
10 **Bed of medicinal plants** (pp.192–193)
11 **Herb spiral**
12 **Chicken enclosure** (pp.236–243)
13 **Roof-mounted wind turbine** (pp.74–77)
14 **Bird box and wildlife area** (pp.132–135)
15 **Woodstore** (p.92)
16 **Window boxes for herbs** (pp.136–137)
17 **Solar photovoltaic panels** (pp.70–71)
18 **Solar thermal panel** (pp.72–73)
19 **Rainwater storage tank** (pp.88–89)
20 **Biodiesel reactor** (pp.94–97)
21 **Straw-bale extension** (pp.50–51)
22 **Air-source heat pump**
23 **Bokashi bin for meat and dairy** (p.116)
24 **Household recycling bins** (pp.102–103)
25 **Water butt to collect rainwater** (p.88)
26 **Pots of herbs and salads** (pp.136–137)
27 **Greenhouse plus heat sink** (pp.130–131)
28 **Wormery for kitchen waste** (p.139)
29 **Automated irrigation system** (p.122)
30 **Wildlife pond** (p.133)
31 **Earth oven** (pp.60–61)
32 **Pumpkin patch** (p.151)
33 **Mushroom logs** (pp.190–191)
34 **Wildflower garden** (p.132, pp.134–135)
35 **Living willow hut** (pp.230–231)
36 **Solar dryer** (pp.170–171)
37 **Seating area**

Keeping bees (5) takes some investment, but can help by pollinating your fruit and veg – as well as providing delicious honey.

A wildlife pond (30) attracts frogs and toads, your allies in the war on slugs.

A large greenhouse (27) has space for exotic plants, from melons to tomatillos, as well as room to start off seedlings under cover.

Harvesting seasonal crops (1) provides healthy kitchen ingredients and reduces your food bills.

Growing from seed (26) is far cheaper than buying mature plants, and in a large garden you can try a wide range of varieties.

The chicken coop (12) should be situated close to the house so it is easy to collect eggs and check on the health of your birds. Close proximity to the home will also deter predators.

Full-size solar photovoltaic panels (17) on a south-facing roof will generate energy to heat water and part of your home.

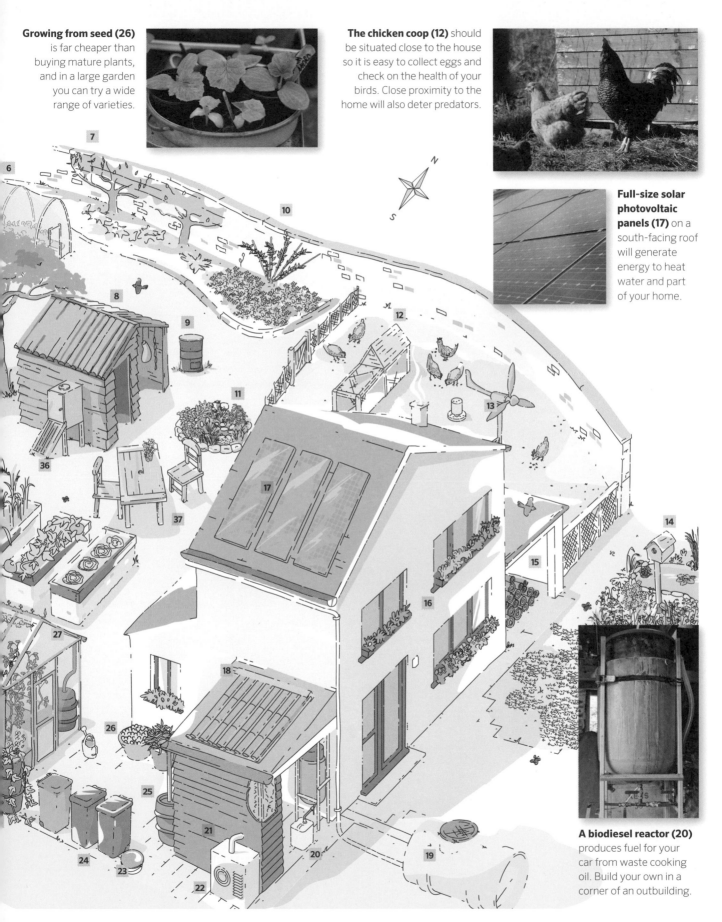

A biodiesel reactor (20) produces fuel for your car from waste cooking oil. Build your own in a corner of an outbuilding.

RURAL ADVENTURE

If you're planning to make the move to a smallholding, the possibilities are enormous, from growing most of your own food to selling your electricity back to the national grid. But do it with your eyes open. You'll find that every day brings different challenges, so stamina and a sense of humour come in handy. Life on a smallholding is hard work but rewarding, and enormous fun.

MAKING THE MOST OF THE LAND

When you're looking for a small-holding, the physical features of the land will determine what you can do with it. The orientation of the plot reflects what you can grow: check to see how much land is facing south and is thereby sunny and warm with lots of potential; for example, a south-facing slope is ideal for planting grapevines (see page 161). If there's a stream running through the plot, then using water power to generate electricity will be an option (see pages 78–83).

The human factor

If you have enough space, time, money, and motivation, you can achieve anything on a smallholding, from growing your own fruit and veg, to rearing your own bacon.

Earning enough money to live means you must plan projects carefully, but you have to be flexible. Priorities are overridden on a daily basis – all it takes is the rain to get in where it shouldn't, or an animal to look off colour. But the work makes us smile: it's satisfying and we're here because we want to be.

Watch out for red tape

In the UK there are a huge number of regulations that may dampen your enthusiasm, from paying an abstraction fee for diverting a stream to applying for planning permission to erect a wind turbine. Thankfully a large number of local councils now have an enlightened approach to eco-friendly projects, but expect some hurdles if you live in a building that is listed for historical reasons, or if your plot is in an Area of Outstanding Natural Beauty (AONB).

There are still some grant funding opportunities available from regional agriculture and environment agencies for sustainability projects. Accessing these valuable grants requires some significant admin work, but there is a current drive for the government to stimulate the renewable energy sector and support micro-generation projects, so it's worth arranging a meeting with your local council to seek advice.

MAXIMIZING ENERGY PRODUCTION

Even in a suburban garden it is possible to generate enough energy to power a home; on a smallholding you may also be in a position to sell some back to the national grid. Analyze your plot to work out which options to go for. Take wind speed readings, measure water flow, and note south-facing roofs on barns and outbuildings as well as on the house (see pages 66–83).

We were keen to experiment with water power, so six months after moving into our Cornish smallholding, our water-wheel was up and running, generating enough electricity to light our home.

Heating

Solar thermal panels for hot water are one of the best investments you can make. In winter, boost their effect by using a wood-fired back boiler to heat up the water to a more comfortable temperature (see pages 72–73 and 92–93).

Wood-burning stoves are the sensible option in a rural area, especially if your smallholding has woodland which you can coppice for a free supply of carbon-neutral fuel.

DEALING WITH WASTE WATER

Most rural properties rely on a septic tank to collect waste water. You can take this one step further and install a reed bed system to treat effluent naturally (see pages 90–91). The best human waste management, though, has to be a compost toilet (see page 87). You will save masses of water used to flush a conventional toilet, and the waste decomposes into a useful compost to be used on fruit and nut trees – not the lettuce!

INCREASING YOUR LIVESTOCK

You need a good few acres to keep cattle and sheep, but goats can be kept in smaller spaces – we once had a pair in a large back garden. Pigs need space to run around and can earn their keep by clearing and manuring uncultivated land (see pages 248–251, 256–257, and 262–265).

Growing fodder for animals

Once you have a few chickens, goats, and a pig or two, it's easy to run up a sizeable feed bill. Reduce it by growing animal fodder crops (see pages 210–213), as well as maize, kale, and fodder beet, try sunflowers.

1. Install a waterwheel if you have a fast-running stream on your land. **2. Pigs rotovate the ground** for you as well as making bacon. **3. A small tractor** and few sturdy wheelbarrows mean you can move equipment and materials whatever the terrain. **4. A decent collection of hand tools** are useful whatever the size of your plot. **5. Large vegetable beds** and fruit trees can cut food bills dramatically.

The smallholding

With more space to play with, you can keep a wide range of animals and grow the crops to feed them. Growing crops under cover on a large scale, combined with successional sowing and crop-storage systems, means you can grow and eat your own food nearly all year round. And you'll have scope for various renewable technologies.

A small tractor can save lots of time when working on a smallholding.

KEY TO PLOT

1   Bamboo for garden canes (p.212)
2   Ram pump to supply water tank (pp.84–85)
3   Grapevines, tea, and hops (p.161, p.213)
4   Spring plus dual-powered pump (p.89)
5   Duck enclosure with pond (pp.246–247)
6   Netted strawberries (p.162)
7   Compost bins (pp.116–118)
8   Water tank to supply polytunnel (pp.128–129)
9   Cold frame (pp.126–127)
10 Greenhouse with heat sink (pp.130–131)
11 Fruit cages (p.164)
12 Beds for crop rotation (pp.120–121)
13 Pig enclosure (pp.248–251)
14 Waterwheel and aqueduct (pp.82–83)
15 Fodder crops (pp.210–213)
16 Turkey enclosure (pp.244–245)
17 Woodland (pp.220–221)
18 Beehives (pp.270–271)
19 Mushroom logs (pp.190–191)
20 Goose enclosure (pp.246–247)
21 Fruit and nut orchard (pp.158–159, pp.162–165)
22 Grid-linked wind turbine (pp.74–77)
23 Geodesic dome (p.128)
24 Medicinal garden and wildlife area (p.193)
25 Sheep and cattle area (pp.256–257, pp.262–265)
26 Pollarded willows (pp.230–231)
27 Chicken enclosure (pp.236–243)
28 Solar dryer and raised bed (p.170, p.138)
29 Woodstore (p.92)
30 Recycling bins and wormery (pp.102–103, p.139)
31 Herb spiral
32 Earth oven and smoker (p.60)
33 Reed bed system (pp.90–91)
34 Rainwater storage tank (pp.88–89)
35 Solar thermal panels on roof (pp.72–73)
36 Ground-source heat pump
37 Grid-linked solar PV panels (pp.70–71)
38 Workshop and hay shed (pp.52–53, p.214)
39 Biodiesel reactor in outbuilding (pp.94–97)
40 Compost toilet (p.87)
41 Root cellar (p.215)

A geodesic dome (23) is the perfect undercover environment for growing plants.

Cattle (25) require a few acres of good pasture to graze, and some types may need a winter shelter.

A polytunnel (8) extends the growing season so that you can pick salad leaves in winter and start off artichoke seedlings ready to plant out for an early crop.

Pigs (13) will clear uncultivated land, but they need a sty to sleep in at night.

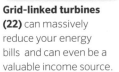

Grid-linked turbines (22) can massively reduce your energy bills and can even be a valuable income source.

Beekeeping (18) is seasonal work. In summer, keep a regular eye on your bees – it only takes an hour a week – and harvest the honey.

THE HOME

IN THE HOUSE

No matter what sort of a house you live in – be it terraced, semi-detached, or detached; in the city or countryside – you'll find it's possible to reduce your energy usage and save a lot of money in the process. Whether you plan to build your dream home or simply renovate an existing property, this chapter will help you understand how your home actually works and suggest possible ways to make it more energy-efficient. We'll even fire your imagination with some radical options for green new builds. For us, though, the kitchen is the heart of a sustainable household, so setting it up to be a practical working space and as waste-free as it can be is a big step towards self-sufficiency.

TAKING STOCK

Most homes were designed before the idea of sustainable living was even thought of, and only the newest homes built in the last decade are designed to truly conserve energy. However new or old it is, by making a few changes to your lifestyle you can make your house more environmentally friendly and save lots of cash by reducing your use of gas, electricity, and water.

TAKE AN ENERGY AUDIT

Assess your energy consumption and bills by taking an audit of different areas in your house, using this checklist. Then read this chapter and see where savings can be made.

EXTERNAL AND BUILDING INFRASTRUCTURE
- Lights
- Taps
- Roof and loft
- Windows and doors
- Chimneys
- Guttering and rainwater
- South-facing walls, roof, and windows

SITTING, DINING, AND BEDROOMS
- Energy usage
- Standby usage
- Lights
- Windows and curtains
- Doors
- Radiators

BATHROOM
- Energy usage
- Standby usage
- Lights
- Windows and curtains
- Doors
- Taps
- Toilets
- Showers
- Bath
- Hot water
- Grey water
- Radiators

KITCHEN
- Appliance energy usage
- Standby usage
- Lights
- Windows and curtains
- Doors
- Taps
- Fridge and freezer
- Extraction
- Hot water
- Grey water
- Radiators

MAKING THE MOST OF YOUR HOUSE

If you are about to build a new property, ensure the performance of the structure is super-efficient. For the rest of us, it's a matter of getting the best from the house we have.

Old properties that were built hundreds of years ago were made to last, using quality materials available at the time. They have amazing insulation properties that have stood the test of time. Unfortunately, this can also make them more expensive to maintain, but any renovations will no doubt last another lifetime.

Your most important, and free, source of energy is the sun, so it's worth checking where it rises and sets, and for how long it falls on windows, walls, and patio doors, as well as the garden. You will then know where to install solar panels and thermal systems, or make windows larger to capture more of this free heat (see pages 70–73).

REDUCING YOUR CONSUMPTION

Your best starting point is to review all your utility bills to see where your cash is going. Draw up a simple chart, noting the cost of each quarterly or monthly bill for heating, electricity, water, and transport, as well as the unit price. This will highlight areas where savings can be made. Reducing your consumption is always the first step, before you start any big "green" projects.

The following ideas will instantly reduce bills and energy consumption.

Turn your heating down one degree and reduce your bill by ten per cent. Pull the curtains at night and use draught excluders on doors. Turn the heating off for a period before you go to bed; a well-insulated house will stay warm. Install reflectors (made by sticking foil to cardboard) behind your radiators, so that they heat the room, not the walls.

To save electricity, use low-energy bulbs and turn lights off when not in use. Switch off your computer, TV, and other equipment at the wall – never leave them on standby, which costs as much as leaving a couple of lights on all night. Switch to a green electricity tariff that uses renewable sources.

For household tasks, only use your washing machine and dishwasher when you have a full load, and dry clothes outside for free. Use pressure cookers – they're fast, very energy efficient, and economical.

Reduce water consumption by taking showers rather than baths, but don't use power showers that use lots of water. Install a "hippo" in the toilet cistern (see page 86).

While out and about, use public transport or walk when you can. When shopping, think about the miles your purchases have travelled and what they are made from. Look for local, seasonal, organic products, or those labelled "fair trade", which help to protect workers in poorer countries.

1. Don't leave taps running when washing; fill the basin instead and use biodegradable or resueable cloths. **2. Solar PV panels** on south-facing roofs can provide power for your home. **3. Hang washing outside** to dry. **4. Locally produced goods** reduce transport costs and pollution, and help support small businesses. **5. Low-energy light bulbs** come in all shapes and sizes, making them suitable for any style of interior.

CONSERVING ENERGY AT HOME

We depend on a huge range of appliances in our homes, from essential items like fridges and washing machines to laptops and sound systems, and they all need electricity. In addition, most of us use lots of energy to keep our homes warm in winter. But adopting energy-saving strategies makes good financial and ecological sense, whatever your house size or lifestyle.

DON'T WASTE HEAT

Heating or cooling your home takes a lot of energy. It pays to have an efficient system, but even more fundamental is to ensure you don't waste the heat you are generating.

One way is to ensure your house is draught-proof. You wouldn't have the heating on and the windows open, and it makes equally little sense to let all the heat escape through lots of gaps. Minimize this loss by fitting draught excluders to doors and windows – there are even systems to reduce the wind rattling through old sash windows. You also don't need to pay for new double glazing to achieve a draught-free, warmer home; thick curtains that you can close every evening are really effective.

Remember, too, that the whole house does not need to be heated to the same temperature. Thermostatic valves on radiators ensure you don't overheat rooms that are seldom in use.

MONITOR ELECTRICITY

Finding out how much energy you are using – or, more importantly, wasting – is the first step to reducing your bills. Wireless energy monitors can give you a live update on how much electricity you are using at any given time and therefore how much you are spending. The result often sparks immediate changes in habits, as you run around the home switching things off.

ENERGY-EFFICIENT APPLIANCES

If you want to save money on your electricity bill, make sure you buy an energy-efficient product when you replace any household appliance. Look at the manufacturer's label to check a product's efficiency. It often pays to spend more for an efficient appliance, such as a freezer, that over its lifetime will generate significant energy savings without compromising on performance.

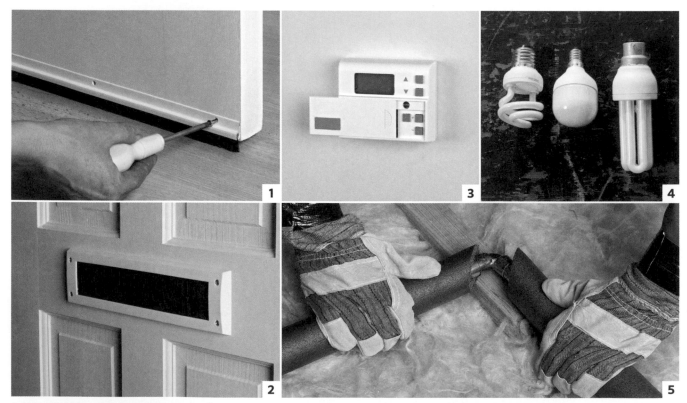

1. Fit a draught-excluding strip to all external doors to keep your home warm. **2. Draught-proof your letterbox** with brush strips.
3. Thermostatic radiator valves allow you to heat rooms to different temperatures. **4. Low-energy bulbs** are available in options to suit all fittings. **5. Lag pipes** in an insulated loft to reduce heat loss.

Comparing running costs

To calculate the running costs of an appliance, multiply its consumption (given in kWh on a label stuck on the side somewhere) by the price you pay for your electricity, and then multiply again by the number of hours you use the appliance.

A plug-in monitor is a handy device when working out running costs. Then, when you go shopping for a new electrical appliance, such as a fridge or washing machine, you can make a proper comparison.

Energy labels rate appliances in terms of energy efficiency. They show how much electricity is used under standard operating conditions. This is measured in kWh per year for fridges and freezers or in kWh per cycle for washing machines and dishwashers. By law, in the UK, an energy label must be shown on all new refrigeration and laundry appliances, as well as on dishwashers, electric ovens, and light-bulb packaging. In the EU, energy labels are used to rate products from A* (the most efficient and least energy used) down to G (the least efficient and most energy used). In Australia, energy labels feature a rating out of six stars for quick assessment and also show annual energy consumption based on average usage. In the USA, the Energy Star label means that a product has met, or has even exceeded, the government's stringent energy-efficiency requirements.

Laundry and dishwasher labels also have ratings for washing, spin and/or drying performance. The A–G ratings are similar to the main energy-efficiency ratings, and are based on standard industry testing. Remember, an A rating is the most efficient.

CHOOSE LOW-ENERGY LIGHT BULBS

At Newhouse Farm our lighting circuit was powered by our waterwheel, but that doesn't mean we weren't careful about conserving our own energy. A decade ago, we were amazed by the difference it made when we switched to using low-energy light bulbs rather than old-fashioned incandescent ones.

Many homes still have ceiling-mounted halogen lights that give off heat similar to old-fashioned incandescent bulbs. Any heat emitted by a light bulb is wasting energy that could otherwise be used to create light. Fortunately, halogen bulb production has recently been banned due to their inefficiency and safety concerns, and sale of remaining stock is being phased out by manufacturers.

There are now low-energy compact fluorescent lamps (CFLs) that fit into any light socket; some of these are even compatible with dimmer switches. Replacing just one 100W incandescent bulb with a CFL that provides the same light for just 20W will reduce the energy usage of that light fitting by 80 per cent.

LED savings

Light emitting diodes (LEDs) are the most energy-efficient option. An LED light bulb uses about 4W to produce as much light as a 50W halogen bulb. What's more, they can last for as long as 25 years. Check out the sums:

- 10 lights fitted with halogen bulbs in a kitchen use 50W each. If they're on for three hours in the morning and five in the evening, they're using electricity for eight hours every day.
- 500W for eight hours is 4kWh units of electricity a day. (A kWh is a kilowatt hour, or 1,000 watts for an hour.)
- If you pay 15p per kWh, you are spending 60p per day, or £219 per year to light the kitchen.
- Replacing them with LEDs at approximately 4W each means they are using 40W for eight hours, or 0.32 kWh per day.
- That means you will pay just under 5p per day, or £17.52 per year.

Replacing your halogen light bulbs with LEDs will therefore save you more than £200 per year – and that's just in the kitchen!

BUYING GREEN ELECTRICITY

We don't produce 100 per cent of our electricity through micro-generation at home, so to support the green energy market, we opt to buy our electricity from green energy providers. It is worth shopping around to find the right deal but also explore where their electricity actually comes from. Some companies produce the entirety of the electricity that they sell from their own solar and wind farms, so you can be sure that your home is completely powered by renewables. In this simple way, however little energy you use, you can make a major environmental impact.

QUICK FIXES

These ideas don't cost much to implement, but will make a sizeable difference to your bills.

Install thermostatic radiator valves and zone the heat.

Change incandescent bulbs to low-energy bulbs, CFLs, or LEDs.

Buy a wireless electricity monitor to keep track of how much energy you are using. They provide an exact reading and show you where savings can be made.

Lagging your hot-water tank, and the pipes around it, will pay for itself in a couple of months – thereafter you are saving oodles.

If your boiler is more than 15 years old, replacing it with a new efficient boiler can generate significant energy savings for you. In the UK, building regulations state that all newly installed gas and oil-fired boilers must be condensing models, which save energy by extracting heat from the exhaust gases.

USING PASSIVE SOLAR GAIN

The sun is a fantastic source of free heat that can be harnessed very simply to warm our homes, using a technique called "passive solar gain". We refer to it as passive because there are no moving parts, switches, motors, or control systems; we let nature do all the work. The idea is to collect and store heat when the sun is shining, and to radiate it back after the sun has gone down.

COLLECTING THE HEAT

The sun's heat energy passes easily through glass and warms the room beyond, which means windows make ideal solar collectors. For maximum capture of solar energy, a house needs large south-facing windows. In a new build this is easy to achieve, as the house can be sited accordingly. Even if it can't be oriented precisely north-south, with careful positioning, you will still be able to harness a good percentage of the sun's energy (see below).

If you want to adapt an existing house, there are several options. You can install bigger windows on south-facing walls, or replace the glass in existing windows (see box opposite), and consider laying a stone floor in south-facing rooms. A black floor is the most efficient at absorbing and storing heat, but any solid floor can act as the main heat sink – the material that stores warmth.

Warm air rises because it is less dense. By using this effect – thermo-siphoning – and installing a system of vents, you can move air warmed by the sun around your house. The most efficient way to do this is with a heat recovery ventilation (HRV) system (see pages 42–45), though it's not strictly passive as it uses a pump.

Capturing the sun's heat

Using solar energy is a constant balancing act between capturing enough heat during the day and preventing it from escaping at night, while ensuring the house doesn't overheat in summer.

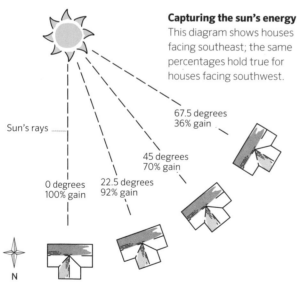

Capturing the sun's energy
This diagram shows houses facing southeast; the same percentages hold true for houses facing southwest.

Sun's rays

67.5 degrees
36% gain

45 degrees
70% gain

22.5 degrees
92% gain

0 degrees
100% gain

N

The ideal house faces due south

Summer solstice

Spring and autumnal equinox

Winter solstice

Seasonal variation
As the earth tilts towards the sun in summer, the sun climbs higher in the sky and rises and sets well to the north of east and west. In winter, the earth tilts away and the sun sinks lower in the sky, rising and setting well to the south.

The ideal position for passive solar gain

A house that faces due south will be most efficient at capturing the sun's energy. But even if your house is 45 degrees off south, you will still be able to harness 70 per cent of the possible solar gain.

Keeping cool in summer

A roof overhang shades windows in summer but doesn't affect heat gain in winter when the sun is lower in the sky. Deciduous trees similarly allow light through in winter and block out summer sun.

OPTIMIZE YOUR WINDOWS

The type of glass in your windows determines how efficiently you can capture solar energy, and it is assessed in two ways:

Solar control refers to how much of the sun's energy passes through the glass and into the room. The solar heat gain coefficient (SHGC) compares the total amount of energy that passes through the glass to the amount of solar energy actually striking the glass. An SHGC of 0.86 means that 86 per cent of the solar energy hitting the window enters the room. The lower the SHGC, the less energy entering the house.

Thermal control refers to the insulating value of glass. It measures the rate at which heat is lost from a warm room to the cool air outside. Up to 60 per cent of heat loss in a well-insulated home is through the windows, but thermal control can be improved by coating the surface of the glass with a microscopic layer of metal or metallic oxide. This is known as a "low-emissivity coating" (see box below). The disadvantage is that the coating reduces the SHGC of the glass, so that less energy is absorbed in the first place.

Connect a conservatory to an HRV system to pump air warmed by the sun round the house during the day. Create a heat sink with a dark stone floor, and it will move the stored heat in the evening, too.

Storing the heat

The materials for a heat sink range from concrete or masonry walls and floors to tanks of water (see page 130). In the average room, the heat sink will be the floor. Walls are a less likely option, unless your house has solid stone walls or eco-walls designed to store heat (see page 48). A material's capacity to store heat is known as its "thermal mass"; the more heat your heat sink stores, the bigger its thermal mass.

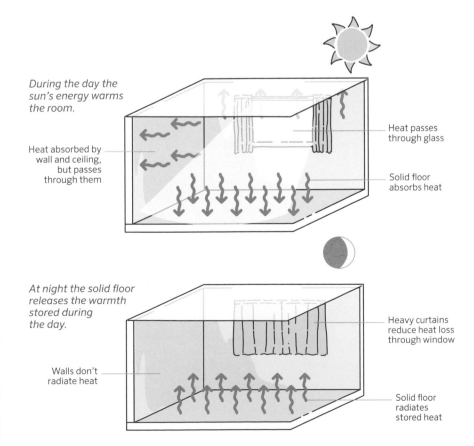

During the day the sun's energy warms the room.

Heat absorbed by wall and ceiling, but passes through them

Heat passes through glass

Solid floor absorbs heat

At night the solid floor releases the warmth stored during the day.

Walls don't radiate heat

Heavy curtains reduce heat loss through window

Solid floor radiates stored heat

KEEPING HEAT IN

For windows that capture heat in the winter and retain heat inside the house as the air outside cools, you need a low-emissivity coating on the inside pane of glass in a double-glazed unit.

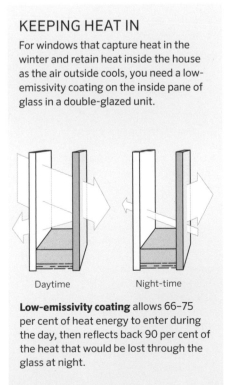

Daytime

Night-time

Low-emissivity coating allows 66–75 per cent of heat energy to enter during the day, then reflects back 90 per cent of the heat that would be lost through the glass at night.

IMPROVING HOME INFRASTRUCTURE

Our houses vary hugely in design, size, and the materials they are made from. In the UK and Europe some of the housing stock is ancient and, in many cases, the buildings are a long way from being efficient to run and heat. That said, they have stood the test of time and it is usually much greener to slightly modify an older building than sinking lots of energy into a new build.

IMPROVING THERMAL PROPERTIES

Harnessing the potential of your home so you can live as sustainably as possible means using what you have to the best of your ability. Not surprisingly, the first thing to do is reduce the amount of energy it takes to run your home. Improving its thermal properties is the logical solution to reducing energy consumption. There are two main areas: insulating, with the minimum amount of thermal bridging; and improved airtightness – draughts mean lost heat.

THERMAL BRIDGING

The purpose of insulation is to slow down the movement of heat; the higher the conductivity of a material, the more quickly heat can move through it. Insulation materials are rated by their resistance to heat transfer, the R value. The higher the R value, the lower the heat transfer.

When different materials with different thermal conductivities make up a wall or ceiling, then the heat will flow through the different materials at different rates. Areas that have poor insulation or where the materials allow lots of heat to flow through them quickly are called "thermal bridges".

The loft hatch tends to be a thermal bridge. Walls and, more specifically, the junctions between walls and the ceiling need to be carefully thought about, to ensure they are airtight.

If you decide you need to add more insulation, you must also sort out all the thermal bridges and draughts or your investment will be wasted. Remember that the first layer of insulation is the most effective. The law of diminishing returns dictates that each additional layer of insulation is less effective than the previous layer.

SUPERINSULATION

It is possible to insulate a property so that you should not need to provide heating at any time of the year. All the warmth you require can be produced by other sources in the house. For example, humans produce about 100 watts of heat per day. Even your cat and dog give off heat – and how about the surplus heat given off every time you boil the kettle?

If you want to go down the path of superinsulation, heating your property predominantly by sources of intrinsic heat (that is, waste heat generated by appliances and the body heat of the occupants), with very small amounts of back-up heat, you must address every area that could possibly be losing heat. For example, solutions include:
• Very thick insulation under floors, on walls, and inside the roof space.

INSULATION MATERIALS

It is fair to say that any insulation is good insulation. However, there is a big difference in the raw materials used to make it and this difference is reflected in the price, with eco-friendly options at the top of the scale.

If you can afford the most expensive, environmentally friendly insulation, that's great. But if you are put off by the price, we would encourage you to get whatever you can afford and install it as soon as you can. Any insulation is better than no insulation. Of all the eco-projects you can possibly undertake at home, insulating provides the best return and it will save you the most energy.

Insulation options
Recycled materials that can be turned into insulation include single-use plastic bottles spun into fleece, recycled denim, newspaper waste, and sheep's wool that is too coarse for knitting and weaving. Wool is naturally fire-resistant (no chemical treatment necessary), won't go mouldy, and is completely safe to handle – no masks or gloves needed – making it eminently suitable for DIY.

Choose the right insulation: from left, expanded foam for pipework, recycled plastic fleece, a jacket for a hot-water tank, foil reflectors for radiators, and a large roll of sheep's wool loft insulation.

- Detailed insulation where walls meet roofs, foundations, or other walls.
- Airtight construction, especially around doors and windows.
- A heat recovery system to provide fresh air (see page 44).
- No large windows.

Fitting superinsulation to an older house may involve building new exterior walls to allow more space for insulation. You must install a vapour barrier to prevent condensation and possible mould and mildew. A vapour barrier should be no further out than a third of the R value of the insulated portion of the wall. This way, the vapour barrier will not usually fall below the dew point, and condensation will be minimized. Alternatively, an interior retrofit will preserve the external look of the building, but room sizes will suffer.

AIRTIGHTNESS

It's surprising how draughty the majority of homes are. Unless houses have been built to exacting standards and tested, there is a very good chance the airtightness will be less than desirable. The "build tight – ventilate right" concept has been promoted in the UK since the 1980s, and building standards for new homes specify maximum air permeability to prevent uncontrolled air leakage. See below for guidance on how to assess your home, and the areas that will most likely need draughtproofing.

Assessing airtightness

Reducing air permeability is a cost-effective way of improving energy efficiency. You can book a specialist company to run a controlled building air-permeability test to measure the performance of your home, but you probably have a good idea of where the problem areas are. Fortunately, they can easily be resolved with simple, inexpensive forms of draughtproofing. Here are the main weak spots in a house.

Air leakage not only wastes energy, it allows warm moist air to flow into cooler areas, resulting in condensation.

Draughtproof crucial areas
Use brush strips and foam-backed tape to seal windows and doors. Make sure holes are filled where pipes and cables pass through the walls.

Gaps in buildings tend to occur during construction, for example between floors and walls.

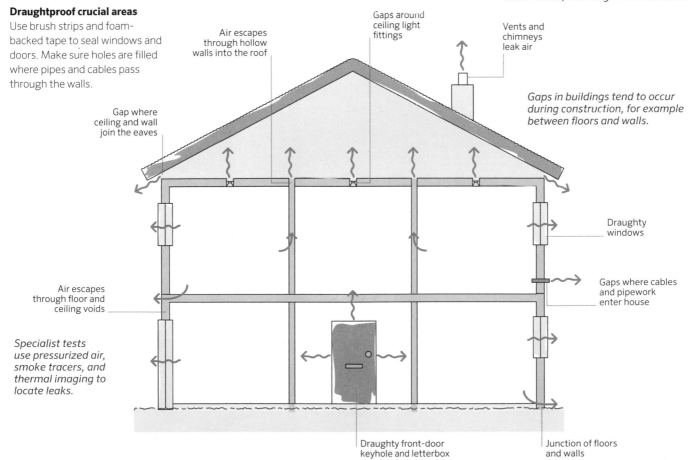

Gaps around ceiling light fittings

Vents and chimneys leak air

Air escapes through hollow walls into the roof

Gap where ceiling and wall join the eaves

Draughty windows

Air escapes through floor and ceiling voids

Gaps where cables and pipework enter house

Specialist tests use pressurized air, smoke tracers, and thermal imaging to locate leaks.

Draughty front-door keyhole and letterbox

Junction of floors and walls

The principles of heat recovery

The idea behind the heat recovery ventilation (HRV) system is very simple. Your house has warm areas, warm wet areas, and cooler areas. All you have to do is get rid of the moisture and take some of the heat from the warm bits and pass it to the cold bits. You may not find it as warm as central heating, but the temperature should be comfortable.

How it works

A system of ducts connects the extractor vents from the warm areas to a heat recovery unit. A roof vent draws in fresh air, which is warmed and forced through another set of pipes to supply warmed fresh air to the cooler areas. Meanwhile, the stale air is vented to the outside. Heat recovery is very efficient: well over 90 per cent of the heat should be kept in the house.

A heat recovery unit installed in a loft. The pipes running to and from the unit are lagged to prevent heat loss.

How warm is warm?

In the past, people would not have expected their homes to be warm, but this is the 21st century and we want to be comfortable. There's an ongoing battle in many homes as to the exact, technical definition of "comfortable". The simplest answer is to start by putting on another layer of clothing.

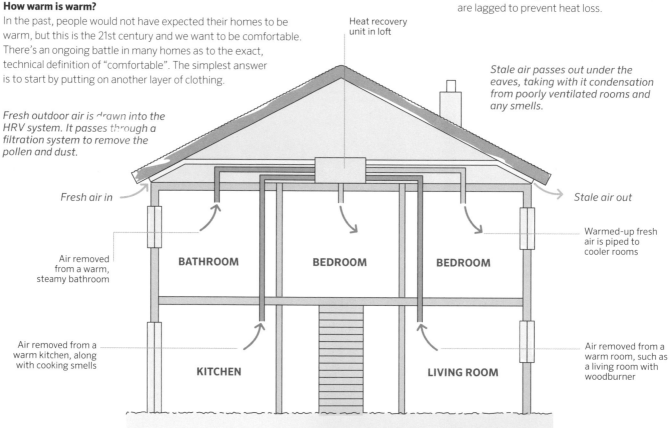

Heat recovery unit in loft

Fresh outdoor air is drawn into the HRV system. It passes through a filtration system to remove the pollen and dust.

Stale air passes out under the eaves, taking with it condensation from poorly ventilated rooms and any smells.

Fresh air in

Stale air out

Air removed from a warm, steamy bathroom

Warmed-up fresh air is piped to cooler rooms

BATHROOM   **BEDROOM**   **BEDROOM**

Air removed from a warm kitchen, along with cooking smells

Air removed from a warm room, such as a living room with woodburner

KITCHEN   **LIVING ROOM**

MOVING HEAT AROUND THE HOUSE

Extraction vents and pipes around your woodburners downstairs (see pages 92–93) can distribute their heat upstairs to the bedrooms. But a system can work using other sources; for example, the heat generated in a busy kitchen to warm a sitting room.

TRADITIONAL HOMES

Living in a historic building, such as a chateau in France, has its pluses and minuses. Older properties have lots of character, and there is a sense of privilege in being a custodian of that heritage. But the owners of historic properties are likely to be restricted in how much they can disrupt the fabric of the building. Preservation orders mean that permission is needed to make any changes, which affects how they can improve a building's thermal properties.

Airtightness vs breathing

Airtightness as a way of conserving energy appears contrary to an old building's need to "breathe" to prevent damp. When we arrived at Newhouse Farm, our former property, damp was certainly an issue. An earlier owner had installed double glazing, so the rattling sash windows you would expect in an old building (which would have provided some ventilation) were nearly all missing. We gradually replaced the plastic PVC windows with wooden, double-glazed ones, with special sash seals to reduce the draughts.

So our method of letting the house breathe, without losing heat, was to install a whole house ventilation and heat recovery system, which was a triumph (see opposite). The vital air-flow system stopped damp building up, while the heat recovery unit minimized heat loss and actually distributed heat from warm rooms to cooler areas of the building.

Accepting what you've got

With an old building it is necessary to just accept it for what it is. Stone and slate have been used as building materials for centuries, and trying to calculate their R value can be meaningless. A thick rubble wall will have very good insulation value, provided it is solid. However, all old rubble-stone walls have voids and it's by no means certain what material

1. Stone buildings in a Cotswold village, UK. Stone has good thermal insulating properties and can withstand harsh weather conditions. **2. Wooden shingle blends** into the landscape and keeps homes warm in cold winters and cool in warmer weather. This is a good example of sourcing building materials from the surrounding area.

SETTING RENOVATION PRIORITIES

Anyone renovating a home faces a lot of decisions. Here's how we prioritize.

Draw up a list of jobs and rate them as urgent or important. Arrange them using this diagram to work out which jobs are both urgent and important – a leaky roof falls into this category. Do those jobs now to the best of your ability. Then tackle important tasks before they become urgent. If you let them drag on until they're in the top corner, you'll spend your time firefighting rather than working to a plan.

IMPORTANT	DO WELL ENOUGH TO SATISFY YOU NOW	DO NOW TO THE BEST OF YOUR ABILITY
	DON'T WORRY AND DON'T FEEL GUILTY ABOUT IGNORING	MAKE A PLAN AND DO PROPERLY AS SOON AS YOU CAN
		URGENT

is inside the core. Even if you were permitted to improve the performance of the walls on the outside, you would undoubtedly change the character of the building. As the rooms of many older buildings are so small, it is not really viable to insulate on the inside. Sometimes it is best to accept that older buildings were not meant to be so warm that you could walk around in your underwear.

GREEN BUILDINGS

We are wary of statistics, but it is obvious that learning to build and live sustainably is becoming more of an issue and it has been shown that buildings consume some 40 per cent of the world's total energy and about 16 per cent of our water.

As the earth's resources become scarcer it is important to re-assess the ways that we use energy and materials. Environmental impact and, very importantly, running costs will soon be critical factors in designing new buildings.

If you decide to build a sustainable home, we reckon the only limitations you have are your imagination, money, and local planners. Each of these will be an issue, so you may as well reconcile yourself with this from the start.

We can learn from traditional methods. Do we have to build from materials that have been shipped halfway round the world, using large amounts of energy in the process? Or could we take the trouble to find locally produced materials that could do the job as well, if not better?

Here are some ideas:

Recycled materials consume substantially less energy than newly fabricated materials. Recycled steel and copper, for example, boast major environmental benefits, while recycled glass and rubber are excellent for flooring, tiling, and countertops.

Choose natural materials that do not rely on high-energy processing. For example, try UK-manufactured sheep's wool insulation, bamboo, or cork.

Source wood responsibly, ideally from trees grown locally.

Avoid timber preservatives and certain paints, removing the need for additives made from toxic chemicals.

Specify durable timber that does not need decorating – rather than re-painting every few years.

Consider a green roof (see box opposite). It is insulating, attractive, and reduces your carbon footprint.

Designing for efficiency

In new construction, the cost of extra insulation and the wall framing needed to contain it can be offset by removing the need for a dedicated central heating system, for example.

But there is a more fundamental calculation required. Whole-life costs also need to be taken into account. If you have a choice of two similar-sized properties and one has bills that will consume about 5–10 per cent of your net income (this will obviously fluctuate from year to year due to supply problems, natural disasters, or geopolitical events), and the other has negligible bills, which one would you rather live in? We have an instinctive aversion to giving lots of hard-earned cash to the utility companies, so that question is a no-brainer for us!

Think about shapes

We've become very used to houses being boxes. Why? The obvious reason is

1. **The first environmentally friendly housing development** in the UK – Beddington Zero Energy Development (BedZED). Homes are made from reclaimed or natural materials. **2. A zero-carbon house** in Kent, UK, needs no fossil fuels for heating. Its vault shape gives it high thermal mass. **3. Earth shelters** in New Mexico, USA, are made from rammed earth and car tyres (see pages 48–49).

SEDUM ROOFS

Green roofs have been standard practice in many countries for thousands of years. Their insulating properties are used in the cold climates of Scandinavia to help retain heat and in African countries to keep buildings cool.

What you should know
Sedum roofs are a mixture of flowering dwarf succulent plants grown on matting for easy installation. They are low maintenance, but not no maintenance, and should be checked at least once a year in spring, when you can give them a dose of fertilizer. The orientation of your roof will affect the plants' growth rate – they need four hours of sun a day– and if you have a south-facing roof they might need an irrigation system. Consider these points before installing a green roof:
• Your existing roof should be sound and watertight, with no guttering leading onto it. Running water will erode the sedum matting and standing water will kill the plants.
• The roof surface must be smooth to avoid creating air pockets underneath the sedum matting.
• The building will have to be strong enough to support the weight – a sedum roof can weigh more than 40kg per sq m (88lb per sq yd). You may find your roof will need reinforcing.
• The roof will be a lot heavier after snow or rain, and any subsequent flexing of the building could cause windows to break.
• Consult an architect or structural engineer if in any doubt.

A sedum roof can be installed on a flat or sloping roof, but the incline should ideally be less than 25 degrees. You'll need access to the roof from time to time to weed it and water if necessary in dry spells.

they're easy to build. But for an object to have the smallest surface area – which will minimize heat loss – it should be spherical. We are not suggesting everyone try living in an igloo (although a geodesic dome would make an excellent conservatory), but there are many creative architects out there – it's all a matter of deciding what you are prepared to live in.

Materials are now available to enable you to build any shape of property you fancy. It's slightly more problematic getting a kitchen fitted, but that's a minor issue. If you are truly starting with a blank piece of paper you can get as close to nature as you want. But, even if you are only considering building an energy-efficient extension, you can still explore all sorts of less popular materials – straw bales for example (see pages 50–51), car tyres, or an earth shelter (see pages 48-49).

Then you can turn your attention to the interior of your home. One area where there is a readily available, eco-friendly choice, is paint.

ECO-FRIENDLY DECORATING
Conventional paints can release certain compounds into the air that are known to harm both human health and the environment. Eco-paints do not contain toxic chemicals such as benzene, toluene, formaldehyde, and mercury. They are completely free from volatile organic compounds (VOCs) – the solvents that carry the pigment and then evaporate as the paint dries, and which can cause headaches, sore throats, and eye irritation. Even their production process has a smaller carbon footprint.

Another advantage is that eco-paints are "breathable" and allow moisture to evaporate from walls. Combined with a natural breathable plaster on the walls, this can reduce condensation and humidity levels.

Organic eco-paints use natural pigments for colour and natural solvents such as citrus oil to carry the pigment. Other eco-paints use casein, which is a milk protein, and come in powder form so you just add water and stir.

BUILDING AN EARTH SHELTER

Earth shelters are buildings that use earth stacked against the external walls to provide an effective thermal mass that will conserve heat when it is cold and keep the temperature cool in warm summers.

This is a sustainable form of construction that relies on well-designed passive solar systems (see pages 40–41) and often uses renewable materials for the fabric of the building.

COULD YOU LIVE IN ONE?

Let's tackle the perceived problems with living in an earth shelter right up front. The question on many people's lips is: are they damp and cold? It's a fair cop: some below-ground designs could be damp, like a basement. The answer is to put in a damp-proof course and drainage, and the sun's energy will keep the shelter warm. Passive solar gain can be used to heat every room with the help of a heat recovery ventilation (HRV) system (see page 44).

CONSTRUCTION TYPES

In its most basic form, an earth shelter could be a cave in a hillside. But we're not suggesting going back in time and living like cavemen; building practice has evolved so that designs now incorporate many forms of modern sustainable architecture in an efficient, earth-covered shell. Like any new building, earth shelters need planning permission. And if you aren't an experienced builder, you will need help with construction.

The earthship

In this design, the majority of the living space is below ground, with a bank of

Off-grid earthship

This is our favourite style of earth shelter. It has to be built into a south-facing hill or slope, so that the northern side of the house is highly insulated by the natural shape of the earth and the south side acts as a source of solar energy (vice versa if you're in the southern hemisphere). Many designs are completely off-grid; that is, self-sufficient in electricity and water.

TYRE WALLS

Old tyres are easy to come by and make solid walls with great load-bearing qualities. They're also weather-proof and fire-resistant.

Shovel earth into the tyre one scoop at a time. Pack it in by bashing it with a sledgehammer. Keep adding more earth, working evenly round the tyre. Do this in situ, as you build up the tyre wall.
Stagger the tyres like bricks. Fill the gaps between them with earth.

Using the soil's assets

The soil's density keeps the temperature inside the shelter at a constant 8°C (46°F) or so. That's a bit too cool to be comfortable, which is why you need passive solar heating. Earth also has excellent soundproofing properties.

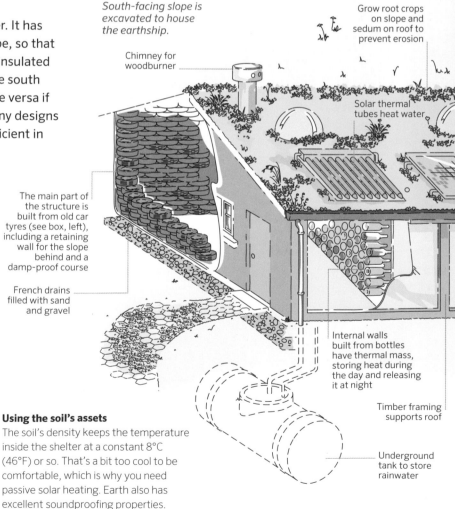

South-facing slope is excavated to house the earthship.

Chimney for woodburner

Grow root crops on slope and sedum on roof to prevent erosion

Solar thermal tubes heat water

The main part of the structure is built from old car tyres (see box, left), including a retaining wall for the slope behind and a damp-proof course

French drains filled with sand and gravel

Internal walls built from bottles have thermal mass, storing heat during the day and releasing it at night

Timber framing supports roof

Underground tank to store rainwater

windows along one wall (see below). Its earth roof means you don't lose any growing space on your plot: planting up your roof can even help prevent erosion.

Earth berming

Earth berming involves heaping earth against a building's external walls and sloping it away from the house.

You can't simply add earth berms to a conventional house. Typically, you'd need a purpose-built heavy-duty timber-framed structure or stonework, stacked with thick layers of soil. Imagine a giant molehill with a cavity of rooms within it. The roof can be covered completely with earth or heavily insulated with your chosen material, from sheep's wool to rock wool or composite insulation. With this type of earth shelter there's less risk of damp compared to building below ground level, and there's little digging apart from foundations. Neither are you confined to a specific landscape.

Going underground

An underground home is a major project, usually best for people living in very hot climates. It involves excavating a central courtyard area, often called an atrium, to supply adequate ventilation and light to the rooms leading off. Unless the ground is really hard, internal structural supports will be a crucial element of the construction.

DESIGN FACTORS

The topography of the land will decide what you build. An earthship must have a south-facing slope, while earth berms and underground homes are suitable for flat sites.

Find the natural water table. If you dig down below it, not only will you have damp problems, but you may end up with flooding.

The soil around the shelter should ideally contain plenty of sand and gravel for drainage. It's a good idea to install French drains – effectively ditches filled with gravel and sometimes pipes – to channel water away from the foundations.

With the huge weight of earth resting on the building, it's crucial that the initial structure is strong enough to support it.

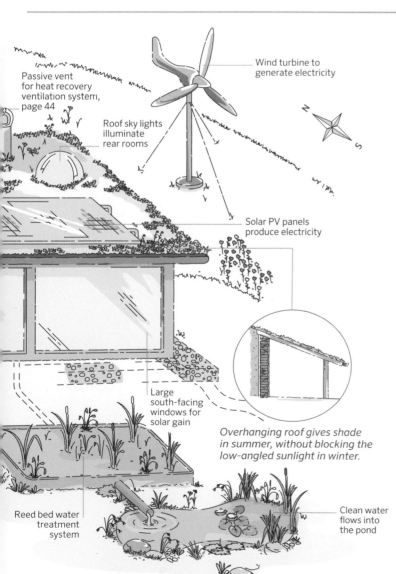

Wind turbine to generate electricity

Passive vent for heat recovery ventilation system, page 44

Roof sky lights illuminate rear rooms

Solar PV panels produce electricity

Large south-facing windows for solar gain

Overhanging roof gives shade in summer, without blocking the low-angled sunlight in winter.

Reed bed water treatment system

Clean water flows into the pond

By working with the landscape to build an earthship, you end up with a structure that blends comfortably into the surrounding area, as well as one that has a smaller ecological footprint.

Earth berms can be incorporated into more familiar houses. These traditional Icelandic homes have stone foundations, timber framing, and a covering of thick turf for insulation, and they can be built on the flat.

BUILDING WITH STRAW BALES

Straw bales are among the most ecologically sound building materials available, and using them for construction is a highly sustainable option. They have great load-bearing strength and effective insulation properties, are non-flammable, and they look good,too. On top of these benefits, the building process is simple, satisfying, and forgiving, as well as being creative and fun.

WHY BUILD WITH STRAW?

Building structures with straw bales is empowering because you can create a structure even if you are inexperienced or unfamiliar with conventional building techniques. Bales are also cost-effective per unit when compared to bricks and mortar, and ecologically sound – the buildings can be composted at the end of their life, which can be over 100 years.

Of course, there are disadvantages. Splashback from rain bouncing up from the surrounding ground onto the base of the walls may cause damp problems, but can be avoided by raising the first straw bales above ground level. High humidity can also be a serious issue in wet regions and in exposed sites; in these cases, straw bales benefit from an outer skin of wooden cladding to protect the walls from wind-driven rain and water damage. However, we feel the positive factors far outweigh the negative points.

MATERIALS

Straw is made from the dead plant stems of grain crops like wheat, barley, rye, or oats. Bales are fairly easy to find, and are often available at local farm shops. They should be dry when you buy them and must be kept under cover if there is a risk of rain. The bales should be well compressed, too; if it is difficult to get your fingers under the string around them, then it is tight enough. Due to their density and the lack of air in them, bales are non-flammable when built, but it's still wise to make sure the storage area is a fire-free zone.

PERFORMANCE

The thermal insulation properties of straw bales are excellent; indeed, they are better than the current UK building regulations require. So straw bales could keep your new home or extension nice and cosy and working efficiently, but only if you make sure you insulate the floors and any gaps in the walls, too. They also have superb soundproofing qualities.

PRINCIPLES OF BUILDING

Straw bales can be used to build anything from two-storey houses and barns to small house extensions. Straw bale building is divided loosely into two methods, which are not mutually incompatible.

For load-bearing walls that can withstand the weight of the roof, bales are placed on top of each other like giant building blocks. They are pinned together with coppiced hazel, and windows and doors are placed inside wooden structural box frames. The flexibility of this style also makes it suitable for curved and circular walls.

Infill involves first building wooden frames for the walls and the roof to provide structural support. The wall spaces are then filled in with bales. The benefit of this method is that you can store the bales beneath the roof out of the rain while you are working. Infill is the best option for creating large open spaces such as warehouses or industrial units, but requires good carpentry skills.

DESIGN FACTORS

Aim to make the walls the length of a number of whole bales with no areas smaller than half the length of a bale, especially by windows and doors.
Position doors and windows at least one bale's length away from load-bearing corners.
Build the foundations so they are the width of the walls.
Avoid using metal pins in the walls as it may encourage condensation from the warm moisture-laden air inside the house.
Use a breathable limewash or lime-based paint to finish your building and help protect it against the weather.

1. **Load-bearing walls** have excellent lateral strength which protects buildings from wind damage and even earthquakes.
2. **Wooden wall frames** are generally more expensive, but provide greater stability for windows and doors than load-bearing walls.

The structure of a straw bale wall

Before starting to build with straw bales, you need to create a solid and stable base that distributes weight evenly over the ground beneath. The base does not need the same degree of solidity as foundations for bricks and mortar, since straw weighs about 60 per cent less. We recommend self-draining foundations made of rubble, stone, or tyres, which can be reused if the building is dismantled.

Working with bales

It is good practice to "dress" a bale before positioning it. Tighten the strings if they are loose, and make the ends squarer with a hammer claw by dragging it along either side of the string. Then pull and tease the straw until the end looks square. If you need to adjust the length of a bale, restring both halves before cutting the original strings. Use a baling needle, which is a giant needle with two holes at one end for the two strings to be threaded through the bale. To curve a bale, simply place one end on a log and jump on top of it until you have achieved your desired curve; but make sure that the strings stay in place.

A wooden frame built with bale-sized spaces between the uprights is the perfect support.

LIME RENDER

Lime render is one part quicklime powder to three parts damp sand (or 1:2 for the first coat, so it is stickier) and is best made two months in advance of using. Mix with a shovel and wear goggles. It gets very warm, so mix it regularly.
Before applying, trim the walls, ironing out unwanted bumps.
Apply a thin coat by hand, wearing thick gloves, and rub it in well. Leave for 1–2 days to dry.
Wet the walls with a mister and apply two more coats, preferably mixed with chopped straw. Rework over cracks, misting and squishing it firmly into any gaps. Protect from direct sunlight, heavy rain, and frost until it has dried.

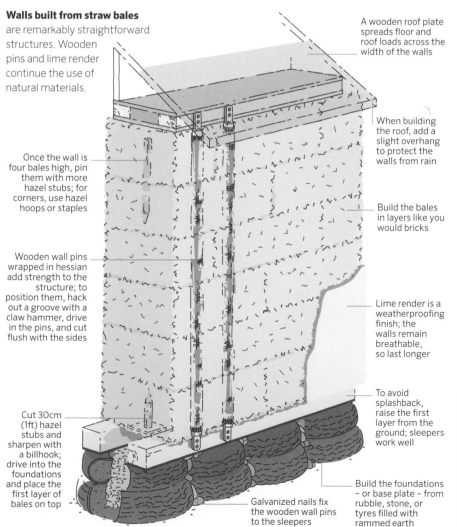

Walls built from straw bales are remarkably straightforward structures. Wooden pins and lime render continue the use of natural materials.

A wooden roof plate spreads floor and roof loads across the width of the walls

When building the roof, add a slight overhang to protect the walls from rain

Once the wall is four bales high, pin them with more hazel stubs; for corners, use hazel hoops or staples

Build the bales in layers like you would bricks

Wooden wall pins wrapped in hessian add strength to the structure; to position them, hack out a groove with a claw hammer, drive in the pins, and cut flush with the sides

Lime render is a weatherproofing finish; the walls remain breathable, so last longer

Cut 30cm (1ft) hazel stubs and sharpen with a billhook; drive into the foundations and place the first layer of bales on top

To avoid splashback, raise the first layer from the ground; sleepers work well

Galvanized nails fix the wooden wall pins to the sleepers

Build the foundations – or base plate – from rubble, stone, or tyres filled with rammed earth

Lime render can be coloured with pigment powders or painted.

SETTING UP A WORKSHOP

A workshop can save you money. You'll be able to make things instead of buying them and repair objects yourself, rather than sending them away to be fixed at great expense. As well as standard hand tools, we reckon every smallholder needs a cordless power drill, an angle grinder, a collection of hammers, and some welding equipment. We also recommend setting up a charging station for cordless power-tool batteries, with clearly labelled areas for "charged" and "charging" so you can always find a fully charged battery when you need one.

KEY TO WORKSHOP

1	Dustpan and brush
2	Overalls
3	Leather apron
4	Paper towels
5	Selection of rawlplugs
6	Power points
7	Eye irrigation kit and Steristrips
8	Nails
9	First-aid box
10	Screws
11	Pens/pencils
12	Window for natural light
13	Voltmeter
14	Big battery charger
15	Power-tool chargers
16	Power tools
17	Cutting disks
18	Goggles
19	Radio
20	Vertical bench drill
21	Portable work bench
22	Bin
23	Plumbing parts
24	Lathe
25	Bench metal grinder
26	Painting/decorating equipment
27	Drill bits and screws
28	Charged batteries
29	Ammo boxes for hand tools
30	Strong vice
31	Compressor
32	Drawers for small items
33	Welding mask
34	Welder
35	Wire brush

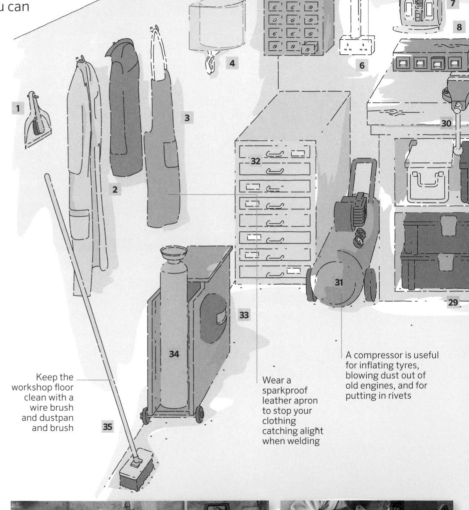

You can't have too many power points: the more you have, the quicker you get a job done

An eye irrigation kit lets you wash out grit or dust

Keep the workshop floor clean with a wire brush and dustpan and brush

Wear a sparkproof leather apron to stop your clothing catching alight when welding

A compressor is useful for inflating tyres, blowing dust out of old engines, and for putting in rivets

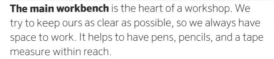

The main workbench is the heart of a workshop. We try to keep ours as clear as possible, so we always have space to work. It helps to have pens, pencils, and a tape measure within reach.

Generous floor space gives you room to work under cover.

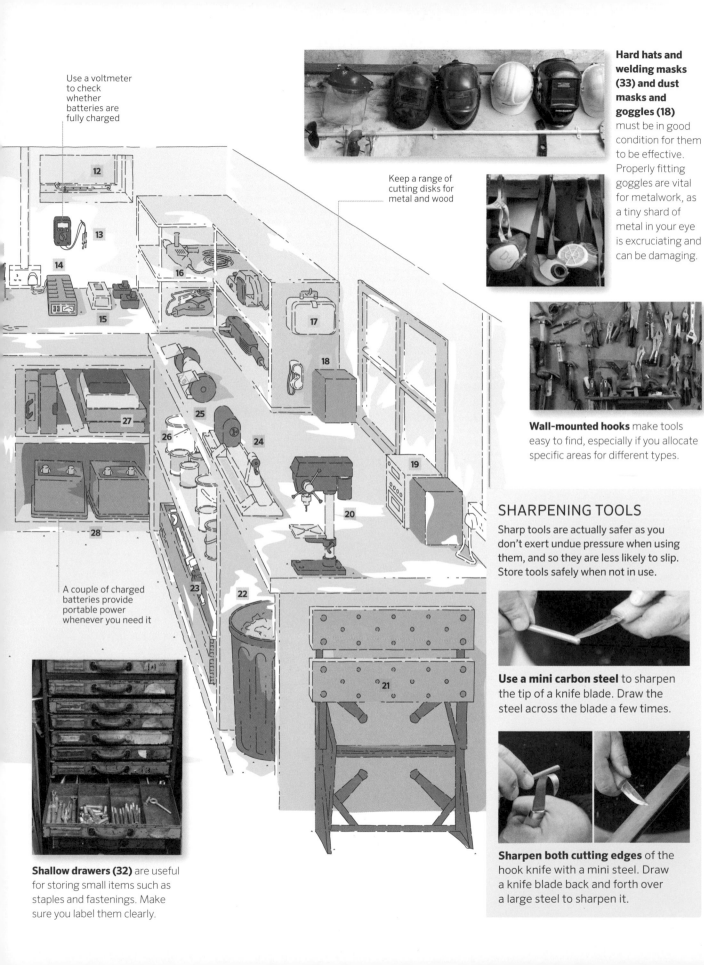

Use a voltmeter to check whether batteries are fully charged

Keep a range of cutting disks for metal and wood

A couple of charged batteries provide portable power whenever you need it

Hard hats and welding masks (33) and dust masks and goggles (18) must be in good condition for them to be effective. Properly fitting goggles are vital for metalwork, as a tiny shard of metal in your eye is excruciating and can be damaging.

Wall-mounted hooks make tools easy to find, especially if you allocate specific areas for different types.

SHARPENING TOOLS

Sharp tools are actually safer as you don't exert undue pressure when using them, and so they are less likely to slip. Store tools safely when not in use.

Use a mini carbon steel to sharpen the tip of a knife blade. Draw the steel across the blade a few times.

Sharpen both cutting edges of the hook knife with a mini steel. Draw a knife blade back and forth over a large steel to sharpen it.

Shallow drawers (32) are useful for storing small items such as staples and fastenings. Make sure you label them clearly.

WORKING WITH WOOD

You can approach working with wood as science, using a carpenter's discipline. Now that we have learned the basics, we prefer to experiment with this amazing material, for pleasure and to save money.

We found that being a perfectionist doesn't always get the job done. Instead we employ a very liberal attitude to woodworking – also known as "near as dammit"!

CHOOSING WOOD

Buying wood can be confusing, but it's worth knowing what to look for, as bad wood can lead to all sorts of problems later. Keep an eye open for defects and try to buy wood that is straight. These guidelines will help:

If wood is warped or bowing it has probably been badly stacked for some time. This may have introduced stresses into the wood that make it harder to cut.

Look for wood where the grain runs straight through the wood, rather than diagonally across it. This reduces the chances of it warping as it shrinks and seasons.

Avoid buying wood with any "shakes" or splits. End splits often occur when the wood is dried too quickly, making it difficult to work – dimensions can shrink and joints may open up. Sealing the ends of timber with waterproof paint can prevent splitting.

Ingrown bark can ruin the appearance and reduce structural strength.

Knots in the wrong places make wood harder to work.

Plywood

This is not good-looking wood but it is cheaper than solid timber and is less prone to shrinking or warping. It's made from sheets of wood laid so that the grain runs in opposite directions in each sheet, then bonded with adhesive.

Marine plywood is a much more expensive version, but has proved its worth to us in both durability and convenience. It's made with stronger waterproof glue and was developed for use with boats, so is ideal for outdoor projects. It's easy to cut and can be varnished or painted.

HAND TOOLS

A properly equipped tool box or, even better, workshop (see pages 52–53) is essential for woodworking. Here are our recommendations for tools you can't do without and some advice on how to use them.

Saws

You want a hand saw to be straight but flexible – it should spring back into place quickly. A taper-ground blade won't stick in the wood. For greater control, hold a hand saw so that your forefinger is extended down towards its "toe". Use the knuckle of your thumb to align the

TYPES OF WOOD

Softwoods come from coniferous trees and hardwoods from deciduous trees – the definition has nothing to do with their relative strength. Each wood has its own particular applications.

Softwoods
Cedar is a good, relatively light wood for furniture, and for building greenhouses and sheds.
Larch is tougher than many other conifers, with a straight-grained timber. Ideal for joinery and fencing.
Pine can be used for light construction work and to make housing for livestock.
Spruce is a lovely pale wood for interior joinery and making boxes.

Softwood

Hardwood

Hardwoods
Ash is excellent for replacement handles for tools or for green woodworking (see pages 222–223).
Beech is close-grained, hard, strong; used for cabinet-making and finer indoor furniture, and for chopping boards.
Cherry is often used in furniture making and for tobacco pipes. Our old cherry tree blew down in a gale and we intend to turn its lovely wood to make bowls (see pages 222–223).
Oak is durable and strong, with a lovely grain – we use it for flooring and furniture. It would be our top choice for an outdoor building project.
Walnut, while fairly hard to come by, is a beautiful wood for carving. Its colour ranges from golden brown to mud red. Ideal for furniture and, due to its often wavy grain, for making gun stocks.

saw against your mark and gently make a few short strokes, gliding past your thumb until you establish a "kerf". Once the cut is made, use the full length of the blade with slow, steady strokes. Near the end of the cut, support the off-cut to avoid splintering the wood.

Planes

Being able to use a plane means you can take salvaged or reclaimed wood and turn it into something smart enough to make indoor furniture. It takes more effort but saves money on buying expensive prepared timber.

To use a bench plane, hold the handle with your forefinger extended in the direction you are going to plane. Place your other hand on the round knob to provide downwards pressure. Stand beside the work bench with your rear foot pointing towards it and your front foot parallel to it. As you start planing, put some pressure on the toe of the plane and, as you finish the length of the stroke, transfer the pressure back to the heel. This avoids rounding off the ends. Plane a flat surface in two different directions, finishing with the grain along its length.

Chisels

Different gouge types suit different jobs. Most chisels have a bevelled edge. Respect the sharp edge. We store our chisels in a wall-mounted rack rather than loose in a tool box as it keeps the tools sharp and our fingers safe.

Using a chisel is a matter of allowing the sharp edge to do the hard work for you. With very little pressure and at the right angle, a chisel can slice through wood like a warm knife through butter. Use your body weight to drive it forwards. If you need some extra force, use the ball of your hand on the end of the handle. For deep cuts, for example a mortice joint, use a wooden mallet. For

1. **Use your thumb knuckle** to align a saw as you start to cut.   2. **A bench plane** smooths rough surfaces. Check your progress by laying the plane on its side to see how flat the wood is.   3. **A sharp chisel** makes neat work of a tenon (see page 57). 4. **Mark wood accurately** with a steel rule.

really delicate work, use the side of the mallet to tap the chisel.

Tape measure and spirit level

A phrase that has stood the test of time with us is: "Measure twice, cut once". Checking your measurements a second time before you cut a piece of wood can save hours of work. Having said that, we are by no means slaves to the spirit level. Often when you are simply knocking something together outside, these tools are just something else to tidy away at the end.

Set squares

These will improve your accuracy and overall carpentry skills. Use one to ensure a joint meets perfectly at 90 degrees, to mark a right angle on a piece of wood,

and even to work out a line at 45 degrees without any fuss. A 45-degree angle line is also known as a mitre.

Hammers

One of our mottoes is: "If in doubt, get a bigger hammer!" We have the largest selection of hammers that we know of and we use every single one.
Cross-peen hammers help when you're holding nails between your finger and thumb and starting them off. Then you can swivel and use the striking face to drive them in.
Pin hammers are for small staples, tacks, and oddly enough, pins.
Claw hammers come in different shapes and sizes. Our favourite is a steel-shafted claw hammer with a non-slip handgrip. The head is moulded onto the shaft.

The hammer's unbreakable and comfortable to use.

A sledge hammer is good for both fine adjustments and gentle persuasion.

Power drills and screwdrivers

Cordless electric screwdrivers have revolutionized woodwork. However, there are still some basic guidelines. Most important of all is to always match the screwdriver tip to the slot in the screw. If you don't you will damage your screwdriver or the slot, score the surrounding wood, or get really annoyed. Always screw or unscrew in line with the screw – imagine an invisible line that goes through the screwdriver and aligns with the screw itself.

With power tools and cordless drills use your battery responsibly if you want it to last a long time. We use ours until they are completely flat and sound like a gramophone playing at slow speed. Then we charge them up for the full amount of time required. It's worth having more than one battery so that you can use and charge them on a rotational cycle.

Some key drill bits worth buying are:

Dowel bit with a central lead-point and two spurs either side, which prevent the bit from sliding or being deflected by wood grain.

Countersink tip which cuts a tapered recess into wood so you can hide the head of a screw below or flush with the wood's surface.

Flat bit for boring larger holes.

PROJECT Make a window box

Making a window box is a great way to practise the basics of woodworking. You can make it out of almost any durable wood and design it to fit a sloping windowsill exactly. Decorate the front with driftwood for a bit of fun.

YOU WILL NEED

Ruler and pencil	Durable wood or plywood
Saw, hammer	Brass or galvanized screws, plus glue (optional)
Spirit level	
Drill plus bits	Driftwood plus tacks

1. Measure your sill. Deduct twice the thickness of the wood from the length and cut two pieces of wood to this length to form the front and back. Make them around 13cm (5in) high to hide plant pots inside the box, without stopping too much light entering the window.

2. Lay a spirit level on the windowsill from the inner to the outer edge. Raise it until it is level, then measure the height from sill to spirit level. **3. Cut each end piece** deeper by this amount at the front, to match the sloping sill. **4. Measure the base** and cut a piece of wood to fit inside the sides, front, and back pieces.

5. Drill 13mm (½in) holes for drainage along the base at intervals of about 10cm (4in) before screwing the window box together. **6. Drill pilot holes** and screw all of the joints together. Use wood glue as well if you want to. **7. Tack driftwood** to the front of the box. **8. Secure** window box firmly in place with an angle bracket fixing.

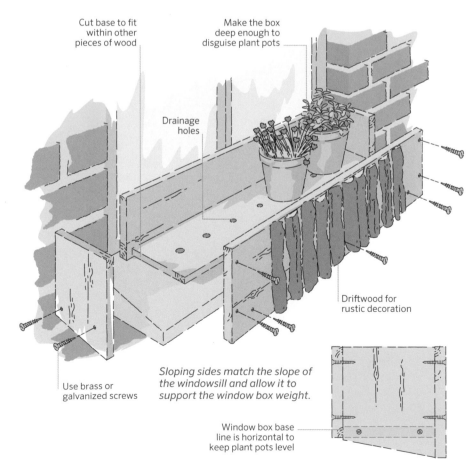

Cut base to fit within other pieces of wood

Make the box deep enough to disguise plant pots

Drainage holes

Driftwood for rustic decoration

Use brass or galvanized screws

Sloping sides match the slope of the windowsill and allow it to support the window box weight.

Window box base line is horizontal to keep plant pots level

SCREWS AND NAILS

Label boxes or keep fixings in clear jars. Always have a large supply.

Countersunk woodscrews are for strength and they also look good. Make a pilot hole before screwing in, to avoid splitting the wood. Countersink to recess the head.

Roundhead screws need pilot holes before drilling. We keep a supply of these in stock to replace screws on old household fixings so that we preserve the original look.

Coach screws are excellent for outdoors. Galvanized ones last longer. Make small pilot holes and tighten with a spanner.

Posidrive screws are one of the inventions that has most enhanced our ability to build things. They don't need pilot holes. Use for all projects.

Staples are indispensable if you keep poultry. Use them to fix wire netting on to chicken runs.

Roofing nails are twisted and won't pull out, even in a gale.

Flat-head wire nails are standard nails for fixing wood together. Use galvanized ones for outdoor work.

Masonry nails are made of toughened steel. Be careful of sparks and splinters when hitting them.

Lost-head nails are so-called because you can bury the head of the nail into the wood. Ideal for floors.

Clout galvanized nails are for fixing battens and slates to roofing.

Wallbolts are for when you want a really strong, load-bearing fixing.

PROJECT Basic joints

Use these two techniques to make a neat, strong joint between two bits of wood. Dowel kits from hardware shops include the lengths of dowel plus the right size drill bits. Make sure that your pieces of wood are cut exactly with a set square so that they will fit flush together. The next stage in joinery for beginners is making mortice and tenon joints. The mortice is a slot in one piece of wood and the tenon on the other piece fits neatly into it. Use this joint for more professional-looking projects.

YOU WILL NEED
Metal ruler and pencil
Clamp
Drill plus bits
Set square
Wood glue
Chisel and mallet
Saw

DOWEL JOINT

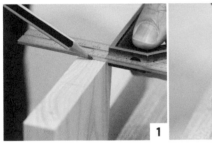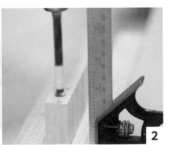

1. Measure and mark the centre of one end of the wood with a metal ruler and a pencil. **2. Mark half the depth of the dowel** on the drill bit with tape as a guide. Drill holes for the dowels. Insert dowels and marry them up with the end grain of the other piece of wood, marking the centres. Drill dowel holes in the other piece of wood.

3. Brush dowels and joint with wood glue, and fit the dowels into the holes in the opposite piece of wood. Clamp until the glue sets. **4. The finished joint**.

MORTICE AND TENON JOINT

1. Mark out the shape of the mortice. Lay the other piece of wood across and draw either side. Then draw a second set of lines inside: the slot needs to be smaller than the width of the second piece of wood. **2. Chisel out** the mortice bit by bit, using a mallet. Mark out the length of the tenon and run the lines over the end of the wood using a set square.

3. Cut out the tenon with a sharp saw. Tidy it up with a chisel.
4. Insert the tenon into the mortice. Adjust the fit with a chisel if necessary. Add glue for extra strength.

ECO-KITCHENS

The kitchen is the heart of many homes – it's where we have our morning cup of coffee, preserve garden produce, enjoy family meals, and entertain guests. But the kitchen is also where we use the most energy, and conserving it by preparing and cooking food with care is one of the easiest ways to reduce the bills. We also minimize food waste and ensure none ends up in landfill.

AVOIDING FOOD WASTE

The first step to creating an eco-friendly kitchen is to avoid any food waste and prevent any from reaching a landfill site. This means eating products before their use-by dates and buying in bulk, which reduces packaging and transport costs to and from the shops.

Sort the contents of your pantry by food type and arrange products like the aisles of a supermarket so that you can easily see what you have and what needs replacing. In our pantry we have one clearly marked shelf for open condiments and another labelled "Closed for storage", which prevents us from opening too many jars at the same time and any going off. You may also find that pull-out baskets are useful for fresh produce.

We have invested in sealed containers to store bulk buys, such as sugar, flour, and oats. If you don't have space to store large amounts of food, then you can try to set up a share scheme with friends or family, which allows you to save money on bulk deliveries but have smaller quantities to store.

Remember, too, that "best before" dates on products simply refer to the time when food is at its best, so it will be okay to eat after that date. The only foods you should avoid after their best before date has expired are eggs. If more perishable food is close to its use-by date and you are going away for a few days, or you have other meal plans, then freeze it.

Another great way to reduce food waste is to be creative and make meals with leftover food. Try established leftover recipes or experiment with your own. You could also learn some new skills such as preserving and fermenting (see pages 168–173, 176–177, 196–201, and 224–225) to use up food if you've over ordered or have a surplus from the garden. Some of our favourite zero-waste recipes include: using stale biscuits for a crumble topping, making chips and crisps from vegetable roots or peel and carrot tops, blitzing leftover greens into a hummus, or adding cucumber ends to water for a refreshing drink packed with essential electrolytes.

RECIPE Sprouting seeds

These are a simple way to boost the nutritional content of meals. You can sprout your own seeds on a kitchen windowsill, then use them to add texture and flavour to stir-fries and salads. Our favourites include radish, broccoli, sunflower (shell them first), and chickpea.

YOU WILL NEED
Seeds
Glass jar with strainer (or use a separate sieve)
Kitchen paper

1. First, soak the seeds in water for 24 hours. Then strain the seeds and place them in a well-lit position. **2. Rinse the seeds** every day with clean water, and drain off the water. They will sprout 4 or 5 days later. **3. Rinse,** pat dry with kitchen paper, and store in a fridge – eat within 2 days.

1. Slow cooking on a woodburner is perfect for a warming stew and to boil a kettle for tea.   **2. Buy flour in bulk;** it has less packaging and you will always have some to hand to make fresh bread.

LOW-ENERGY COOKING

Most ovens are powered by gas or electricity. Electricity is the most inefficient option, as energy is converted from fossil fuels to electricity and then back to heat; taking energy directly from gas to heat is more efficient. However, the most sustainable choice is an electric oven that is powered by renewables on a green tariff.

Another green fuel option is wood. Wood is a carbon neutral fuel and modern, DEFRA-approved woodburners are highly efficient (see pages 92–93). Although we find that using a woodburner for cooking only makes sense on colder days, it is at these times when a warm hot pot is just what we want. When cooking on a woodburner make sure that you use a pan or pot with a thick base, and start nice and early, as cooking times are much longer. When the woodburner is fired up we also use it to boil a kettle and to heat water for the washing up.

Earth ovens (see pages 60–61) and barbecues also burn wood as fuel. However, we only really use our outside oven in the warmer months.

Pressure cookers

Old-fashioned pressure cookers used to be a bit scary, but modern versions are absolutely safe and very easy to use. They are far more energy efficient than conventional pots and pans because the higher internal temperature and pressure they exert cooks food in a third of the time. In addition, more of the vitamins and minerals in foods are retained when cooked in a pressure cooker. If you want to live a greener life and save a bit of money at the same time, then a pressure cooker is a must-have eco-gadget for the kitchen.

Kelly kettles

Ideal for a quick cuppa while you are working outside, a kelly, or volcano kettle has a double-walled chimney that you fill with water. Then fill the fire base with twigs and dry grass, and light. Heat goes up the chimney, and the large surface contact area with the water leads to a quick thermal transfer and faster boiling.

QUICK TIPS
Try these simple tips to reduce food waste and energy usage.

Shopping

Plan meals in advance so that you know what you are going to eat and when. You'll save money by avoiding the temptation to buy more food than you need when out shopping.

Order food in bulk. We order ours from an organic wholesaler, allowing us to enjoy the benefits of organic produce at a cheaper rate. It also has less packaging.

Buy from local shops rather than supermarkets. This reduces your transport costs and the food may also contain fewer preservatives.

Make a shopping list; check your pantry before you go, and only replace items that are finished. Stick firmly to your list.

Opt for recyclable packaging by choosing juice and condiments in glass rather than plastic bottles. Glass can be recycled infinitely and you can reuse jars at home to make more preserves or chutney.

Choose long-life products like tinned beans, dried fruit, nuts, pasta, noodles, rice, and grains, which are unlikely to go off before you use them.

Cooking

Just cover veg with water; the less water you use, the quicker it will boil. Don't throw excess water away after cooking; allow it to cool and use it to make stock, or feed it to your plants.

Cover pans as the lids prevent heat loss. Also ensure pots and pans cover the coil or gas rings to maximize their efficiency.

Try using a solar dryer for cooking in the summer (see pages 170–171).

Try one-pot cooking – stews and soups that take one burner to cook and produce less washing up.

Allow food to cool before freezing or refrigerating, as previously cooled food won't need as much energy.

Don't fill kettles and only use as much water as you need.

Keep ovens door closed – opening them to check on food wastes energy.

Make stocks and bone broth from roast chicken carcasses to make the most of the chicken's nutrients and flavour. Boil the carcass with a tablespoon of cider vinegar to help extract all the calcium and goodness from the bones.

PROJECT Make an earth oven

With its simple clay structure and use of carbon-neutral wood fuel, a traditional earth oven is an eco-friendly option when it comes to cooking. Besides, few tastes compare to the stone-baked flavour it creates. Earth ovens can be used for everything from bread and pizzas to fish pies. The cost of building one is next to nothing and the process, while a bit messy, is a lot of fun.

YOU WILL NEED

Tape measure	Clay – dig your own if possible
Pen	Sand
Garden riddle	Newspaper
Rolling pin	Bricks or breeze blocks
Knife	Stone slabs
Trowel	Wood

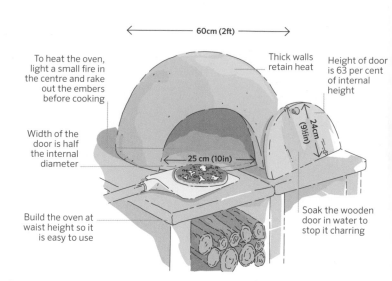

← 60cm (2ft) →

To heat the oven, light a small fire in the centre and rake out the embers before cooking

Thick walls retain heat

Height of door is 63 per cent of internal height

24cm (9½in)

Width of the door is half the internal diameter

25 cm (10in)

Build the oven at waist height so it is easy to use

Soak the wooden door in water to stop it charring

Each oven has its own unique cooking temperature. Light the fire inside the oven, leave the door open, and wait a few hours. When the soot burns off the inside, remove the cinders with a shovel and start cooking.

BUILDING THE STAND

1. Build a firm support. We used some spare breeze blocks. Keep it level and build the structure up to a comfortable working height for cooking. **2. Prepare a solid floor** for your oven. It must have a smooth surface. We used a couple of paving slabs. **3. Mark the centre** of the oven. Make it as large as possible. Ours is about 60cm (2ft) in diameter. **4. Draw two circles**, one 7.5–10cm (3–4in) inside the first, to show the thickness of the walls. Note down the radius of the inner circle.

PREPARING THE CLAY

5. Riddle the clay to remove pebbles and debris if you dug your clay from the ground. **6. Lay a big tarp** on the ground and mix your clay – best done with bare feet. **7. Add sharp sand** (about a bucketful) and some water if the clay is very dry. Mixing takes time and effort. Don't slip over!

8. Keep turning and mixing the clay. Test it to see whether it is ready to work with by making a clay sausage and holding it with half in your palm and the other half dangling over your hand. If the clay bends but does not break, it's ready to use.

MAKING THE OVEN BASE AND WALLS

9. Roll out a circular layer 1cm (½in) thick on the internal circle of the base. Trim the edges with a trowel. Wet the clay, then smooth it with your hands. This will serve as a smooth base for sliding whatever's being cooked in and out of the oven. **10. Cover** the circle with a layer of moist newspaper to stop any sand sticking to it. **11. Pile on moist sand** and sculpt the shape of the earth oven, making a dome that is 5–7.5cm (2–3in) taller than the internal radius of the oven. Measure the height of the sand dome, which will be the interior height of your oven. Multiply this by 0.63 to get the height of the door.

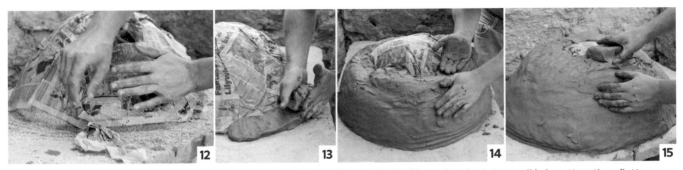

12. Cover the sand dome with wet newspaper to stop the clay sticking to it. **13. Shape** the clay into small briquettes, then flatten and squash them into place. Start at the base and work round and up, to cover the dome. **14. Use** the width of your hand as a rough measurement: the layer of clay should be around 7.5–10cm (3–4in) thick. **15. Try to push** the clay against itself, not against the mound of sand, as you join each briquette. Cover the entire dome with clay, making sure it is still the same thickness at the top as the bottom. Wet your hands and smooth the surface of the finished dome.

MAKING THE DOOR

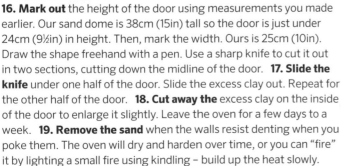

16. Mark out the height of the door using measurements you made earlier. Our sand dome is 38cm (15in) tall so the door is just under 24cm (9½in) in height. Then, mark the width. Ours is 25cm (10in). Draw the shape freehand with a pen. Use a sharp knife to cut it out in two sections, cutting down the midline of the door. **17. Slide the knife** under one half of the door. Slide the excess clay out. Repeat for the other half of the door. **18. Cut away the** excess clay on the inside of the door to enlarge it slightly. Leave the oven for a few days to a week. **19. Remove the sand** when the walls resist denting when you poke them. The oven will dry and harden over time, or you can "fire" it by lighting a small fire using kindling – build up the heat slowly.

FINISHING TOUCHES

Repair any cracks that emerge on the oven as it dries, and fit a door.

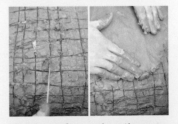

Wet the surface and gently score it with a cross-hatch pattern (left). Apply more clay to the cracked area (right). Repeat if cracks appear after the oven is used. **Make a wooden door**. Use a paper template to get the shape right. However, it doesn't need to be a perfect fit.

GREENER CLEANERS

Surprisingly, the most environmentally damaging objects in our houses tend to be found hiding under the kitchen sink or lurking in the cleaning cupboard. Among all those bottles and cans there are all sorts of harmful combinations of toxic bleaches and chemical-based polishes and disinfectants that can be seriously damaging both to humans and to the environment.

ECO ALTERNATIVES

Several years ago we switched from using artificially produced cleaning products to eco alternatives. Many of the cleaners we now use are made from simple home-made recipes that originate from old-fashioned cleaning techniques. As well as being less harmful to humans, animals, and the environment, they are also cheaper. We are reducing packaging waste as we've stopped buying new bottles and packets over and over again.

We also find that, having stopped using conventional products that always seem to promise to "kill 110 per cent of all germs", we have become noticeably more sensitive to the noxious smell of commercial cleaners. These days we much prefer the aroma of our natural alternatives, many of which are outlined here for you to make and enjoy yourself.

BICARBONATE OF SODA

Also known as sodium bicarbonate, baking soda, or simply bicarb, this naturally occurring substance is extremely cheap and will not harm you or the environment.

To deal with a smelly carpet, sprinkle baking soda over the carpet. Add dried crushed lavender or basil leaves and leave for 30 minutes. Vacuum thoroughly afterwards.

To clean work surfaces, apply baking soda and scrub with a damp cloth or sponge. To remove ingrained dirt or stains, use a brush.

To clean the oven, sprinkle the mess with some salt and then mix 2 tbsp of baking soda with water in a cup to make a thin paste and apply to the oven surface. Use an old toothbrush or bristle brush to scrub it off.

For clogged sinks and shower drains, try to avoid using conventional drain cleaners in your home because they contain particularly nasty ingredients. If you have a blocked drain, pour in a cup of white vinegar plus one cup of baking soda. Let that sit for a few minutes (don't panic – it's normal for it to bubble like Vesuvius). Then flush the mixture down with a kettle of boiling water.

SOAP NUTS

A soap nuts is actually a fruit that contains a natural surfactant that works like a normal detergent to break down dirt and grease in water. They leave laundry clean but can also be used for cleaning floors or even as shampoo. The aroma is less pronounced than chemical cleaners we're used to, so we'd recommend adding your favourite essential oils to your laundry to provide a fresher smell.

DISTILLED VINEGAR

Also known as clear or white vinegar, distilled vinegar is another great cleaner for the home as well as an effective deodorizer. Any kind of vinegar will do, but we recommend distilled as its smell is less likely to linger than other types. This is the only issue that we have with using

1. Bicarbonate of soda is a very efficient cleaner. Sprinkle it over a surface and then wipe off with a damp cloth. **2. For a blocked drain,** combine bicarbonate of soda with distilled vinegar in the proportions described above. **3. Spray a mix** of vinegar and water (see opposite page) onto windows to clean them. **4. Wipe half a lemon** over your chopping block to clean and deodorize it.

vinegar – the smell reminds us of being in a fish and chip shop. However, mixing in lavender essential oil and/or lemon juice is a very effective way to mask the smell.

If you are using a vinegar cleaner on tiles or marble, always be sure to rinse the cleaned area well because it can continue to react with any lime-based product.

To clean windows, mix 120ml (4fl oz) vinegar with 4 litres (1 gallon) of water and use to wipe down your windows. For extra shine, try rubbing the window glass dry with a couple of sheets of screwed up newspaper.

Use as a fabric conditioner. A splash of white vinegar in the final rinse leaves your clothes, sheets, and towels soft and absorbent. It has the added benefit of breaking down the detergent more effectively, thus keeping the washing machine drum clean. It is especially useful for family members who have sensitive skin or allergies. Don't worry

– any smell of the vinegar goes when the clothes are dry.

For a basic furniture polish, mix 60ml (2fl oz) vinegar with 175ml (6fl oz) olive oil. Alternatively, mix 60ml (2fl oz) lemon juice with 120ml (4fl oz) olive oil. Wipe down your furniture with a soft cloth soaked in the solution for excellent eco-cleaning results.

LEMON JUICE

This everyday liquid is a mild, yet very effective, green cleaner. You can use lemon juice as a non-toxic way of removing grease and also as an effective antibacterial in many parts of your home.

To remove stains, squeeze some neat lemon juice over any tough stains on your chopping board, let the juice sit there for about 10 minutes, then wipe it away.

To freshen and deodorize your microwave, place a few slices of lemon in a small bowl of water and heat it for

30–45 seconds. Use the hot liquid to wipe down all the surfaces in the microwave.

For everyday toilet cleaning, sprinkle some lemon juice and baking soda into your toilet and leave it to fizz for a few minutes, and then scrub it clean with a toilet brush.

HERBAL CLEANERS

Choose one fresh herb, such as lavender, juniper, or thyme. Simmer a handful of the leaves and stems in a pan of water for about 30 minutes. Add a little more water – the less you add, the stronger the solution and the more powerful its properties. Strain the liquid and pour it into a bottle with a dash of natural soap. This herbal solution cuts through grease and smells good too.

RECIPE ## Natural hand scrub

Keeping your hands clean can be expensive, especially using squirty hand cleaners, so we started making our own scrub to keep next to the washroom sink. Our recipe is basically a salt and oil scrub, infused with soothing calendula, which we have found to be excellent at removing garden grime. Use it regularly and you'll find that it is also a great moisturizer for the skin.

YOU WILL NEED
Rock salt
Calendula flowers
Calendula oil
Small jar

1. To make the scrub, put a 5cm (2in) layer of rock salt in the bottom of a jar.
2. Mix the salt with some dried calendula flowers and then cover with calendula oil. **3. To use,** scoop a small amount of the scrub and use it like normal hand soap. Wash off with warm water. If you don't like calendula, try this recipe with your preferred blend of dried flowers and oil.

ENERGY AND WASTE

Having reduced the amount of energy you are using, why not start generating your own? Any investment you make now will help future-proof you against price rises that you have no control over and all the hard-earned cash that used to go to the utility companies will be yours to spend on the good life. Even the most technophobic will discover there is something very empowering about not being reliant upon others for your energy, water, and waste-disposal needs.

UNDERSTANDING ELECTRICITY

Before you decide what renewable energy systems you are going to install, it's definitely worth getting an overview of the basic science governing electricity. So here's the techie bit – this isn't an excuse to skip ahead! Have a go at reading these pages. They will help you to work out what your needs are and what renewable options are possible on your plot.

KEY TERMS

Amps measure the flow of electrons forming an electric current.

Conductors, as the name suggests, conduct electricity. Good conductors, such as metals, have loosely bound electrons so there is little resistance.

Insulators are materials such as rubber that don't readily conduct electricity as their electrons are not easily displaced.

Power is a measurement of how much energy you are using each second, measured in watts.

Volts measure the "potential difference" between two points on a conductor.

GENERATING POWER

You can use renewable technologies to generate electricity in one of two ways. One option is to generate alternating current (AC) electricity mechanically by moving magnets past wires (see below) – this is what a wind turbine or the axle of a waterwheel does. Another method that we use on a domestic scale absorbs energy from the sun into solar photovoltaic (PV) panels, generating direct current (DC) electricity. The energy can be used immediately, for example, to light your home or run a pump, or can be stored in batteries.

STORING ENERGY

In a battery, a chemical reaction occurs and then electrons travel through a wire from one terminal to the other, and the result is DC electricity (see below). Batteries are particularly useful for off-grid systems as they are fairly cheap and power can be stored and used when needed. If you wish to store electricity that you have generated in a battery bank, use leisure or "deep cycle" batteries – the sort used on caravans or boats, or to power electric fences. They are designed to deliver less current for a longer period of time. Once flat,

---

THE ESSENCE OF ELECTRICITY

At the centre of every atom is a nucleus, which is divided into even smaller particles called protons and neutrons. Around the nucleus, electrons orbit.

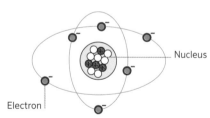

Nucleus

Electron

The protons within the nucleus have a positive charge and the orbiting electrons are negative. When there is a balance between the two, the atom is neutral.

Usually atoms have fixed numbers of electrons, which cannot be moved away from the atom. But if there is an excess of free electrons on some atoms in one site (for example, in one of the silicon-based layers of a solar panel), and a deficiency of electrons on other atoms in a neighbouring site (such as the adjoining layer of the solar cell), and there is a route between them, then the electrons will start to move from one location to the other. This flow of electrons is called a current. The "potential difference" between the two sites is known as voltage and the electricity generated is measured in volts.

Direct current (DC) electricity

When a battery is connected to a circuit, some of the material in the battery starts to dissolve and some of the atoms lose electrons. This means there is a deficiency of electrons at one of the terminals. This potential difference sets the electrons in motion and the current flowing. The current that a battery produces is called direct current, or DC – this is

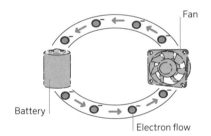

Fan

Battery

Electron flow

Direct current can be used for small appliances that are often associated with camping, such as a 12V kettle or a fan.

easy to remember because DC electricity travels from one battery terminal *directly* to the other terminal.

Alternating current (AC) electricity

AC is what the mains supply provides and so it's good to be able to produce it at home, either to sell back to the national grid or to feed into the existing electrical circuits in your house.

they're intended to be recharged. They cost about the same as car batteries, but the latter are unsuitable for most renewable applications as they aren't meant to be fully discharged then recharged once flat.

AC/DC CONVERSIONS

You might imagine that one form of power would work for everything, but there are good reasons for having the capability to switch from AC to DC and vice versa. Rectifiers and inverters are devices that enable such conversions. **Rectifiers** change an AC supply into DC. On our farm we used a rectifier to charge up a set of batteries using the AC current produced by our spinning waterwheel. We then stored the energy that the wheel generated in batteries that could be drawn upon in the evening when we used the most indoor lighting.

Inverters work the other way to rectifiers, converting DC to AC. They are pretty much essential if you want to continue to enjoy a 21st-century lifestyle while making full use of green technology.

There are a number of inverters available, with significant differences in quality. It has to be said that some inverters can cause lights to flicker and have been known to destroy complex electronic equipment such as computers.

So it's worth opting for the more expensive, most useful version. This is the pure sine wave inverter, which ensures the electricity you produce matches the electricity distributed in the national grid.

When choosing an inverter, you need to know what sort of appliances you want to power from it. Ask the installer and, if in doubt, check the details with your electrician.

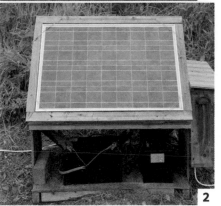

1. **Our grid-linked wind turbine** allowed us to sell any excess electricity we created back to the national grid. 2. **The energy created** by these solar PV panels is stored in leisure batteries for such things as pumping water from our spring and powering our greenhouse heat sink.

To understand AC electricity, it helps if you think back to school, as you may well remember a diagram of a magnet with lines of flux linking the two poles (see right). These lines of flux illustrate the invisible power in a magnet. Now imagine moving a magnet near a coil of copper wire. The magnetic field passing through the wire will cause the electrons to move in the coil and so generate a current. The stronger the magnets, the more lines of flux cut the copper coil, and the more electricity you generate. You can make more electricity either by moving the magnets past the coil faster or by bringing them closer to the wire.

Mount a set of magnets on a rotating disc and you have a system that makes AC (see right). All you have to do to create the electricity is put in energy to make the magnets spin. In this way you could harness the power running in a stream or the energy contained in a breeze blowing over an exposed hill.

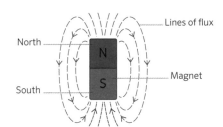

Lines of flux — North — Magnet — South

Lines of flux between the north and south ends of a magnet are invisible.

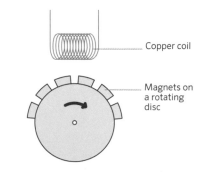

Copper coil

Magnets on a rotating disc

Magnets rotating past a coil of copper wire generate alternating current (AC).

GRID LINKING

You must use the skills of a certified electrician to ensure you are set up correctly so that you can be linked to the national grid. The systems must have a safety mode that switches them off if the grid connection fails. It's easy to imagine what would happen if you were gaily putting electricity into the grid while a maintenance team were working on a problem with their mains supply switched off. Also, renewable energy systems must synchronize to the grid. If they are not in synch, the combined effect will produce dirty AC.

RENEWABLE ENERGY OPTIONS

Before you start harnessing the elements to heat and light your home, it's worth considering your motivation as this will determine what systems are best for you. You will also need to assess your site as you may not have suitable resources. Finally, you should decide on how self-reliant you want to be. Do you want to remain connected to the national infrastructure as a safety blanket?

WHY DO YOU WANT TO USE RENEWABLE ENERGY?

If you simply wish to reduce your impact on the planet and money is not an issue, then all you need to ensure is that the energy it takes to set up your renewable energy technology, known as the "embodied energy", will be less than the energy you save during its lifetime.

However, for the majority of us, financial savings tend to be just as important as environmental savings – so, the systems you choose must make financial sense now and for the future, as well as help the planet. When doing your sums, you may be concerned about the length of the payback time, but remember that your investment can also future-proof you against energy price rises and possible shortages. So add peace of mind into your calculations. There is also the advantage that well-chosen renewable energy systems will add to the value of your property, making them a better investment than many interior refurbishments.

WHAT ARE YOUR RESOURCES?

Your opportunities depend on different factors that relate to your plot. Carry out a site survey and ask yourself some key questions before hiring a certified installer to put in an expensive system. For example:

• **Do you have south-facing roofs** or surfaces on which to place solar thermal systems or solar photovoltaic (PV) panels?
• **Do you get winter sun?**
• **Can you use the wind?** To decide this you will need to know the following: Where are the prevailing winds? What is the average wind speed? And will your local planning authority give you permission to erect a turbine?
• **Do you have a water source?** If so, how far will you have to move the water for it to be useful? Is there sufficient flow/drop to harness the water energy to turn it into electricity? (See pages 78–83.) You will also need to apply to DEFRA for a licence to abstract water.

When you are considering what is possible on your plot you should also think about what you would really *like* to do. If you have always wanted a wind turbine or a waterwheel, and either or both will work on your site, then go for it. However, it's still better to focus on one energy source than to try using loads of different ones. You need to prioritize because ultimately cost will dictate what you can achieve.

HOW SELF-RELIANT SHOULD YOU BE?

There are some big advantages to using the national grid as a buffer for your renewable projects. You have something to fall back on if any of your systems fail and you can sell any surplus to the grid. There is also something comfortingly mainstream and normal about still being connected to the rest of the country.

That said, grid connection does mean you won't have a truly independent system and unless you are disciplined and use no more than you produce, you will still be facing regular bills and price rises when energy costs increase.

1. The grid-linked turbine is positioned on the brow of a hill to take advantage of the wind.
2. Solar PV panels can be a great sustainable energy source for urban and suburban homes.

Renewable energy systems

At Newhouse Farm we decided to remain connected to the grid, but not to be fully dependent upon it, so some systems were grid-linked while others were stand-alone.

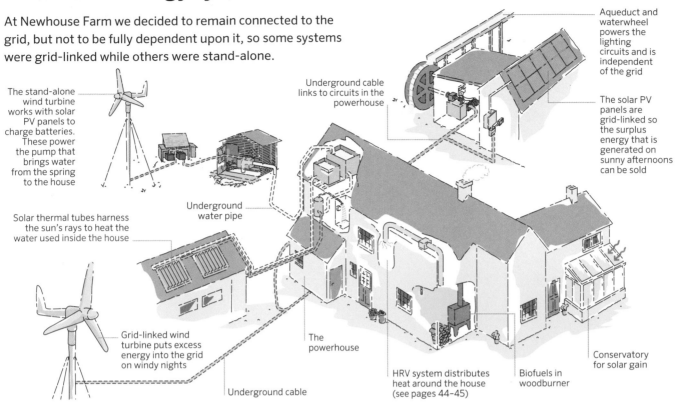

The stand-alone wind turbine works with solar PV panels to charge batteries. These power the pump that brings water from the spring to the house

Underground cable links to circuits in the powerhouse

Aqueduct and waterwheel powers the lighting circuits and is independent of the grid

The solar PV panels are grid-linked so the surplus energy that is generated on sunny afternoons can be sold

Solar thermal tubes harness the sun's rays to heat the water used inside the house

Underground water pipe

Grid-linked wind turbine puts excess energy into the grid on windy nights

The powerhouse

Underground cable

HRV system distributes heat around the house (see pages 44–45)

Biofuels in woodburner

Conservatory for solar gain

OFF-GRID LIVING

If you wish to break free and come off-grid, you won't be able to use the existing infrastructure for electricity, gas, or water as a back-up to your systems. You will be taking full responsibility for your own energy and water supplies, and while that is a heady prospect, the reliability of the systems you're putting in place will become a key issue. Start by addressing a list of "what ifs?". What if...

• it's not sunny for days/weeks?
• it's not windy for days/weeks?
• it's neither sunny nor windy for days/weeks?
• the water flow from your source reduces during the summer?
• you have to take a system out of service for planned maintenance?
• something malfunctions in a system and you need to fix it?
• your stored/harvested water is not replenished for a day/week/month?

A huge storage capacity for excess energy may seem to be the answer, but providing masses of storage that you may only use occasionally is a very expensive option. Instead, focus on what you consider to be essential and then be prepared to compromise if nature temporarily leaves you a bit short. For example, you may consider that it's essential to flush the toilet but that baths are a luxury. So if your rainwater tank starts to get low, you will choose to limit baths and use the shower instead.

There are plenty of other examples that demonstrate that limiting your energy usage is not necessarily a hardship. How about limiting the amount of time the television is on each day; or using the kettle on the woodburner rather than switching on the electric kettle? You can also give up the tumble dryer and use a washing line instead.

CALCULATE YOUR ENERGY USAGE

Before you invest in a renewable energy system, you need to work out how much energy you use so you can calculate how much you need to make. To find out the amount of energy an appliance uses, multiply its power by the number of hours it is turned on for, or use a plug-in energy monitor. The standard unit to measure this is kilowatt hours (kWh) – a kilowatt is 1,000 watts – and it is the unit of electricity you get billed for.
• A 2kW heater that is run for 1 hour will use 2kWh of energy.
• A 100W light running for 5 hours will use 0.5kWh.
• A 3kW electric oven run for 3 hours will use 9kWh.

If your ambitions lie firmly in the realms of living "off-grid", using only home-generated power, it's important to first reduce the amount of energy you are using (see pages 36–39).

USING SOLAR PV PANELS

Whenever the sun shines, it offers us a powerful source of energy in the form of photon rays that can be converted instantly to direct current (DC) electricity by photovoltaic (PV) cells. These can be used for all sorts of jobs, from powering low-voltage water pumps to charging a mobile phone. Solar panels can be fitted to most roofs or set up on the ground in a south-facing sloped field.

HOW SOLAR PANELS WORK

A solar panel is made of a lattice of layers of silicon-based cells. When the sun's photons hit a layer of silicon that has free outer electrons around the atoms, these flow to the layer beneath, which contains atoms with electrons missing. This flow causes a small current. With so many cells being linked together, the panel produces a useable voltage of DC electricity. This current is then changed to AC through an inverter if it is being grid-linked (see page 67), or remains as DC current if it is going straight into a battery.

POWER CAPABILITY

As a rough guide, a south-facing solar panel installation in the UK with a 3kW peak capability should provide enough energy over the year to meet the estimated average household demand of 3,300kW.

Solar PV systems are specified as having a "peak power capability", because they depend on how much sun shines on the panels. Surprisingly, the systems do still make electricity on overcast days, though obviously less than on sunny days. It is also worth mentioning that your specific microclimate is key when making any decisions about power capability. After all, a solar panel simply might not be the right option if you live in a shady valley.

CHOOSING SOLAR PV PANELS

There are all sorts of types of solar PV panels available to buy, with more efficient options coming to market all the time. This increased efficiency means a small area of roof can provide enough power to be useful for an average home.

Conventional solar PV panels are now relatively straightforward to buy. Our advice is to shop around to make sure you're buying the latest, most efficient technology. Try to calculate the payback time to put your purchase in perspective and to incentivize yourself. We'd recommend working with an established solar PV fitter who knows the industry. Also beware anyone who tries to sell you solar PV for a north-facing or shaded roof.

You can also buy solar roof tiles, which are used as part of the material for the build of your roof. While they are a lot more expensive than conventional panels, according to the Energy Saving Trust, they can increase the value of your property by as much as 10 per cent.

Another option you could potentially consider is combined solar photovoltaic and thermal (PVT) panels. These panels improve upon the efficiency of standard solar PV panels (which are typically around 15–20 per cent efficient) by

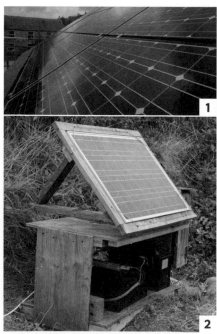

1. **Solar PV panels** on the roof of our barn. 2. **Adjustable ground-mounted panels** enable you to track the sun during the day. 3. **A small panel** powered the fan in our greenhouse heat sink (see pages 130–131).

SITE CHECKLIST

How much electricity do you use in your house? Can you reduce it?
Do you have a south-facing roof?
Does your roof get winter sun?
If you are hoping to install panels on the roof of your house, have you checked with the local authority if planning permission is needed?
Are there any grants available in your area?
Size does matter – how much space do you have for your panels?
What is the peak output in kW of the system you have been quoted (i.e. how much electricity can the panels generate in full sunshine)? How long will it take for the system to cover its costs?
Have you considered a combined system for operating a small off-grid energy solution?

converting excess energy into heat for the home. This technology is not currently widely available in the UK, although it has had some success in the US.

Whichever type of panel you choose, you can expect to get good value for money. PV panels are often the best-value energy systems to link to the grid, and their lifespan is impressive as no moving parts are required to convert the solar energy into electricity. The glass is also self-cleaning.

STAND-ALONE SYSTEMS

Solar PV panels are extremely useful for powering a stand-alone system that is remote from a mains supply, such as a pump for a watering system. For a stand-alone system, it is important to be sure that your system can provide enough energy to power whatever you intend to use it for. If you are using your pump for two hours a day, work out how much power you will use (see right) and install enough panels to supply it. As long

as your panels are a suitable size and you have sufficient batteries, you should seldom have an occasion when there is not enough power to work the pump because you will have enough stored energy for these "just-in-case" situations.

We used solar PV panels to run a pump near a spring to provide water for the farmhouse. It wouldn't have been acceptable for us to run out of water just because there had been no sunshine, so we opted for a dual wind-sun system in addition to the charging batteries. The wind turbine acted as a back-up when there was less sunshine.

There were times (and these were surprisingly frequent) when nature provided us with so much energy the batteries would become full and risk being damaged if they continued to be charged. To overcome this, we connected a charge controller to the PV panel. It monitored the battery voltage and when the battery was full, the charge controller simply turned the panels off.

WORKING OUT HOW MANY BATTERIES TO USE

The amount of energy that a battery can supply is specified in ampere hours (Ah). So, a 12-volt battery that is specified as 100Ah can theoretically deliver 1 amp for 100 hours or 100 amps for 1 hour. Multiplying the volts and amps gives you the power (watts) the battery will produce. For this battery, then, it is possible to have:
• 12 volts x 1 amp = 12 watts for 100 hours OR
• 12 volts x 100 amps = 1,200 watts for 1 hour.
Back in the real world, these numbers don't quite add up as you can't expect to get more than 80 per cent capacity from your battery, i.e. 80Ah. Also, the smaller leisure batteries are not designed to deliver masses of current so you shouldn't really drain more than about 10-15 amps (that's a device of about 120–180 watts). The pump at our spring needed 60 watts to function so we divided 60 watts by 12 volts, which gives 5 amps. As we used our pump 2 hours a day, a fully charged battery therefore lasted us about 8 days.

Important elements of a PV system

The position of your panels makes a big difference to how much energy is obtainable. South-facing is ideal and as the angle of incidence of the sun varies by season and time of day, the optimal pitch is 40° from vertical.

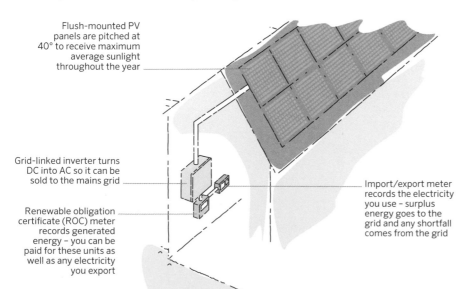

Flush-mounted PV panels are pitched at 40° to receive maximum average sunlight throughout the year

Grid-linked inverter turns DC into AC so it can be sold to the mains grid

Renewable obligation certificate (ROC) meter records generated energy – you can be paid for these units as well as any electricity you export

Import/export meter records the electricity you use – surplus energy goes to the grid and any shortfall comes from the grid

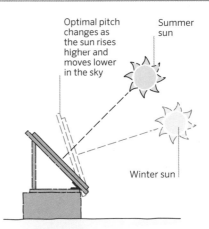

Optimal pitch changes as the sun rises higher and moves lower in the sky

Summer sun

Winter sun

Adjusting pitch
Optimal pitch is the angle from vertical that PV panels should be positioned to receive maximum sunlight at solar noon. This changes according to the season and where you are in the world. You can adjust the pitch manually yourself, or instead invest in solar tracking mounts, which maximize electricity production by automatically adjusting to the sun's changing position in the sky.

USING SOLAR THERMAL SYSTEMS

Solar thermal generation is one of the simplest means of harnessing the sun's energy. It works by the direct transfer of energy; no power is generated, which means there's no mucking about with electricity. Solar thermal systems rely solely on the power of the sun to warm water and neatly provide you with plenty of pollution-free hot water for your home – as long as the sun shines.

HOW IT WORKS

Solar thermal units, or solar collectors, are the most widely used type of solar power. In its simplest form, a unit comprises a black heat-absorbing plate, bonded to a series of pipes that allow water to flow through them. It is housed in an insulated box with a glass cover to protect the unit from cooling winds. Sunshine passes through the glass and heats up the plate and the resulting heat is transferred to the water flowing through the pipes, which is then pumped to wherever it is to be used.

KEY ELEMENTS

The solar collector is the key part of the system as it captures the solar energy. There are several different types that vary in price and efficiency, ranging from a plastic system (the cheapest and least efficient) to a copper system or even heat pipes inside evacuated tubes (the best and usually most expensive). Evacuated tubes are double-walled glass tubes with a vacuum between the layers to stop heat escaping. They have black-painted copper fins attached to a heat pipe inside them, which transfers heat up to a manifold, or "junction", where cold water passes over it and is instantly heated up to a higher temperature. As you decide which type of solar collector you wish to purchase, remember that in this context size really does matter, as you want to capture as much solar energy as possible.

The solar hot-water tank is usually double-insulated, so when you heat water it is still warm and available for a shower next morning. There is a sensor at the top of the tank that allows you to confirm the temperature of the water.

The solar water feed is placed at the bottom of the tank so when it's sunny you get masses of free hot water. And when you do have to pay to heat the water, for example by using the boiler, you only heat the top portion rather than the whole tank.

The controller is the brains of the outfit. It makes the simple decision that if the collector is hotter than the tank, it turns the pump on to circulate the hot water until the whole tank has heated up; if the collector is cool it turns the pump off.

Our thermal set-up

We installed our solar thermal system on an outbuilding for a good south-facing position. To overcome the losses we expected because of the water travelling a great distance from outside to the hot-water tank, we installed a slightly bigger system than was theoretically required and made sure that the pipes were heavily insulated.

System efficiency

Our system averaged out to the equivalent of turning an electric immersion heater on for 3–4 hours a day in the summer and about 30 minutes per day in the winter. It was a revelation to discover that we could still get enough solar energy to help heat our water even in the middle of the winter.

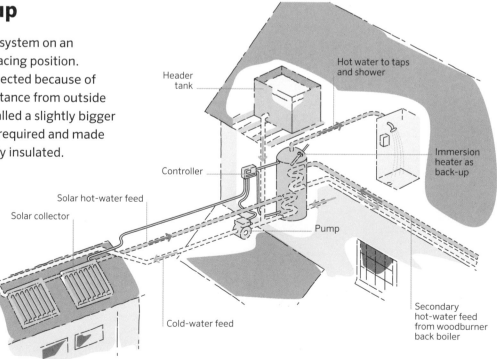

Header tank

Hot water to taps and shower

Controller

Solar hot-water feed

Solar collector

Cold-water feed

Pump

Immersion heater as back-up

Secondary hot-water feed from woodburner back boiler

SITE CHECKLIST

How much do you pay for your hot-water bill?

Do you have a south-facing roof or shed?

Do you live in a conservation zone or have a listed building? If so, you must first secure planning permission from your local planning authority. If you are not granted permission, another option would be to use a ground-mounted system.

Are there any grants available in your local area?

Have you researched different manufacturers and checked what they quote as the peak kWh performance per sq m? Compare this to other companies to find an efficient system at the right price.

1. Camping solar showers are black bags that are filled with water and left in full sunlight – they will heat up in a couple of hours. **2. Our solar collectors** comprised evacuated tubes with heat pipes connected to the manifold, through which water flowed and was heated.

DIY solar shower

It is possible to make your own garden solar shower for next to nothing using a recycled radiator, a large black plastic dustbin, and a variety of leftover materials. An ideal location for the shower would be a disused outhouse or large garden shed with a south-facing roof for mounting the collector to receive maximum sunlight. When you're ready to start using your shower, fill the reservoir with enough water so that it comes above the level of the top pipe from the collector, then leave it to heat up. Empty the reservoir after each use to prevent the water from turning into an unhealthy bacterial soup; refill using a garden hose.

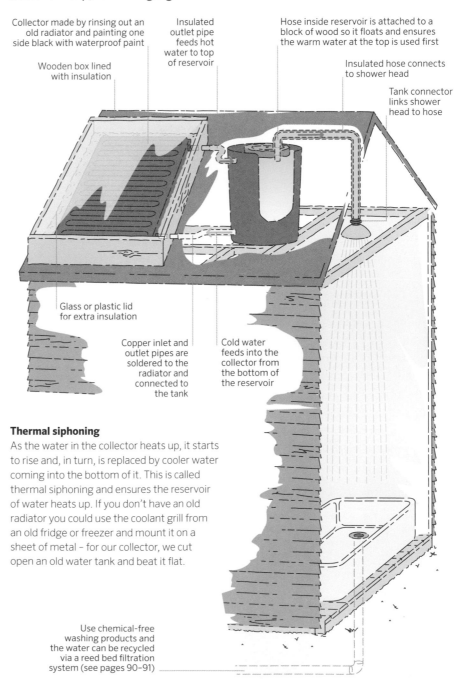

Collector made by rinsing out an old radiator and painting one side black with waterproof paint

Wooden box lined with insulation

Insulated outlet pipe feeds hot water to top of reservoir

Hose inside reservoir is attached to a block of wood so it floats and ensures the warm water at the top is used first

Insulated hose connects to shower head

Tank connector links shower head to hose

Glass or plastic lid for extra insulation

Copper inlet and outlet pipes are soldered to the radiator and connected to the tank

Cold water feeds into the collector from the bottom of the reservoir

Use chemical-free washing products and the water can be recycled via a reed bed filtration system (see pages 90–91)

Thermal siphoning

As the water in the collector heats up, it starts to rise and, in turn, is replaced by cooler water coming into the bottom of it. This is called thermal siphoning and ensures the reservoir of water heats up. If you don't have an old radiator you could use the coolant grill from an old fridge or freezer and mount it on a sheet of metal – for our collector, we cut open an old water tank and beat it flat.

HARNESSING WIND ENERGY

Wind power ultimately derives from the action of the sun warming the earth and creating weather patterns. Here in the UK we benefit from strong winds, mainly caused by depressions coming in from the Atlantic Ocean. A wind turbine uses that wind to create electricity: the wind turns the blades, which rotate a series of magnets past coils of copper wire, which generate electricity.

WIND ON YOUR PLOT

Jutting out into the Atlantic Ocean, Cornwall is a windy place, and at Newhouse Farm, we harnessed this free energy with a grid-linked turbine that provided mains electricity for the house and a stand-alone turbine that we used to charge batteries in order to power a pump for the spring. The term "stand-alone" simply means that this turbine wasn't connected to the national grid.

The stand-alone turbine works with a solar PV panel (see pages 70–71) as part of a dual system. This underlines the first thing to bear in mind when considering a wind energy option: wind is not classed as "firm power" as it does not always blow. As a result, wind power has to be part of a mix of generation capabilities, or else you have to store the energy you harvest in batteries (see pages 66–67) for times of low or no wind.

There are two other important things about wind that you should know if you are considering installing a wind turbine. **Wind speed matters** – lots. The power you can get from a wind turbine is proportional to the wind speed cubed (i.e. speed x speed x speed). You can get hold of information about average wind speeds in your area by doing an internet search and then use these figures to see if it's worth your while installing a wind turbine. Even if you don't get a lot of wind, with the right kind of turbine it's still possible to generate electricity.

Local wind turbulence is another crucial factor. The key to a good output from the turbine is to have "smooth" wind (known as "laminar") with no turbulence (see below). A wind turbine will never be efficient in turbulent air as it will be trying constantly to change direction. The rotor speed also fluctuates, which will reduce the power output of the turbine and increase the wear and tear.

With a bit of preparation you will be in a good position to make that most crucial decision of where to locate the turbine. Get it right and whenever the wind blows you'll be rubbing your hands with glee at all that free power. Get it wrong and you will get disappointing results and waste a lot of time, money, and effort.

COMING TO GRIPS WITH TURBULENCE

Turbulence is caused by obstacles on or near the ground, which change the wind from a smooth flow from one direction into vortices. These cause the wind to rapidly change speed and direction, which is not a good thing for a wind turbine (see below left). As a result, a turbine on the coast will probably get "better air" compared to one in the middle of a city, where there are many more obstacles causing turbulence. But local deviations in the strength and laminar quality of the wind can be huge, even out in the countryside where you'd think air flowed freely. The illustration below, right, shows how obstacles can effect turbulence. As a general rule, to avoid turbulence, plan for a turbine to be 6–9m (20–28ft) higher than any obstacle within 100m (330ft). If in doubt, higher is always better.

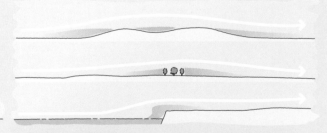

The turbulence caused by an obstacle begins twice its height before reaching the obstacle and carries on for 10–15 times the height after the obstacle, peaking at twice its height.

The grey areas above show where turbulence is created when the wind blows over hills (long areas of turbulence) and over a cliff, where the turbulence is relatively short.

1. The stand-alone turbine stores the energy it creates in a battery. **2. The area around the base** of grid-linked turbines can still be used as pasture land and for grazing livestock.

HOW TO BUY

Not so long ago, commercially bought wind turbines were very expensive and if you wanted a lot of power for your pound, then the only real way forward was to build your own.

Growing environmental awareness, cheaper foreign imports, and increased competition have created a huge market for wind-generation products. This means it's much more economical than it once was to buy an off-the-shelf system and have it installed. On the other hand, building your own brings huge rewards and is a lot of fun. You'll find it's relatively easy to source all the components you need (see pages 76–77).

A quick internet search will reveal most of the commercial turbines that are available. The market is constantly changing, so to make a good buy you need to check out all the claims made by turbine manufacturers, especially claims about their rated power. Manufacturers will claim different power outputs that are based on different wind speeds, so they become difficult to compare. For example, a turbine rated 500W at 8m (26ft) per second is almost the same as one that claims 1,900W at 12.5m (40ft) per second.

The best thing to do is to ask for the power curves based on how you will use the turbine, for example to charge a 24V battery. Power curves show the power supplied by the turbine at different wind speeds. This will then allow you to compare like with like.

HOUSE-MOUNTED SYSTEMS

There are systems that mount onto the side or the roof of your house. For people in urban and suburban locations this may be your only option and, if so, it is still possible to obtain reasonable performance using one of these systems. This isn't the way we'd go, however, and you need to be aware of a few issues relating to roof-mounted installations:

Turbulence can have a negative effect. Just imagine the kind of turbulence around a huge obstacle like a house – a roof-mounted turbine is located in this zone of turbulence.

Wind turbines create noise. If a turbine is mounted directly on to your house, the sound may be very noticeable and could become a source of irritation, not only for you but also for your neighbours.

SITE CHECKLIST

Do you understand how the wind blows on your plot?

Have you found out the wind speed for your area?

Have you established the direction of the prevailing wind?

Have you identified topographical features that form obstacles and create turbulence? Have you worked out how they might be overcome?

Have you looked into planning permission? Virtually all wind turbines need this and the following factors need to be investigated before planning permission is granted: proximity to power lines, airports, roads, and railways; shadow flicker; noise; electromagnetic interference; siting and the landscape; ecology, archaeology, and listed buildings; and disturbance during construction.

Components of a domestic wind turbine

If you are to harness the wind, you will require all of the items that are discussed on these pages. If you are intending to build your own turbine, most parts can be bought on the internet at very reasonable prices. This allows you to customize the turbine to your own needs, for example, if your location isn't very windy, you may need bigger blades.

Flat farmland can be especially good for siting wind turbines as it has good laminar air flow.

BLADES

A very clever chap called Betz calculated that, even with the best designed blade in the world, you can only ever harness 59 per cent of the wind's energy. So, if you can harness only some of the energy, it is important that you have good blades that are as big as possible.

Designing blades is a little complicated and the choices vary: cheaper blades have a constant angle of attack and width (rather like a plank), while complex ones vary in width and look twisted. The better the blade, the more efficient your turbine. The turbine will also be less noisy, especially in high winds, and it will be easier to start the blades spinning.

ELECTRICAL GENERATOR

It may not seem like it when you look at a wind turbine, but the blades rotate relatively slowly with only a couple of hundred rpm, even in high winds. Because of this low rpm, there is really only one type of electrical generator to use: a permanent magnet alternator (PMA), which can start generating useful power at as little as 50 rpm and can have an efficiency of 90 per cent. A few points worth noting:

Buy a multiphase alternator. This produces continuous power resulting in much reduced "cogging". Too much cogging can make it hard to start turning the blades.

Consider efficiency. Check out the power curves of any PMA that you are thinking of using.

Check for rectifiers. If you intend to use the turbine to charge batteries, you will need to change from AC to DC (see opposite). Some manufacturers of PMAs and wind turbines have rectifiers built in, while with others you have to buy or make your own at extra cost.

HIGH-SPEED PROTECTION

It is essential that all wind turbines have an emergency shutdown device to prevent them being damaged in high winds. If a turbine over-spins, due to either heavy weather or a malfunction, there is a risk of mechanical damage. If a blade breaks off, it can travel a long way and potentially injure someone.

Furling systems

The term "furling system" describes a system that turns the rotor of the turbine away from the direction of the wind, either horizontally or vertically, to prevent damage. Furling not only protects the turbine from over-spinning, but will also reduce excess loading on the tower and its supports. Check out these designs:

Variable pitch blades form the best but most complex system. The blades automatically adjust their pitch, depending on the wind speed.

Tail furling involves mounting the turbine and tail to the side of the yaw bearing and hinging the tail onto the turbine body.

Tilt-back furling hinges the turbine so it can tilt backwards.

Blade flexing allows the blades to bend back and twist during high winds.

The parts of a wind turbine aren't numerous, but there are choices to be made for the blades, inverter, and generator. Tail furling (**1**), flaxing blades (**2**), cheap/noisy blades like planks (**3**), well-designed, efficient blades (**4**), electrical generators (**5**), electrical generator with integrated blades (**6**), pure sine inverter (**7**), charge controller (**8**), cheap quasi-sine inverter (**9**), switch-over switch (**10**), and home-made monitoring capability (**11**).

Shutdown systems

These are systems that can be used to manually shut down the turbine for maintenance or weather protection.

Electrical shutdown is the most popular method of shutdown. One advantage of using a PMA is that if the AC outputs are shorted, the PMA becomes very difficult to turn and so stops the blades from turning. A change-over switch is installed to prevent damage to the PMA.

Mechanical shutdown forces the turbine out of the wind or applies a brake to stop the blades turning.

TOWER

There are two main types of tower for installing wind turbines:

A tilt-up tower can be assembled on the ground and then raised into the vertical position. It also has the advantage that it can be tilted for maintenance. However, this type of tower usually requires guys to keep it steady and will therefore take up a larger area.

A fixed tower is assembled in situ, usually using a crane, and the turbine is fixed on to it later. This type of tower can be free-standing and relatively trouble-free. To maintain it, however, you have to climb up the tower or move it to the ground.

Whichever system you go for, it must be extremely strong. A 2.5cm (1in) steel pipe may feel strong if you have 1m (3ft) of it, but it is very easy to bend 10m (33ft) of steel pipe. Some people recommend that you should spend at least as much on the tower as you do on the turbine, and we'd be inclined to agree with them.

ELECTRICAL SET-UP

There are two main types of electrical systems that usually apply to wind-generation installations.

Direct grid-tie systems are usually only an option on larger turbines. There are two main components: the turbine and a grid-tie inverter. These systems can link to the electricity grid in the same way as the solar PV system (see page 71).

Battery storage retains the electricity produced by the turbine in batteries until it is needed. It can, if necessary, then be converted by a suitable inverter to 240V AC to run mains equipment. The advantage of battery storage is that the electricity can be stored and then used, even if the wind stops blowing. Most installations use lead-acid batteries, which is why the voltage of most systems is in multiples of 12V. Forklift or submarine batteries also work well if you can find them. If you intend to store energy in batteries, you will have to ensure the battery bank is close to the turbine.

BATTERY STORAGE SET-UP

Turbines are usually optimized to match the voltage of the battery bank that you will be using, such as 12V or 24V. The turbine usually produces AC electricity, which must be converted before it can be used to charge batteries. Lead-acid batteries always "hold" the voltage generated by the turbine to their own voltage until they are fully charged. When the voltage increases dramatically, it could easily "boil" the batteries (which will ruin them and creates an explosion hazard). To prevent this happening, use a charge controller and a dump load. Heating elements are the best dump loads. Don't use light bulbs or motors as they have a large start-up current, which could damage the charge controller.

The elements of battery storage
The rectifier converts AC electricity to DC, which can then be used to charge batteries.
The battery bank holds the electricity generated by the turbine.
The charge controller monitors the charge state of the batteries and when they are full diverts the incoming electricity into a dump load.
The dump load takes the excess electricity and so must be able to cope with the maximum amount of power produced by the turbine.
An inverter converts DC power from the batteries to AC and will be needed if you want to power the mains electricity, which runs on 240V AC.

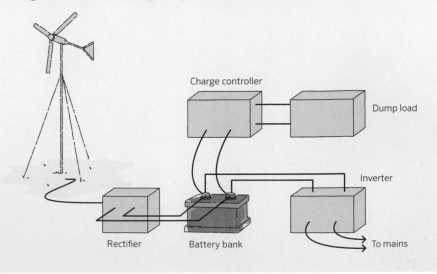

Charge controller

Dump load

Inverter

Rectifier   Battery bank   To mains

HARNESSING WATER ENERGY

Anyone who has carried watering cans back and forth to their crops knows that they are heavy, and therein lies a blessing and a curse. As water flows downhill it has lots of energy which can be harnessed, but to transport it to a convenient location, such as a water tank, takes a fair bit of effort as well as additional energy, which we have to provide.

WATER POWER

The most efficient way of harnessing the energy in moving water is to use a water turbine, which converts water power into rotational power at its shaft. This is then converted to electrical power by a generator.

Waterwheels are covered on pages 82–83, so here we are looking at a hydroelectric system. This is a series of interconnected components, with water flowing in one end and electricity coming out of the other (see opposite page).

GETTING STARTED

Before you can begin planning your system, or estimating how much power it will produce, you will need to make a few essential measurements:

The flow determines how much water is available to use.

The head is the vertical distance between where you plan to start capturing the water and where you intend to use it.

These are the two most important facts you need to know about your site. To measure them, see pages 80–81. You simply cannot move forward without these measurements as they determine everything about your hydroelectric system, including the pipeline size, turbine type, rotational speed, and generator size. Even costs are impossible to estimate without flow and head measurements, as you need them to calculate the potential power output generated by your hydroelectric system.

Once you have obtained your measurements, do an online search for domestic water turbine companies to find those that can either build and install the whole thing for you, or, if you have some engineering experience, supply the components for you to make one yourself.

THE TURBINE

This important piece of equipment is the heart of the hydroelectric system. Its efficiency determines how much electricity is generated. There are two main types:

Reaction turbines run immersed in water and are used in low-head/high-flow systems. Makes include Francis, Propeller, and Kaplan.

Impulse turbines operate in air and are driven by one or more high-velocity jets of water. They're typically used with high-head systems and use nozzles to produce the high-velocity jets. Examples of these turbines include Pelton and Turgo.

There are many types of turbine and making the right selection requires considerable expertise. A Pelton design, for example, works best with a high head, whereas a Crossflow design works better with a low head but high flow.

SITE CHECKLIST

Have you sought permission for your water power system from the Environment Agency? And have you purchased an abstraction licence? You will need this permission if you are re-directing more than 20,000 litres (4,400 gallons) of water per day.

Have you measured the flow and head (see pages 80–81)?

Have you discussed your system calculations with potential installers to see if you have satisfied the minimum requirements?

1. This Pelton turbine is an example of an impulse-type turbine driven by high-velocity water jets. **2. Even small streams** may be able to create enough power to run a turbine.

A hydroelectric system

The major components of a hydroelectric system are a water diversion, a pipeline for creating pressure in the water, a turbine and generator for generating electricity, a tailrace for exiting water, and transmission wires. You need to divert water from a stream by building the diversion where water enters the pipeline that feeds your turbine.

Water diversion

At the start of the system you need to build a water diversion. This is a deep pool of water that creates a smooth flow and removes dirt and debris before it reaches the pipeline.

The pipeline

Water is transported to your turbine via a pipeline that forces the water into a limited space and creates pressure as gravity takes it downhill. Its diameter, length, and route affect the overall efficiency.

The turbine and generator

Your system's efficiency depends on the turbine's design, especially the water entry and exit points. Controls ensure the generator constantly spins at the correct speed. Both the turbine and generator must include an emergency shutdown.

The tailrace

This directs water back to the water course. If part of the tailrace is a sealed pipe below the turbine, it can increase the head.

An expert must match the turbine design to your flow and head. This will ensure that the turbine works efficiently and generates the maximum amount of power.

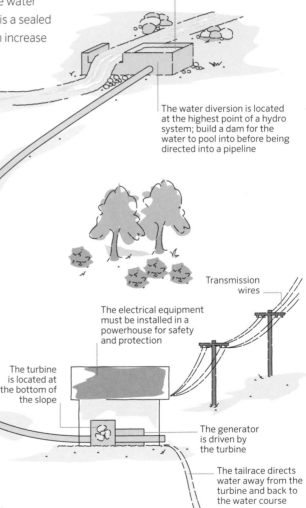

Deep pool of water creates a smooth, air-free inlet to the pipe

The water diversion is located at the highest point of a hydro system; build a dam for the water to pool into before being directed into a pipeline

Avoid using too much water, as plants, fish, birds, and other wildlife rely on the stream for survival.

The pipeline focuses all the water power at the bottom of the pipe where the turbine is connected

Transmission wires

The electrical equipment must be installed in a powerhouse for safety and protection

The turbine is located at the bottom of the slope

The generator is driven by the turbine

The open stream dissipates energy as it travels downstream

The tailrace directs water away from the turbine and back to the water course

Taking measurements for head and flow

Before you set up your hydro system, you will need to measure the flow and head of your water to determine the viability of the system (see page 78). Flow is expressed as volume (litres or cubic feet) per second or minute. Stream levels change through the seasons, so it is sensible to measure flow at different times of the year. Greater flow means greater power. The head is measured as a vertical distance in metres or feet. The higher the head, the greater the pressure – and the greater the available power.

A fast-flowing stream can generate a great deal of energy if you know how to harness it.

MEASURING THE FLOW

METHOD A: MEASURING THE TIME IT TAKES TO FILL A CONTAINER

1. Build a temporary dam with a single outlet pipe. Place a container of a known volume at the end of the pipe. Use a stopwatch to time how long it takes to fill.

2. Divide the container volume by the seconds. For example:
15-litre bucket
Time to fill = 3 seconds
Flow = 5 litres/second.

Temporary dam forces the water through a single outlet pipe

The time it takes to fill the container is the "flow"

Small stream method
This is an ideal way to calculate the flow of very small streams.

METHOD B: MEASURING WITH A FLOAT

1. Measure the average depth of the stream. Lay a plank across the stream and measure the stream depth at 30cm (12in) intervals. To work out the average depth, add all the measurements together and divide by the number of measurements you have made.

2. Calculate the cross-sectional area of the stream. Multiply the average depth you just worked out by the width of the stream. For example, if the average depth is 0.2m (8in) and the width is 1.5m (60in), then the cross-sectional area is 0.3sq m (3.23sq ft).

3. Measure the speed of the stream by marking off a 5m (5yd) length of the stream that includes the point where you measured the cross-section. Place a weighted float that can be seen clearly in the stream a good distance upstream of your measurement area, and then use a stopwatch to time how long it takes to cover the length of your measured section. The stream speed will probably vary across its width, so record the times for different locations and calculate an average. To work out the speed, divide the distance travelled by the time it took. For example, if it took 10 seconds for the float to travel 5m: speed = 5/10, or 0.5m per second (equivalent to 18in per second).

4. Calculate the flow. Multiply the speed by the cross-sectional area.
Flow = 0.5m per sec x 0.3sq m
= 0.15 cubic m per second.
This is equivalent to 150 litres (33 gallons) per second.

5. Account for friction. Because a stream bed creates friction with the moving water, the bottom of the stream moves a little slower than the top. This means the actual flow is less than our calculation. To get a more accurate rate, we need to multiply the result by 0.83:
flow = 0.15 x 0.83 = 0.1245 cubic m per second. This is equivalent to 124 litres (27 gallons) per second.

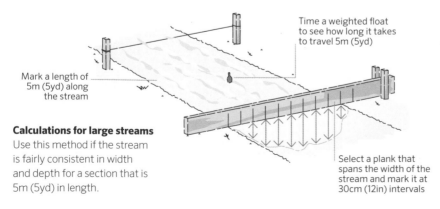

Mark a length of 5m (5yd) along the stream

Time a weighted float to see how long it takes to travel 5m (5yd)

Select a plank that spans the width of the stream and mark it at 30cm (12in) intervals

Calculations for large streams
Use this method if the stream is fairly consistent in width and depth for a section that is 5m (5yd) in length.

MEASURING THE HEAD

METHOD A: USING HORIZONTAL PLANKS AND VERTICAL POLES

1. Decide your pipeline intake point, then measure 1m (3ft) from it and hold a measuring pole vertically at that point. Place a horizontal plank from the intake point to the vertical pole.

2. Measure the height on the vertical pole from the horizontal plank to the ground and note it.

3. Repeat to the bottom of the hill. Ensure each plank starts from where you took the last measurement.

4. Add up all the measurements taken from the vertical pole. The result is the "head".

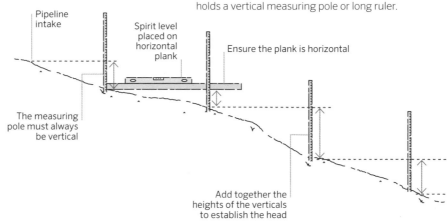

Find a friend
This job is easier with two people. One holds a horizontal plank with a spirit level on it; the other holds a vertical measuring pole or long ruler.

Pipeline intake

Spirit level placed on horizontal plank

Ensure the plank is horizontal

The measuring pole must always be vertical

Add together the heights of the verticals to establish the head

METHOD B: MEASURING PRESSURE

If you measure the pressure in a hose, you can calculate the elevation change of your system. This method relies on the fact that each vertical foot of a head creates 0.433 pounds per square inch (psi) of water pressure (10 vertical feet would create 4.33psi).

1. Run a hose (or hoses) from the intake site to the proposed turbine location.

2. Attach an accurate pressure meter to the bottom end of the hose and completely fill the hose with water. Read the gauge on the meter. If the pressure were to read 4.33psi, the head would be exactly 10ft; if it reads 6.49psi the head would be 15ft.

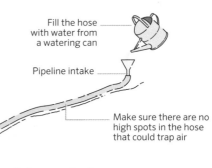

Fill the hose with water from a watering can

Pipeline intake

Make sure there are no high spots in the hose that could trap air

A pressure meter attached to the end of the hose gives an accurate psi reading of pressure

Using a garden hose
If the distance is short enough, you can use one or more garden hoses to measure the head.

HOW MUCH POWER IS AVAILABLE?

Now you can roughly estimate the power output from your system. Note, however, these figures don't account for the inevitable losses from your system, so actual power output will be considerably less.

Theoretical power (W) = flow (litres/sec) x head (metres) x gravity (9.81sq m)

If, for example, a stream has a flow of 150 litres per second and a head of 10 metres, the maximum theoretical power is:

Theoretical power = 150 x 10 x 9.81 watts
= 14,715 watts (14.7kW)

What is brilliant about this figure is that you can see just how much electrical power your hydroelectric system could theoretically make.

14.7 kW for one hour = 14.7kWh.
In a day, this totals 14.7 x 24 = 352.8kWh.
In a year, this totals 128,772kWh.

Assuming an average house uses 3,300kWh, the stream in our example could theoretically generate enough power for more than 40 houses.

Accounting for loss of power
The friction of the water moving through the pipeline and turbine, and the energy loss in the drive system, generator, and transmission lines, account for some loss of power from your hydroelectric system. For example, a home-sized system generating direct AC power may operate at about 60–70 per cent "water-to-wire" efficiency, measured between turbine input and generator output.

BUILDING A WATERWHEEL

Waterwheels are a long-established way of harnessing water power. In fact, British industry was powered by watermills for hundreds of years. The principle is simple: the wheel comprises a series of angled buckets and as the flow of water enters each bucket it pushes the wheel round before pouring out again, and this turning of the wheel can be converted into electrical power.

THE WATERWHEEL SYSTEM

As with water turbines, a waterwheel system must make the most of the quantity of water flowing and the drop it falls through, or "head" (see pages 80–81). Having said that, it is still possible to operate a waterwheel system on a stream without any head, using an undershot wheel design (see box, opposite).

The water source (the flow) can be a lake, river, stream, a constructed millrace or, as in our case, a spring-fed stream that is diverted along an aqueduct. The diversion or sluice is usually the main civil works to be undertaken. Remember, you may need Environment Agency approval and have to pay for an abstraction licence (see page 78).

Aim to have the largest wheel possible to create the maximum drop. The distance from the centre of the shaft to the inside edge of the buckets determines the torque. So, the longer the distance, the more power it creates.

It is also worth considering making the wheel wider to fully utilize the flow of water rather than making the buckets deeper.

To harness the power from the rotating shaft, you can use a direct belt drive, as used to drive machinery; a gear box to increase the shaft speed so it is possible to drive a generator; or a system of cogs and chains or pulleys and belts, which can achieve the same increases in shaft speed as a gear box.

Our waterwheel system

Our waterwheel was designed to provide the power for all the lights at Newhouse Farm. Our wheel was only 4m (13ft) in diameter and had very little flow, so the power it produced was limited to between about 40W under normal conditions and 250W when the stream was in spate.

Converting water energy to electricity requires the following: a rectifier (**1**), ammeter (**2**), charge controller (**3**), dump load (**4**), and pure sine inverter (**5**).

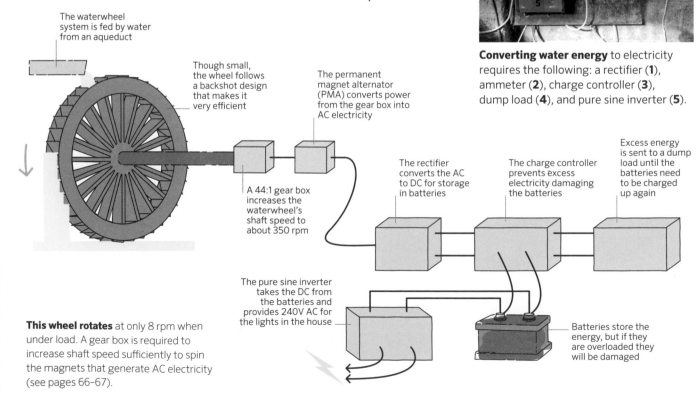

The waterwheel system is fed by water from an aqueduct

Though small, the wheel follows a backshot design that makes it very efficient

The permanent magnet alternator (PMA) converts power from the gear box into AC electricity

A 44:1 gear box increases the waterwheel's shaft speed to about 350 rpm

The rectifier converts the AC to DC for storage in batteries

The charge controller prevents excess electricity damaging the batteries

Excess energy is sent to a dump load until the batteries need to be charged up again

The pure sine inverter takes the DC from the batteries and provides 240V AC for the lights in the house

Batteries store the energy, but if they are overloaded they will be damaged

This wheel rotates at only 8 rpm when under load. A gear box is required to increase shaft speed sufficiently to spin the magnets that generate AC electricity (see pages 66–67).

MAXIMIZING EFFICIENCY

It sounds a bit trite but the one fact you need to remember is that the angle of the bucket should be 114°. You might assume that such an exact figure has a long and complex mathematical derivation, but actually it has been arrived at empirically over hundreds of years, as engineers experimented with waterwheel design. Having unveiled the great secret of waterwheels, it must be said that there are still many more decisions to make and the efficiency of your system will ultimately depend on optimizing the use of the water you have at your disposal. For example, the type of wheel you choose or are able to install, depending on your site, can mean the difference between 90 or 20 per cent efficiency.

WHAT TO DO WITH THE POWER

At Newhouse Farm we stored our energy in batteries (see opposite), so that we could use more than 40W for those periods when we needed lights – mainly early morning and evening. If we generated at 40W for the full 24-hour period, then we could use 160W for a 6-hour period, which is equivalent to about 16 low-energy light bulbs – more than enough for our lighting needs!

As you can see from the details about our waterwheel system, we didn't produce a great deal of electricity. But if you have sufficient power, it may be worth thinking about linking your output power to the national grid. To achieve this you would need to include a grid-linked inverter in your system.

1. If you need to include an aqueduct, it must be as level as possible so you don't lose the energy stored in the water until it is being used to turn your wheel. **2. The waterwheel** is a remarkably simple design, but it isn't always as efficient as it could be (see right). **3. The gear box** is crucial for getting the permanent magnet alternator (PMA) spinning and efficiently generating electricity.

WHEEL TYPES

Your major decision is what sort of wheel to build or install. This depends on the available water flow and head, the possible route for the aqueduct, and the site for your wheel.

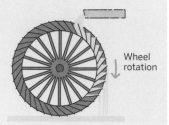

Wheel rotation

A backshot wheel is the most efficient and can convert nearly 90 per cent of the water energy into useful power.

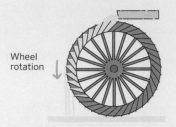

Wheel rotation

Overshot wheels can achieve up to about 70 per cent efficiency when converting water energy into power.

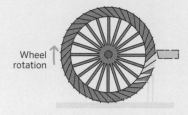

Wheel rotation

Breastshot wheels are useful when there's not much head but only achieve up to 50 per cent efficiency.

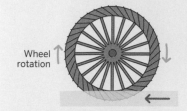

Wheel rotation

Undershot wheels are used when there's no head and are very inefficient (only up to 20 per cent). They're like paddles that are driven by the water flowing under the wheel.

USING A HYDRAULIC RAM PUMP

A ram pump is an ingenious device that uses the water's own pressure to move it to where you need it most – and no electricity supply is required. With a bit of plumbing know-how, you can put one together yourself. We installed a ram pump in our stream to pump water uphill to a bulk container behind our polytunnel, where we used it to water the plants.

WHAT IS A RAM PUMP?

A hydraulic ram, or ram pump, is a water pump – powered by water. When water is flowing down a pipe and is suddenly stopped, the energy in all that moving water is reduced to zero with a thump. This is sometimes called the "hammer" effect. A ram pump utilizes this effect to force a portion of the water that powers the pump upwards, to a point higher than the original water source.

THE MECHANICS

A ram pump is cheap to build, easy to maintain, and very reliable. It has only two moving parts: a spring- or weight-loaded waste valve – or "clacker" valve – and a delivery check valve.

Water flowing into the waste valve eventually forces it to close. Then the hammer effect opens the delivery check valve. This pushes some water into the delivery pipe. Because this water is being forced uphill, the flow slows down. When the flow reverses, the delivery check valve closes and the whole cycle repeats itself.

A pressure vessel cushions the shock when the waste valve closes and improves efficiency.

You also need a delivery pipe to take water to your storage tank, and a drive pipe to bring water from the water source. The pump should run indefinitely unless air gets into the drive pipe or the valves get blocked.

HOW A RAM PUMP WORKS

To power the pump, the water source has to be running water that is higher than the pump. The higher the "head", the higher the water can be pumped – up to 10 times higher than the head. Site the pump so that it has a decent head of at least 2m (6ft 6in).

The flow rate of the source water also affects how much water can be pumped and how fast. A ram pump can be used to supply a household or an entire village with minimal maintenance.

Source of running water

Head of water above ram pump

Drive pipe

Hydraulic ram

Delivery pipe

Water storage tank must be higher than water source

Height water can be pumped

The length of the drive pipe affects how often the waste valve opens and closes – the stroke period.

RAM PUMP COMPONENTS

• Water flows from the drive pipe into the inlet **(1)** and out of the open waste valve **(2)**. The delivery check valve **(3)** is closed.
• The water picks up speed and kinetic energy and forces the waste valve to close.
• The "hammer" effect raises the pressure and forces open the delivery check valve.
• A pressure vessel **(4)** containing air reduces hydraulic pressure shock when the waste valve closes, prolonging the life of the pump. The pump will work without it, but less efficiently, as the vessel also creates a more constant flow rate.
• Water flows through the tap **(5)** into the attached delivery pipe and is forced uphill.

A ram pump works on an endless cycle of opening and closing valves, powered solely by the pressure of water.

Make and install a ram pump

Building a ram pump entails joining different widths of copper pipe. The simplest way is to use reducers and make compression joints. To carry the water to and from the pump we used polyethylene piping. The choice of materials and pipe sizes is flexible, depending on what size valves and joints you have.

YOU WILL NEED

Copper piping
2 brass T-joints
Brass one-way valve
Brass spring-loaded
  one-way valve
Delivery check valve

Brass end cap
Adjustable spanners
Filter and mesh cage
Pliers
Tap
Polyethylene piping

ASSEMBLING THE RAM PUMP

1. Make a compression joint. Slide the end nut over the end of the pipe. Slide the olive on to the end of the pipe. Push the pipe into the connector. Hand tighten the nut onto the connector. **2. Tighten compression joints** with two adjustable spanners, twisting connector and nut in opposite directions. **3. Use the same technique** to join the pressure vessel – a length of 28mm (1in) copper pipe sealed with the end cap. **4. Join the delivery check valve** in the same way.

ATTACHING THE PUMP TO THE WATER SUPPLY

5. Attach a filter to the drive pipe to stop debris clogging the system. Ensure it is submerged – we tied ours to a post under the water. Fit an additional coarse mesh basket to stop larger bits of debris sticking to the filter. **6. Run the drive pipe** from the water source to the pump. **7. Connect the drive pipe** to the pump. Water should flow freely through the waste valve. **8. Fix the ram pump upright.** We built a frame from spare timber and lashed the pump to it.

9. Unscrew the waste valve to adjust the spring. Cut the spring down using pliers if it won't close, or add a stronger spring if it won't open. **10. Connect** the delivery hose. **11. Push in** the "clacker" element of the valve with a finger, to charge the ram and start normal operation.

CONSERVING WATER

We take a plentiful supply of clean water for granted in the developed world, but in recent years droughts and a less-than-efficient capture and distribution infrastructure have led to hosepipe bans in some areas. Combined with rising water rates, this has made us aware that drinking water should be valued and used more sparingly. Here are some ideas for conserving this precious resource.

USING LESS WATER

Treating water so it's safe to drink when it comes out of the tap takes time and energy. Water is heavy, too, so moving it around is significant. To get an

Most homes now have dual-flush toilets, but it is worth converting any with old handle types to save water.

idea, fill an upstairs bath from an outside tap using buckets. Now imagine the "work" or energy needed to move water to your bathroom from the nearest water treatment station, which is miles away.

So before you even think about harnessing your own water supply or using rainwater, try to reduce wastage. Simple ways include:

Take showers not baths, but do not use power showers which can use nearly as much water as a bath.

Dual-flush toilets. Devices can be fitted to older loo handles so the cistern flushes only when the handle is held down. This means you can use as much or as little water as you wish. Or why not pop a water saver into the cistern (see below)?

Turn off the water when brushing your teeth and put in the plug when you start running water in the sink or bath, then adjust the temperature accordingly.

Keep a jug of water in the fridge to avoid wasting water by running the tap until the water's cold enough to drink.

Use a stiff wire brush and a bucket instead of a hose when cleaning outside paths and driveways.

Aerated taps and shower heads add air to the water to give the effect of using more water for less.

Dishwashers and washing machines should be run on full load and economy settings. An efficient machine may cost you more initially but it will save money throughout its life, especially when water and electricity rates inevitably increase.

PROJECT DIY loo water saver

You can buy a ready-made water saver, but it's just as easy to make your own from a single-use plastic bottle. Put it underneath the large cistern float. When the toilet is flushed, the water sitting in the saver isn't used in the flush, saving that much water.

YOU WILL NEED
Single-use plastic bottle
Scissors
Pebbles

1. To make a water saver, cut the top off a 1 litre (1¾ pint) single-use plastic bottle.
2. Add pebbles to the base to weigh it down.
3. Stand it upright in the toilet cistern.

REUSING GREY WATER

Waste water from all sources other than toilets is known as grey water (sewage is sometimes referred to as black water). Many grey-water recycling systems collect and treat waste water from showers, baths, and wash basins rather than the more contaminated water from washing machines, kitchen sinks, and dishwashers. It can then be reused in other systems where you do not need drinking-water quality –

flushing toilets and watering gardens, for example.

Before deciding to install a grey-water recycling system you should compare how much you are likely to generate (which depends on the number of baths and showers taken) with your demand for reclaimed water (which depends on the number of toilet flushes or volume required for the garden). Only then can you calculate the potential savings.

Treating grey water

You can't simply collect grey water in a tank for reuse. It doesn't take a lot of imagination to picture what happens if water containing dead skin, hair, etc., is left to stand for a period of time. Treating the water is essential, and can be based on either physical or biological filters, or a combination. You also need somewhere for a storage tank of an appropriate size. The systems are expensive and need energy to power them.

Compost loo

A typical compost loo has two chambers: while one is in use the other is composting the contents. When the second chamber is nearly full, you can use the compost from the first compartment. For composting to work properly, urine must be separated from solid waste. There are absolutely no bad odours and a clever fly trap takes care of pests.

Instead of flushing sprinkle half a cup of sawdust down the loo and – very important – close the lid.

Composting loo in action
Even after four years of use, the first compartment in our compost loo at Newhouse Farm wasn't full (we made it slightly bigger than necessary – oops).

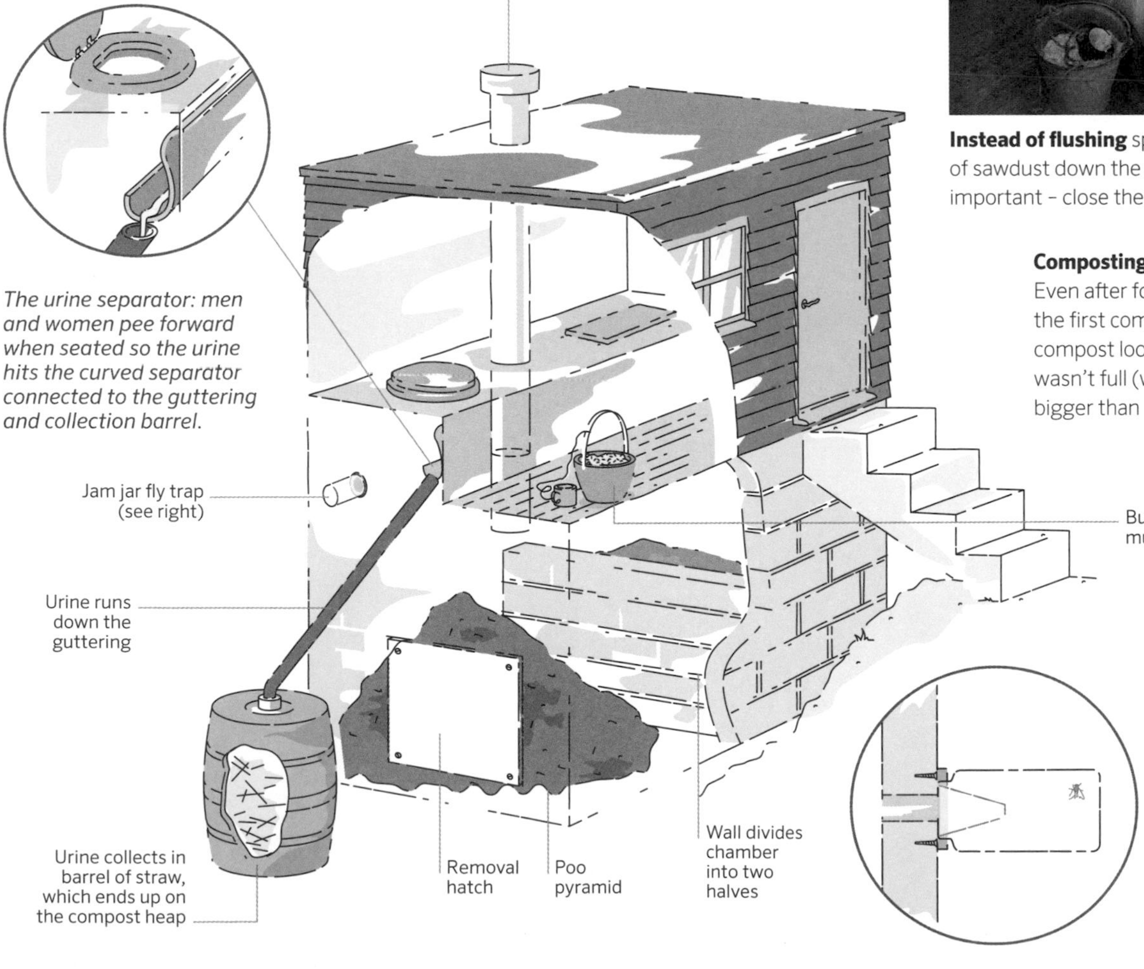

The urine separator: men and women pee forward when seated so the urine hits the curved separator connected to the guttering and collection barrel.

Jam jar fly trap (see right)

Urine runs down the guttering

Urine collects in barrel of straw, which ends up on the compost heap

Air vent

Removal hatch

Poo pyramid

Wall divides chamber into two halves

Bucket of sawdust and mug on a string

The fly trap lures flies towards the light. Once they've gone through the plastic funnel into the jam jar, they can't get back out. The trap won't work unless the loo seat is closed.

COLLECTING YOUR OWN WATER

If there is a spring, stream, or even an old well on your plot, you can be far less reliant on mains water. And even if you don't have your own water source, it's possible to harvest rainwater. Before you look at water treatment systems, decide what you will use the water for. If you plan to use it only to flush the toilets, simple filtration will remove debris. If you plan to drink it, you'll need a purification system.

HARVESTING RAINWATER

In the UK there is no real shortage of water; in fact, the rain tends to be a regular topic of conversation. We measured the surface area of the roof on our house, researched the average rainfall, and calculated that about 207,000 litres (45,534 gallons) of water fall on our roof every year – never mind the outbuildings! When we realized that even our little shed received more than 4,000 litres (880 gallons), the logic was obvious. Even if you don't want to invest in a whole system (see below), you can at least install a couple of water butts.

Unfortunately, rain doesn't fall regularly throughout the year. If you want to use rainwater you will have to store it. A big underground tank is expensive, but fitting a system that isn't big enough to meet your needs is a false economy and means you will end up using mains water as a back-up more often.

On average we use 153 litres (34 gallons) of water per person per day in the UK (in the US it is about 489 litres or 108 gallons), which is shocking when you consider we only drink about 2 litres (3½ pints) each.

WELLS AND SPRINGS

One of the reasons we moved to Newhouse Farm was that it had a spring. We realize not everyone is fortunate enough to have such an accessible water source, but you may be able to find water by drilling down to the water table, or if you have an older property, you may find that you have an old well that has been capped.

Rainwater harvesting system

Before designing a system, decide how much water you need to store. Check your water meter to see how much you use in a day and calculate what size tank to install.

Filtering rainwater

A vortex rainwater fine filter in underground piping removes debris and diverts 90 per cent of clean water to the storage tank. It consists of a fine-mesh filter in a polypropylene housing and needs cleaning regularly.

Rainwater is best used for supplying household appliances rather than for drinking water. It would need more complex filtration and purification systems to make it safe to drink.

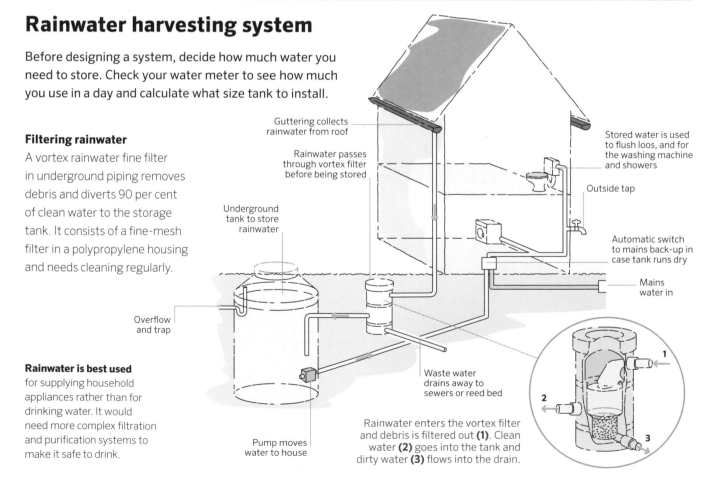

Guttering collects rainwater from roof

Rainwater passes through vortex filter before being stored

Underground tank to store rainwater

Overflow and trap

Pump moves water to house

Stored water is used to flush loos, and for the washing machine and showers

Outside tap

Automatic switch to mains back-up in case tank runs dry

Mains water in

Waste water drains away to sewers or reed bed

Rainwater enters the vortex filter and debris is filtered out (**1**). Clean water (**2**) goes into the tank and dirty water (**3**) flows into the drain.

Types of pump

In a shallow well or spring, you can use a pump that floats on the water surface. But if you don't know the level of the water or don't have easy access to it – for example, in a bore hole or a deep well – use a submersible pump.

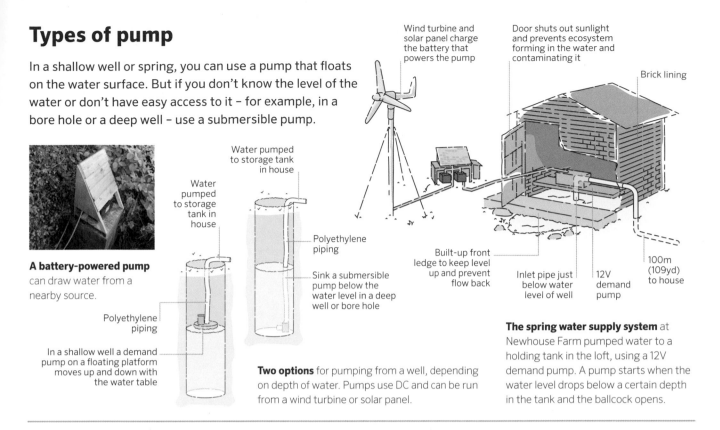

A battery-powered pump can draw water from a nearby source.

Polyethylene piping

In a shallow well a demand pump on a floating platform moves up and down with the water table

Water pumped to storage tank in house

Water pumped to storage tank in house

Polyethylene piping

Sink a submersible pump below the water level in a deep well or bore hole

Two options for pumping from a well, depending on depth of water. Pumps use DC and can be run from a wind turbine or solar panel.

Wind turbine and solar panel charge the battery that powers the pump

Door shuts out sunlight and prevents ecosystem forming in the water and contaminating it

Brick lining

Built-up front ledge to keep level up and prevent flow back

Inlet pipe just below water level of well

12V demand pump

100m (109yd) to house

The spring water supply system at Newhouse Farm pumped water to a holding tank in the loft, using a 12V demand pump. A pump starts when the water level drops below a certain depth in the tank and the ballcock opens.

We had our spring water tested regularly by our local council's environmental health department, just in case it became contaminated somewhere down the line. We used it for everything except for the cold water tap in the kitchen, which was mains water, and we also had a mains water back-up system.

FILTRATION

There are a number of ways to filter water, each with varying degrees of effectiveness and different costs, but all with the same general purpose of improving the hygienic and aesthetic qualities of the water.

Activated carbon filters have been around for a long time and work by absorbing and removing unwanted compounds. They have a very large surface area of highly porous material that attracts and holds chemical pollutants. They are used primarily to improve taste and odour.

Water distillation involves heating the water to boiling point and then condensing the steam. An obvious drawback to this system is that it requires a large amount of energy. Generally, the distilled water is very high quality, but tastes flat as there is less oxygen dissolved in it. You can make a solar still to distill water, using the sun as an energy source.

Sand-based water filters have been used for more than 100 years to treat waste water. They can be employed on a large scale to treat a water supply for a whole community, or they can be scaled down to suit an individual household (see pages 90–91). Most require a constant flow of water to work correctly.

Reverse osmosis filters force water under pressure through a semi-permeable membrane. They allow water through but filter particles such as bacteria, toxins, and salts.

Ultraviolet (UV) filters kill the majority of bacteria and viruses in the water that passes through them. However, they won't remove chemical pollutants. For a UV filter to work effectively, the water must be filtered first to remove any solid particles, so there is nothing for bacteria and viruses to hide behind and avoid being zapped.

In a typical system, UV radiation from a lamp passes through a special quartz-glass sleeve and into the untreated water, which flows in a thin film over the lamp. The glass sleeve keeps the lamp at an ideal working temperature of 40°C (104°F).

UV treatment does not remove organisms from the water; it merely inactivates them. The intensity of the lamp decreases over time, and it needs to be replaced regularly. UV treatment is an effective technique, but the disinfection only occurs inside the filter unit, which means that any bacteria introduced afterwards can be an issue.

USING SAND FILTERS AND REED BEDS

If you are lucky enough to own land that has a spring you can use for drinking water, you will need to filter and then purify it first with a sand filter. How about building your own sand filter? Then you will need to deal with the waste water and sewage your household produces. This is where reed beds come in – they're environmentally friendly, as well as being truly functional.

USING A SAND FILTER

Sand filters require no energy or chemicals to filter water, and need very little maintenance. They filter out small particles, and remove more than 90 per cent of bacteria, making them an excellent primary filter. It's a good idea to use an ultraviolet filter afterwards (see page 89). Sand filters are generally used in tandem with a water storage tank, which produces a constant flow of water.

Before setting up a sand filter and drinking water from a well or spring, ask your local environmental health department to test the water to see if it will be suitable. If their analysis shows traces of metal and chemical pollutants, you may want to use a reverse osmosis filter too. If you intend to use more than 20,000 litres (4,400 gallons) a year, contact the Environment Agency for details of abstraction fees.

USING A REED BED

A vertical-flow reed bed is a sealed, gravel-filled trench with reeds growing in it (see opposite). The common reed (*Phragmites australis*) oxygenates the water, which helps to create the right environment for colonies of bacteria to break down unwanted organic matter and pollutants. The reeds also make the bed attractive to wildlife: birds, amphibians, predators like herons and grass snakes, and colourful dragonflies and butterflies in summer.

Before building a reed bed, you need to notify the Environment Agency and your local planning department. Then install a settlement chamber to separate the solids from the liquid effluent. An existing septic tank is fine, or you can make a new, similar chamber. You'll also need to deal with the solid sludge that separates out, just as with a septic tank.

The downside to reed beds is that they use up lots of space and they take quite a long time to produce clean water.

Final treatment stage

If you are intending to drink the water or give it to your livestock, or if it is issuing into a stream, we recommend a two-stage water treatment system: a vertical-flow bed (see opposite) for the initial treatment, and then a horizontal-flow bed for the final stage of treatment.

The horizontal-flow bed is a lined shallow bed that produces an oxygen-depleted environment to enable bacteria in the reeds' root zone to break down nitrates. The water runs through the bed like a stream and you can plant it with bog plants such as yellow flag (*Iris pseudacorus*) and marsh marigold (*Caltha palustris*). The purified water is clean enough to run into a fish pond or soakaway.

1. Reed-bed filters are based on wetland ecosystems that naturally purify water flowing through them. They can be used as a finishing treatment in sewage plants or on a domestic scale if you have space.
2. The common reed (*Phragmites australis*) grows to about 2m (6ft 6in) high. In wetlands, it forms an important habitat for insects and birds. Buy your plants from specialist suppliers or grow from seed, but don't take them from the wild.

How a sand filter works

Raw, untreated water, known as the "supernatant", passes through the filter from the top to the bottom of the tank, under pressure from the water above the sand. This process takes up to two hours, and, generally, the smaller the grains of sand, the more effective the filter.

Filtering the water

Algae grow on top of the sand and form a sticky net that strains out large particles. Small bits of unwanted organic matter form a green slimy layer in the top 2cm (¾in) of sand, which is eaten by protozoans and bacteria.

Cleaning the filter

Drain and clean the filter every three months. Take off the top 2cm (¾in) of sand; rinse it clean and return it to the filter.

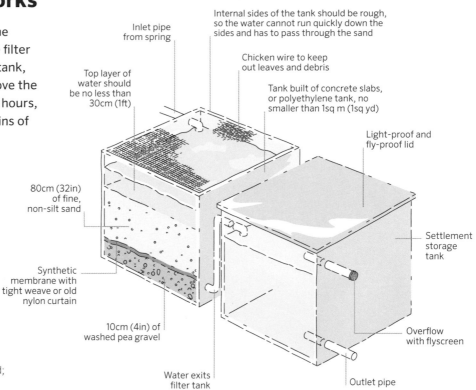

Inlet pipe from spring

Internal sides of the tank should be rough, so the water cannot run quickly down the sides and has to pass through the sand

Chicken wire to keep out leaves and debris

Top layer of water should be no less than 30cm (1ft)

Tank built of concrete slabs, or polyethylene tank, no smaller than 1sq m (1sq yd)

Light-proof and fly-proof lid

80cm (32in) of fine, non-silt sand

Settlement storage tank

Synthetic membrane with tight weave or old nylon curtain

10cm (4in) of washed pea gravel

Overflow with flyscreen

Water exits filter tank

Outlet pipe

How a vertical-flow reed bed works

Vertical-flow beds work by gravity and need a fall of more than 2m (6ft 6in). Their effluent flows onto the surface of the bed and percolates slowly through the layers into an outlet pipe, which leads to a horizontal-flow bed. There is no standing water, so there should be no unpleasant smells.

Digesting the waste

Waste water flows through layers of sand, reed roots, gravel, and stones, and is cleaned by millions of bacteria, algae, fungi, and microorganisms that digest the waste, including sewage.

Getting the size right

Design your reed bed to allow 1sq m (1sq yd) per person in the household, and allow for about 100 litres (22 gallons) flow per person per day.

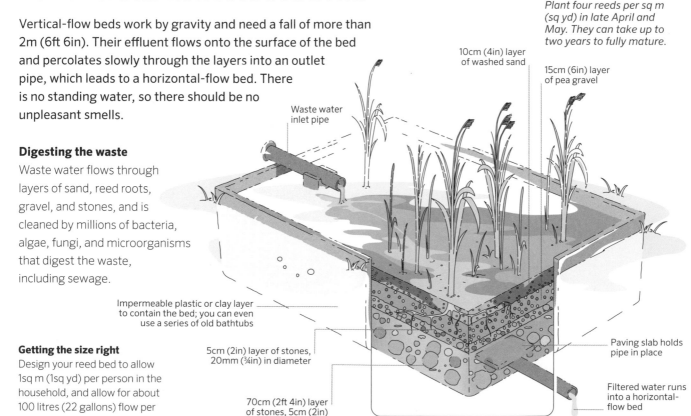

Plant four reeds per sq m (sq yd) in late April and May. They can take up to two years to fully mature.

10cm (4in) layer of washed sand

15cm (6in) layer of pea gravel

Waste water inlet pipe

Impermeable plastic or clay layer to contain the bed; you can even use a series of old bathtubs

Paving slab holds pipe in place

5cm (2in) layer of stones, 20mm (¾in) in diameter

Filtered water runs into a horizontal-flow bed

70cm (2ft 4in) layer of stones, 5cm (2in) in diameter

USING BIOFUELS

The simplest definition of biofuels is any fuel that is obtained from a renewable biological resource. This distinguishes them from fossil fuels, which were laid down millions of years ago and are considered non-renewable. Some examples of biofuels are wood (in its many forms), elephant grass, ethanol, methanol, biodiesel, and even animal waste.

WOOD

As a fuel source, wood is sometimes described as "biomass", which means vegetable mass that is used as a source of energy. Biomass is sub-divided into woody biomass and non-woody biomass, which includes materials like animal waste, high-energy crops (sugar cane, rape, and maize), and some biodegradable by-products from food processing.

The most common type for domestic use is woody biomass, which can fuel a range of appliances, from open fires to ultra-modern pellet-burning boilers and room heaters. By far the best wood is free wood, though it usually takes considerable effort to cut it to size and season it. However, even if you buy your timber for fuel, it is usually significantly cheaper than oil or gas, and, if it is sourced locally, there's the additional environmental benefit in lower transport miles, emissions, and pollution.

Apart from the sun's rays, open fires are probably the oldest form of heating known to man. Everyone knows that an open fire forms a mesmerizing focal point, but it is extremely inefficient, with only about one fifth of the heat generated from the fire being radiated into the room, while the remainder goes to waste up the chimney.

Woodburners

Unlike fires, modern woodburners can be about 80 per cent efficient, which means the vast majority of the available heat warms the room. In real terms, putting three logs on a woodburner for the evening will provide you with a similar amount of heat as putting 12 logs on an open fire. As if that isn't enough reason to invest in a woodburner!

A well-designed burner is very efficient, pre-heating the air that is drawn in prior to combustion so that the flow of gases will ensure that as complete combustion as possible takes place. The high temperature and complete burning that takes

CHOPPING WOOD

You can use either an axe or a hydraulic log splitter to split rounds, but in our view there is no competition. A hydraulic log splitter makes short work of splitting logs and is much safer than wielding an axe (although you must remember to wear goggles at all times). When we lived in Newhouse Farm, our log splitter was ideal for our needs.

Before investing in a hydraulic log splitter, you need to think about a few factors, such as:
Power/size – ours had a seven-ton push, which was enough for us.
Vertical or horizontal design – our log splitter operated horizontally, which suited us perfectly as we would never split logs too big to lift onto it.

If you choose to use an axe, it's useful to have a large block of wood set at a convenient height, a sharp axe, and a sharpening stone to hand. A couple of productive hours of chopping will supply a few months' worth of kindling.

Never hold the wood you intend to chop; instead, hold it steady using another offcut while you cut into it with an axe.

A hydraulic log splitter is relatively expensive, but in the long run it will save you money, time – and your back.

ECOFANS

We use a clever device called an ecofan to circulate the warm air around our woodburner. It uses the Seebeck Effect: when two dissimilar semiconductors (p-type and n-type) at the same temperature are connected together they establish a static electric potential difference. When the bottom of the fan sits on the hot woodburner and the fins at the top are cooler, electrons flow (a current). This current powers the fan's motor, with no batteries or cables required!

An ecofan on a log burner circulates the warm air.

Modern domestic boiler systems can achieve efficiencies of nearly 90 per cent because they are computer controlled to achieve the best results. For wood pellets to compete with fossil fuels in terms of convenience, automatic methods of delivering pellets to the burner have been developed. This greatly increases the time between refills.

Wood-chip boilers

Wood-chip boilers can be as flexible as wood-pellet boilers. Indeed, it is probably easier to source chips locally than pellets, but therein lies some of the issue; fuel quality may vary, depending upon your source.

ELEPHANT GRASS

Growing elephant grass (*Miscanthus*) for fuel is relatively new in the UK. These crops are essentially biological solar panels and batteries as they capture sunlight and store it in a form that can be readily harvested and used as fuel. Companies will plant the grass for you with specialist equipment on non-prime agricultural land, and after a year of fairly careful weed control and a bit of fertilizer, the grass should be established. In a couple of years, it grows up to 3m (10ft) tall annually. Sell your crop to power stations, or commercial combined heat and power (CHP) systems are available.

place in one means less residue collects in your chimney so it needs sweeping less frequently.

Commercial woodburners are usually made from cast iron or steel, but ceramic fires are also available. Look out for the latest DEFRA-approved models (usually labelled "ecodesign ready"), as these should conform to the government's latest regulations on domestic burners.

If you are to use wood as a fuel, it needs to be well seasoned (ideally left for a minimum of two years after cutting in a dry, ventilated store) and it must be free of preservatives, paint, or galvanized nails, as these can all emit harmful gases when burned.

At Newhouse Farm all our heating was provided by four woodburners. The heat was then transferred round the house using a heat recovery ventilation system (see pages 42–45). We only had to empty our ash about once a week.

It's also possible to fuel your central heating with wood by installing a back-boiler attached to certain higher-output woodburners, or by installing one of the new technology wood pellet- or chip-burning boilers. If using a log-fired woodburner with a back boiler, take into account that the burner will not heat the room it is in to the same degree as

it would normally because the heat is being diverted around the pipes and radiators instead. Also be prepared to feed your burner large quantities of wood to make it function properly.

Wood-pellet boilers

There has been a recent increase in the popularity of domestic and small, industrial wood-pellet boilers. The wood pellets are generally the by-products of sawmills, though an increase in demand has led to new plants that produce the pellets commercially. Despite the fact that this processing takes energy, pellets are nearly carbon neutral and much less detrimental to the environment than a fossil fuel-powered system.

1. If you have felled trees on your plot, split them and then leave them to season for a couple of years before burning.   **2. Elephant grass** looks like a crop of maize and yields 10–18 tons of biomass per hectare.

BIODIESEL

Biodiesel is the biofuel equivalent to conventional fossil diesel and it can be made from vegetable oils or, less frequently, animal fats. Biodiesel is not raw vegetable oil, but an ester created by reacting vegetable oil with an alcohol. The scientific name for this is "transesterification" – the reaction of a triglyceride (fat/oil) with an alcohol to form esters and glycerol. At Newhouse Farm we used waste vegetable oil (WVO) – vegetable oil that has been used for cooking – to make our biodiesel.

Pros and cons

There are many benefits to using biodiesel instead of conventional diesel as a fuel:

Biodiesel is nearly carbon neutral: although energy is required to make it, when the vegetable crop was growing it was absorbing CO_2, and when the fuel is combusted roughly the same amount of CO_2 is released.

As biodegradable as sugar, biodiesel is less toxic than table salt.

Biodiesel reduces emissions of tailpipe particulate matter, hydrocarbon, and carbon monoxide from most modern, four-stroke compression ignition engines.

Biodiesel has a very low concentration of sulphur, significantly lower than in traditional diesel and is comparable to ultra low sulphur diesel (ULSD). Sulphur in other fuels can lead to sulphur dioxide being produced and, as a result, acid rain. All this makes it sound like the new "wonder fuel", but there are a few disadvantages to using biodiesel:

Biodiesel starts to coagulate at a higher temperature than fossil diesel, which can be a problem if the fuel is used in cold climates.

On some older diesel engines it has been known to dissolve the rubber seals within the fuel system and engine.

Biodiesel is a great source of food for microbes – add some water and they will have all they need to multiply, resulting in blocked filters.

Commercial biodiesel suppliers now buy most of the available WVO, so a steady supply is becoming more difficult to source.

The all-important reactor

If you're serious about making your own biodiesel, you will need to build yourself a reactor. This is less scary and complicated than it sounds, but it does require a certain amount of engineering know-how and isn't something we can comprehensively cover in this book. What you don't need is a whole lot of specialist equipment; we built ours from a recycled water tank housed in a galvanized metal frame. The maze of pipework underneath is explained on page 96.

Starting the process

Before you begin making biodiesel, it is essential that you first clean and dry your WVO (see box, opposite). Then you can move on to the science. Biodiesel is produced from the reaction of triglycerides (vegetable oil) with

1. Our reactor was made from an old copper water tank topped with a metal dustbin lid. **2. Biodiesel and glycerol** are mixed together when first reacted. They are then left to stand so they separate. **3. The heavy glycerol** is dark compared with the brighter biodiesel. The difference in colour makes it easy to see when all the glycerol has been drained off.

methanol. Unfortunately, if you just mix the two together, nothing will happen: you need a catalyst. In this case, an alkali is needed and those most commonly used are sodium or potassium hydroxide.

To aid the delivery of the solid catalyst into the biodiesel reaction, it is first dissolved in the methanol. The amount of methanol needed is about 20 per cent of the volume of oil used. For example, if you were using 50 litres of oil, then you would need 10 litres of methanol.

There is a standard amount of catalyst needed for the biodiesel reaction – many people agree that it is 3g per litre of oil, although some use 5g without any problems. If WVO is being used, then a titration (chemical analysis) must be carried out to determine the amount of extra catalyst required to neutralize the acidity of the oil. The acidity is caused by cooking in it, which produces free fatty acids (FFAs).

If there are FFAs present, they will react with the catalyst before the biodiesel reaction is complete, which results in a very poor yield of biodiesel, which will be contaminated with unreacted oil. Therefore, you need to add extra catalyst, so that after reacting with the FFAs there is still enough left to make biodiesel.

Carry out a titration

A titration must be carried out on WVO to test for the amount of FFAs present. A titration does not take a lot of time, but it can look a little complicated on paper. The basic idea is to react a small sample of the WVO (which will be acidic as it contains FFAs) with a measured amount of sodium hydroxide (which is a base). When the acid is neutralized, you can assume that all the FFAs have been used up.

Titrations are carried out on a very small scale using accurately measured amounts of chemicals. Steps 1 to 6 on page 97 describe the method in detail.

Local restaurants, school and work canteens, and catering companies are your best bet for sourcing waste vegetable oils, or WVOs. People are usually happy to give you the oil, as it saves them time and money disposing of it, but some do demand a small charge as more of us cotton on to biodiesel.

Any bits of food must be filtered out and water removed by heating the oil. It is very important that all of the ingredients used in the production of biodiesel are as water-free as possible or you will end up with soap. If you are using new vegetable oil, it is already dry.

Pour the WVO through some muslin into a storage barrel to filter out any large pieces of food in the oil.

Allow the oil to settle so that smaller pieces of food sink to the bottom of the barrel.

Draw off the oil from a tap set about a third of the way up the barrel to remove the cleaner oil above the food contaminants.

Dry the oil in your reactor. It is of paramount importance that the oil contains no water. Heat the oil to about 60°C (140°F), then pump it around the system so there's an opportunity for the water molecules to escape. This drying process takes up to 4 hours.

We know that all the FFAs have been used up when the pH rises to about 8.5, which is also the point at which phenolphthalein indicator solution changes from colourless to purple. This is also the point when you stop adding drops of sodium hydroxide.

It is essential that you note how many drops you have used. Each drop from the pipette contains 0.0455 ml or, to put it another way, there are 22 drops in 1ml. Repeat steps 5 and 6 three times and take an average of the result. If one result is totally different to the others, you have probably made a mistake so ignore it and do another one. And that's it. You can then scale up these numbers to work out how many extra grams of catalyst are required per litre of WVO. For example, say the titrating solution

required 2.5ml to use up the FFAs, then an extra 2.5g of catalyst should be used for each litre of WVO. In the example on page 97, we used 1ml of WVO (which is 1/1,000 of a litre) and a solution that contained 1/1,000g of sodium hydroxide per litre. Therefore if 2.5ml of solution is needed for 1ml of oil, then 2,500ml would be needed for 1 litre of oil and 2,500ml of solution would contain 2.5g of sodium hydroxide. Phew!

Once you know the correct amount of sodium hydroxide to add to the methanol for an efficient biodiesel reaction, mix carefully, shaking all the time, and pour into a large container. Then attach to the reactor with a hose, checking for leaks.

Use a reactor to effect a separation

Heat the oil to about 50°C (122°F) and then add the methanol/catalyst solution. Place the lid on the tank so that any methanol that evaporates from the reactor will condense on it and drip back into the tank. After the whole lot has mixed for about an hour, turn off the reactor and leave its contents to settle (see below).

Once the reaction is complete, you are left with a mixture of biodiesel and glycerol. Luckily, the glycerol is heavier than biodiesel and will therefore separate out naturally and sink to the bottom. When the glycerol has settled, drain it off.

Wash the biodiesel

The biodiesel at this stage is by no means ready to use. It will still contain some methanol, sodium hydroxide, soap, and glycerol. These impurities could easily damage parts of a diesel engine and must be removed. Luckily they are all soluble in water, so they can be "washed" out, which we did with a process called "bubble washing", using a fish tank aerator.

Add water to the reaction vessel (we used a hose attached to a funnel so the water went straight to the bottom) in a ratio of about 1 part water to every 2 parts biodiesel.

Lower the fish tank aerator into the bottom of the tank and switch it on to the lowest setting that creates bubbles. Each bubble carries a small amount of water through the biodiesel and when it bursts at the surface, the water drop sinks back through the biodiesel and dissolves the impurities. Allow the biodiesel to wash for 2–4 hours. Turn off the air pump and leave to settle for a further hour. Then drain off the water (use the hose to watch for the colour change again). Each wash takes 2 hours and then the water is drained off. It should take four or five washes until the water comes out clear.

Once the biodiesel has been washed it will be slightly cloudy because it has been saturated in water. Dry it as you did the original WVO (see box on page 95) and avoid breathing in the fumes.

The final product

The biodiesel should now be ready to use. It should be a clear, amber-coloured, pH-neutral nectar. Although it is possible to get about a 98 per cent yield, realistically you can expect 80 per cent plus.

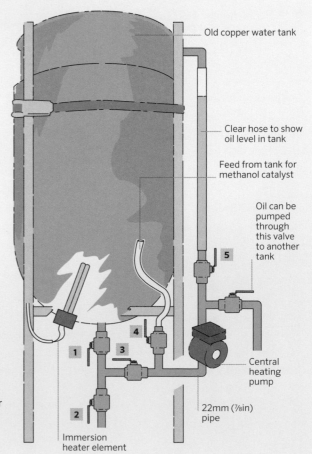

USING THE REACTOR

Our reactor comprised a water tank, immersion heater, and a set of pipes and valves. Be very careful to prevent methanol fumes escaping at the top of the tank. It's best to try to seal the lid and vent the tank to the outside. When making biodiesel, you use a reactor as follows:

Slowly open valve 4 and the methanol/catalyst is sucked in and mixed with the oil.

When the methanol/catalyst tank is empty, close valve 4.

Let reaction mix for 1 hour. Turn off pump and close valve 1.

Some of the mixture will remain in the pump and pipe. Use valve 2 to drain this off and tip it into the top of the tank (you may want to put a bit into a glass jar so you can watch it).

Close all valves and leave the whole lot to separate for about 6 hours.

Drain the glycerol once it has separated out by opening valves 1 and 2. Attach a piece of clear hose below valve 2 and watch for the colour change between the glycerol and biodiesel, which shows that the glycerol has drained off. NB: Oil must always cover the immersion heater element when it is switched on.

If an emulsion forms when washing

Stop the washing and drain the water, leaving the emulsion/biodiesel in the tank.

Close valve 2 and open valves 1, 3, and 5.

Switch on the immersion heater and pump and leave to heat up to 50°C (122°F).

Turn off the heater and pump and close the valves (drain off the pipes and put the mixture back into the tank).

As the mixture cools, the emulsion should break and the water will separate out. Adding a little vinegar to the first wash may help prevent emulsions from forming.

Old copper water tank

Clear hose to show oil level in tank

Feed from tank for methanol catalyst

Oil can be pumped through this valve to another tank

Central heating pump

22mm (⅞in) pipe

Immersion heater element

PROJECT Make biodiesel

Observe all safety procedures when handling these chemicals. Always wear suitable gloves and goggles and read the labels. Wash all equipment and rinse with deionized water before use. Read through the instructions before you start and make sure that you understand them and have everything ready.

1. Mix the sodium hydroxide solution by measuring out 1 litre of distilled water and dissolving 10g of sodium hydroxide into it. Measure out 100ml of the solution and add 900ml of distilled water. Stir and pour this solution into an airtight bottle and label it.
2. Measure 10ml of propan-2-ol and add it to a conical flask.   **3. Using a syringe,** measure 1ml of the waste vegetable oil.   **4. Put it into the conical flask.** Swirl the contents vigorously until the oil dissolves, and then add a couple of drops of the phenolphthalein solution and mix. You must be accurate with your measuring and keep records.

5. To carry out the titration, use a separate pipette to suck up 3ml of the sodium hydroxide solution and add it to the conical flask drop by drop while swirling continuously. Each drop will cause the liquid to turn pink temporarily, and then it returns to pale yellow.
6. When it turns pink and stays pink for 20–30 seconds, stop adding drops. Repeat steps 2 to 6 three times and do the calculations described under "Carry out a titration" on page 95, so you know how much extra sodium hydroxide is needed.   **7. Measure the sodium hydroxide catalyst** and mix it up carefully with the methanol in a large container.

8. Attach the catalyst container to the reactor with a clear hose. Heat the WVO in the reactor to 50°C (122°F) and add the catalyst/methanol mixture. Turn the immersion heater off, but keep the pump running. Follow the process for opening and closing valves described on page 96.   **9. As the catalyst mixes** with the oil and the reaction takes place, the oil turns from clear to cloudy.   **10. Wait for the glycerol** to separate out.   **11. Wash the biodiesel** (see page 96) and repeat the washing until the water comes out clear. Then dry the biodiesel again as described on page 95.

USING BIOFUELS

97

USING AN ANAEROBIC DIGESTER

An anaerobic digester can take any type of organic matter, from food waste to manure and sewage, and convert it into a soil fertilizer and biogas, which can be used as an alternative energy source. Ideal for farms and waste-processing plants, it is also an option for smallholders interested in reducing their food waste output or perhaps even going fully zero-waste (see pages 101 and 116–118).

WHAT IS ANAEROBIC DIGESTION?

Anaerobic digestion is the process whereby bacteria that occur naturally break down organic waste material in an oxygen-free environment. These anaerobic bacteria are part of nature's waste management and are found in soils and deep waters, as well as in landfill sites. Anaerobic digestion converts organic waste into a residue, which can then be used as a fertilizer, and a biogas that can be used as a renewable energy source.

PRODUCING BIOGAS

In the UK we throw away about a quarter of the food we buy. If we add to that the waste from commercial catering and food processing, as well as farm manure and crop waste, it creates an awful lot of material that could be turned into biogas. This organic waste often ends up in landfill sites where it generates harmful methane as it decomposes – methane is 24 times more damaging than carbon dioxide as a greenhouse gas.

You can recycle your waste in a system known as a digester. This creates a controllable supply of biogas, which is a mixture of 40 per cent carbon dioxide and 60 per cent methane, plus other trace elements, and can be harnessed as a renewable energy source. Although anaerobic digestion also releases carbon dioxide, the carbon was absorbed by plants previously and so forms part of a complete carbon cycle. The released gas does not contribute to global warming in the same way as carbon released from fossil fuels, which has been trapped under the ground for millions of years.

TYPES OF DIGESTER

There are commercial anaerobic digesters, designed for farms, which have the capacity to produce 100kW to 1MW of electricity. For our smaller plot, we created a DIY digester (see opposite). You can also use an anaerobic lagoon (see box, below) to produce biogas in a large balloon.

1. The contents of a compost bin heat up, and naturally occurring anaerobic bacteria break down the organic material.
2. The residue or digestate from anaerobic digestion goes back onto the land to improve soil fertility.

ANAEROBIC LAGOON

Like the digester, this anaerobic lagoon creates biogas from your waste organic material; you can buy the materials to make one from specialist suppliers. You add organic matter to a large closed balloon with no air in it – biogas plus air is an explosive combination. When anaerobic digestion takes place, the biogas balloon inflates. Heat enhances and speeds up the reaction.

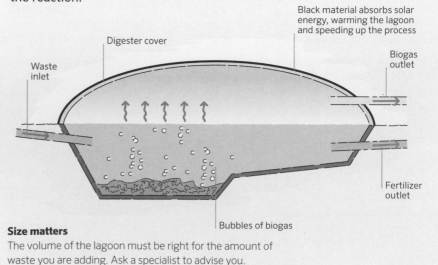

Black material absorbs solar energy, warming the lagoon and speeding up the process

Digester cover

Waste inlet

Biogas outlet

Fertilizer outlet

Bubbles of biogas

Size matters
The volume of the lagoon must be right for the amount of waste you are adding. Ask a specialist to advise you.

DIY digester

We used an old muck spreader as the anaerobic digester in a system that we designed for a farmer friend. He uses it to manage the waste produced on his farm: the biogas is used to power an oil press to make biodiesel, while the fertilizer goes back onto the fields.

You can use biogas to power a generator to produce electricity, or go for a combination of heat and power and run a generator and feed heat back to speed up the digesting process. Collecting solar energy from black radiators is another way of generating the heat needed. Alternatively, simply burn the biogas for heat or use it instead of petrol to power machinery. In India and China, it is very common to build domestic digesters to power gas rings for cooking.

The digestate is an excellent fertilizer and can be applied with a muck spreader.

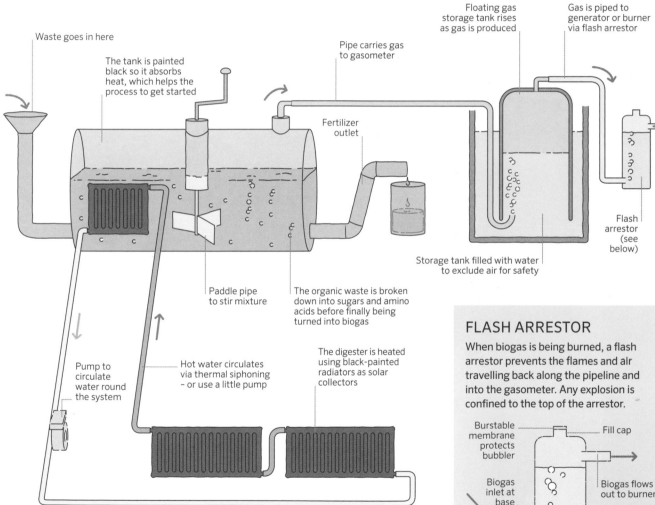

Waste goes in here

The tank is painted black so it absorbs heat, which helps the process to get started

Pipe carries gas to gasometer

Floating gas storage tank rises as gas is produced

Gas is piped to generator or burner via flash arrestor

Fertilizer outlet

Flash arrestor (see below)

Storage tank filled with water to exclude air for safety

Paddle pipe to stir mixture

The organic waste is broken down into sugars and amino acids before finally being turned into biogas

Pump to circulate water round the system

Hot water circulates via thermal siphoning – or use a little pump

The digester is heated using black-painted radiators as solar collectors

Stirring up a reaction
The process can slow down if waste is allowed to separate into layers. It can take up to a month to produce biogas. Stirring the mixture speeds things up. The gas is explosive when in contact with air, so sinking the paddle pipe below the level of the liquid in the digester ensures that no air is introduced.

FLASH ARRESTOR

When biogas is being burned, a flash arrestor prevents the flames and air travelling back along the pipeline and into the gasometer. Any explosion is confined to the top of the arrestor.

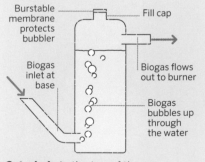

Burstable membrane protects bubbler

Fill cap

Biogas inlet at base

Biogas flows out to burner

Biogas bubbles up through the water

Cut a hole in the top of the cap and place a burstable membrane to limit damage.

REDUCING, REUSING, AND RECYCLING

We are committed to recycling and have got sorting waste down to a fine art. But before we send anything off to be recycled – from old cabling and cardboard boxes to plastic bottles – we try to go one step further and find an alternative use for it around the house or in the garden. After all, one person's waste is another person's treasure.

THE WASTE PROBLEM

Waste has evolved as a result of rapid economic development, but its disposal is not an entirely new problem. Court records show that in Stratford-upon-Avon in the mid-16th century, Shakespeare's dad was fined for "depositing filth in a public street". Later, mass industrialization led to an increase in waste, and in the 1800s dustmen were introduced to collect ash from coal fires.

We're now more aware than ever of the mounting problem of waste, and the pressure it puts on the environment and our planet's finite resources. When we look at a piece of rubbish, we see its whole lifecycle. First, there is the cost and embodied energy it has taken to manufacture the object. Then there's its useful lifespan. If it has been designed for only one use, this is followed by the serious question of, what next?

Any domestic waste that isn't recycled, and instead goes in a black bag of general rubbish, ends up as landfill. Not only are landfill sites now reaching capacity, but they eventually leach toxins into the surrounding soil and farmland. Equally as worrying is the amount of waste pollution that ends up in our rivers and oceans. The waste loop is broken and we all need to take responsibility to fix it.

SINGLE-USE PLASTICS

In order to really make a change to our buying habits, it helps to understand the problems we're facing. All kinds of waste are contributing to the current crisis, but one of the most notorious is single-use plastics.

Thanks to their relatively low cost, plastics have crept into all parts of our lives. Almost everything these days seems to be plastic-wrapped, often unnecessarily. Despite recycling efforts by local governments and incoming legislation against some single-use plastics and micro-plastics (tiny particles that are imperceptible to the naked eye), the reality is that, on a global scale, plastic pollution is having a catastrophic effect on wildlife, especially in our waterways and the oceans.

We all need to do our part to reduce plastic waste and help to be part of the solution rather than part of the problem. By thinking more about how and where we shop, and what single-use items we could cut down on or avoid altogether, we can not only reduce the amount of plastic waste we personally produce, but also help pressure big businesses to swap their plastic packaging for more eco-friendly alternatives.

BUY LESS, WASTE LESS

It might sound obvious, but much of the waste we produce nowadays comes as a result of our shopping habits, particularly

1. Reclaimed beams from an old building made a strong frame for our raised bed (see page 138). **2. Bamboo toothbrushes** can be composted after use, minimizing your single-use plastic waste. **3. Rocks grubbed up** by our pigs were used to build a herb spiral.

when it comes to buying groceries. In our experience, one of the key pieces of advice for reducing food and packaging waste is to plan better. If you check your fridge and cupboards before you go out, and keep to a detailed shopping list, you'll avoid buying unnecessary goods and save yourself time and money.

We always take a range of cloth shopping bags for carrying groceries home, as well as a few smaller reusable bags and containers for delicate herbs, cheese, or deli goods. It's also worth keeping a few sturdy cardboard boxes for carrying heavier items.

GOING ZERO-WASTE

The zero-waste movement is a growing response to the amount of waste (particularly plastic waste) that we

TIPS FOR REDUCING WASTE

The easiest way to cut down on the amount of waste you send to landfill is to avoid accumulating items that you know will soon have to be thrown away. Here are some simple solutions we suggest:

Refusing junk mail by registering online and unsubscribing from mailing lists.
Saying no to free stuff that you don't need. This includes anything from a "free" pen to flyers and gifts.
Buying second-hand or upcycled items (see page 14) rather than buying new.
Reducing the number of shopping trips you make, and buying in bulk whenever you can.
Swapping single-use disposables with re-useable alternatives. We always use a handkerchief rather than a packet of tissues, and stainless-steel water bottles and bamboo straws rather than single-use plastic ones.
Avoiding single-use plastic where possible. If it's unavoidable, make sure you buy plastic that you are certain can be recycled locally.
Keeping leftovers fresh by storing them in longer-lasting stainless steel boxes or glass jars rather than plastic alternatives. Beeswax wraps offer a great reusable alternative to cling film.

each generate in our daily lives. Going zero-waste is exactly what it sounds like: making a commitment to avoid sending any waste to landfill.

Cutting out waste completely requires a radical shift in many people's routines, but it is achievable if you set your mind to it. If you're interested in living a zero-waste lifestyle, we suggest starting by looking at all the waste you currently produce: not just food and packaging, but old clothes, furniture, toys, and so on. Decide if you really need it all, or if there are some items that you could cut down on or avoid altogether. For the rest, see if there are any swaps you could make (see below), and make a commitment to reusing and repairing as much waste as you can. Recycling and composting should be your last resort.

Even if you don't succeed in going fully zero-waste, the changes you do make will help you reduce your overall impact on the environment.

REUSING DIFFERENT MATERIALS

Reusing is a creative, money-saving, and environmentally friendly activity. We reuse all sorts of different materials – it's not only a lot of fun but also an incredibly satisfying process.

Household and office rubbish

Paper can be recycled, but we first use scraps for writing notes, or stick a piece of paper over a used envelope as a label. You can even buy a gadget that turns paper into a "brick" for burning.
Printer ink cartridges in printers can easily be refilled.
Rechargeable batteries can be reused again and again, and save toxic chemicals going into landfill.
Glass jars are excellent for storing your own jams, preserves, dried food, seeds, and chutneys. Clean them thoroughly and keep the right lid next to the right jar.

Building materials

Reusing old timber is hard work. De-nailing, sanding, and planing it can take time and effort, but the satisfaction of turning what could be termed rubbish into furniture, or using it as a building material, makes the effort worthwhile. Plus, old timber is much cheaper than new wood and you can use the off-cuts as kindling in a woodburner. Be aware, however, that it takes much more work to prepare reclaimed timber that has creosote coatings or has been painted.
Copper and plumbing parts are valuable commodities. We have used them for all sorts of projects, from building a ram pump (see pages 84–85) to constructing a solar shower (see pages 72–73). If you don't reuse your copper, make sure you get a good price for it at a scrapyard.
Old cables and electrical components are slightly harder to reuse if you are not an experienced electrician, but replacing and wiring a plug is relatively simple. It's also worth stripping out the lengths of copper inside cables and using it to suspend slug-proof hanging baskets (see pages 20–21).
Ferrous or scrap iron is another great material for reusing in all sorts of projects. We keep a small metal store so that if we need to fix a wheelbarrow or build a spit roast then there is always metal available for us to weld together.

In the garden

Single-use plastic bottles should ideally be swapped for reusable alternatives, but if you do have a few lying around they can be transformed into mini cloches (see page 102).
Glass that's still in a window frame is great for building cloches or making cold frames in the garden (see pages 126–127). Once glass breaks it's harder to reuse, but we have made a heat sink in the greenhouse from broken glass,

maximizing its thermal properties (see pages 130–131). Handle glass with care.

Old guttering makes a perfect tray for sowing pea seeds.

Egg boxes can be composted, but first we use ours for growing seedlings (see page 146).

Rubber tyres can be reused as towers for growing potatoes. Fill the base tyre with compost and add chitted potatoes. When shoots appear, stack another tyre on top and fill with more compost. Add a few more tyres for the perfect pot.

RECYCLING EXPLAINED

We will only consider recycling an item after we've first reused it around the house and garden, because the energy taken to recycle materials is still significant.

We were fortunate enough to be able to follow our recycling bags to the local processing centre, which proved a real eye-opener. Once the lorry arrived on site, our bags were part of an extremely well-choreographed sorting operation. They were emptied on to a conveyor belt and plastic bags were separated by hand before the first of the clever machines did its bit.

A rotating belt with a very powerful electromagnet sucked steel cans about 30cm (1ft) off the conveyor and sent them into a huge cage. Then the main belt passed an area where an electromagnetic vortex flung aluminium cans into yet another cage. Recycling aluminium can save up to 95 per cent of the energy needed to make it from raw materials. Pallets of crushed aluminium

cans are so valuable that they are kept under lock and key.

The plastic bottles were separated into a huge pile and were then fed on to yet a conveyor belt where they were sorted by hand to extract any that were not recyclable. Elsewhere in the plant, paper and card were sorted by hand.

Conventional bags for landfill are also processed at the same site and, as the bags split open, we saw bottles, tins, clothes, and many other items that could all have been recycled. But there's no way to sort through it all once it's reached this stage. Recycling has to be sorted as far as possible in the home.

Get organized at home

For us, a well-organized system and proper segregation makes all the

---

PROJECT # Reuse materials in the garden

Loo roll tubes make great biodegradable pots to start off carrot and other root veg seeds, as you can plant them in the soil, tube and all, without disturbing the roots. You can also give single-use plastic bottles a second life by turning them into mini cloches for individual plants.

YOU WILL NEED

Toilet roll tubes	Seeds
Potting compost	Plastic drinks bottle
Tray	Scissors

BIODEGRADABLE POTS

1. Fill tubes with sieved potting compost after standing them in a tray. Sow one seed per tube. Water gently and leave to grow. **2. Plant out** seedlings when they are growing strongly; bury the tube directly into the soil.

MAKE A BOTTLE CLOCHE

1. Cut the base off a 2-litre (3½-pint) single-use drinks bottle.
2. Place the bottle over a plant, such as a tomato, to make an instant mini cloche. Use the cap as a ventilator and take it off on warm days when condensation collects in the bottle.

difference. Try our top tips (see right) to help you sort out your recycling at home, ready for your local recycling centre. While most houses don't have dedicated recycling areas as part of their design, try to designate a place for sorting recyclable waste that is close to the kitchen, practical to store recycling until collection, and easy for you to carry outside. Try to keep your recycling clean and separate it so that it's easier for the teams at recycling plants to process.

Generally, it's best to send textiles and clothes to charity shops or sell them at car boot sales for reuse, but they can also be recycled if there's no useful wear left in them – just check the rules at your local recycling centre first.

Not all recycling involves processing raw materials. For items like spectacles and shoes that still have plenty of life in

them, look into schemes that will repair your old belongings, ready to be worn by those in need.

Upcycling also offers a fun and imaginative way to give new life to your old belongings, particularly the likes of clothes, home furnishings, and decorative items. More and more artists and craftspeople are making a living by upcycling rubbish into functional, beautiful items. Why not have a go at upcycling your own clothes or furniture?

Some enlightened councils collect food waste separately for processing (see anaerobic digesters, pages 98–99), but the easiest thing to do at home is to compost your raw vegetable and fruit scraps to make a wonderful soil-conditioner (see pages 116–119 for a guide to everything you can and cannot compost).

(see anaerobic digesters, pages 98–99), ... (see pages 116–119 for a guide to everything you can and cannot compost).

RECYCLING TIPS

Recycling can be messy if you don't have a well-ordered system. We have a unit in the kitchen that is labelled, lined with the council's specific recycling bags, and easy to empty. It saves lots of hassle and is a reminder to recycle every time we pass it.

Ready for recycling

Wash tins, cans, and plastics so that they can be more easily sorted down the line and to avoid your recycling area smelling bad.

Plastics need to be sorted into different types – councils recycle some plastics but not others. It is usually easier and more economical for centres to recycle heavier plastics, such as single-use plastic bottles, while plastic film is often rejected. Black food trays are also difficult to process. Wash and flatten bottles and containers before recycling.

Carrier bags can be take to recycling points, often outside supermarkets. Bags made from recycled polythene use one-third of the energy needed for virgin plastic.

Glass can be recycled hundreds of times (unlike plastic, which wears out after being recycled a few times), so try to support businesses selling their goods in this material. Recycle jar lids as for tins.

Paper and cardboard are easy to recycle and reclaimed waste paper now makes up around 63 per cent of the fibre used to produce paper and board in the UK. You can also use it as a mulch or tear it up into smaller pieces for compost in the garden.

Foil-lined drinks cartons and Tetrapaks are hard to recycle, but in some areas you can take them to your local recycling centre.

Flatten cardboard, paper, and cans so that you can fit more in your bin.

1. Reuse old guttering to start pea seeds off under cover. **2. CDs and bottle tops** threaded on string and hung between stakes so that they move in the breeze make effective bird scarers. **3. Turn an old window frame** and some timber into a cold frame. **4. Cardboard boxes** laid flat make a great biodegradable mulch (see no-dig gardening on page 112).

MENDING

We firmly believe in the philosophy "make do and mend" and only throw things away as a last resort. Fixing everyday objects, like clothing and gadgets, will save you lots of money and reduce the polluting waste in landfill. Also remember the old saying "a stitch in time saves nine", which means that it'll be easier to repair an item as soon as you notice it's broken, rather than waiting until it gets worse.

BUYING GOODS THAT CAN BE FIXED

Our throw-away culture and high levels of consumption are made worse by small, plastic, breakable components and over-technical parts that the average person cannot fix. Indeed, many products are now designed with an artificially limited useful life so that it no longer functions after a certain period of time (known as "built-in obsolescence"), forcing you to buy the latest, more fashionable model. Fortunately, consumers are becoming wise to this trick: by buying better quality products less often, it is possible to circumvent this wasteful strategy. If something does break, attempt to repair it instead of throwing it away, or find clever ways to recycle or reuse it (see pages 100–103) to give it a new lease of life. Not only can this cut down on waste, but it can have a positive impact on your pocket too.

REPAIR MADE EASY

We always try to fix something that is broken first rather than reusing or recycling it. Not everyone can fix a computer or car, but there are still lots of things that anyone can mend.

If you've never repaired a particular type of item before, ask family or friends for help – whether that's someone who's handy with a screwdriver or nifty with a needle. For more complex repairs, search

---

PROJECT **Basic clothes mending**

Mending a tear and darning a hole are both easy and take no time at all once you have the knack. If you have worn through some socks or knitwear, or have a tear in jeans or a shirt, try giving them a new lease of life.

MENDING A TEAR

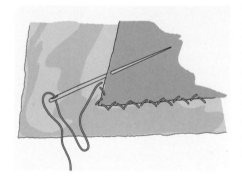

Tack a piece of stiff paper to the underside of the tear with small pins, so that it is temporarily held back together. Using small slanting stitches, push the needle down through one side of the tear and the paper, and then bring back up through the paper and material on the other side of the tear. When you have completely sewn the tear neatly back together, carefully remove the paper.

BASIC SEWING KIT

A fine sewing needle that has a small eye and is sharp.
Small sewing scissors are useful for snipping off the ends of thread.
Different-sized safety pins.
Soft fabric tape measure.
Pins to fix everything in the right place before sewing.
Large pair of fabric scissors that are sharp and cut clean, straight lines. Keep them just for fabric or they will quickly go blunt.
A darning mushroom enables you to stretch a worn sock over the mushroom-shaped wood and gather it tightly at the neck, making it easier to work on a hole in the toe or heel.

DARNING A SOCK

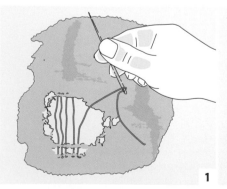

1

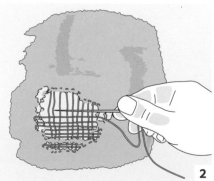

2

1. Use thick thread or thin wool in a matching colour to darn a sock or sweater. First sew around the edge of the hole using a running stitch; the stitches on the back should be equal lengths and half the length of those on the front. Next, sew vertical lines up and down the hole. **2. Fill the hole** in by sewing horizontal lines, weaving the thread from side to side, over and under the vertical lines. Keep darning until the hole is no longer visible.

for your local repair café. These are free events, often held in community centres, where people can bring items in need of repair – such as clothing or gadgets – and work together to fix or modify them.

Reheeling shoes

Breaking the heel on your favourite pair of shoes doesn't mean the death of them: most can be repaired in minutes. If the rubber piece of your heel is wearing down, you can buy a replacement piece – they're usually very cheap. Pull the worn piece out with a good set of pliers, then place a piece of wood inside the shoe at the heel end to reinforce it. Wrap your shoe in fabric to prevent it from getting scuffed, and secure it in a vice if you have one, or hold it steady on a table. With the heel pointing upwards, hammer in the replacement until it is secure.

Saving electronic gadgets

The number of mobile phones and small electronic gadgets that are thrown away every year is quite astonishing. Many are difficult to fix, but sometimes quick reactions and common sense can save them. And if not, remember that broken items can often be recycled.

If you drop a gadget in water, fish it out as quickly as possible and remove the battery immediately. Take out any SIM or memory cards and allow them to dry. These can often be saved, even if it's too late for the rest of the device.

If any of the components are covered in wine, sugary drinks, or milk, it's important to clean them as the residue can corrode the circuitry. Use a rag or paper towel dipped in methylated spirits, nail polish remover, or some very strong alcohol, and dry with a hairdryer (set to cold) or a compressed air can. You can also submerge items in a bowl of uncooked rice grains. Leave for 24 hours and with luck you will have saved your gadget from a watery grave.

PROJECT Repair a bicycle puncture

This is a very simple repair and you can save quite a lot of money if you take five minutes to fix a tyre yourself rather than take the bike to a repair shop. You will need a puncture repair kit and, ideally, a bucket of water, which makes the whole process much easier – a deep puddle at the side of the road will do in an emergency.

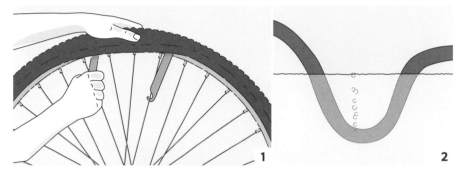

1. Remove the inner tube using levers. Then pump it up and listen for any hissing noise, or place the tyre near your face and rotate it until you feel the air escaping. **2. Alternatively, dunk the tube** in a bucket of water and find the puncture quickly by locating the area from which bubbles are escaping.

3. Mark the hole with chalk from your repair kit. Sand the area with sand paper, and dry the tube. **4. Apply a thin layer of glue** and wait until it starts to dry, then place a suitable patch of rubber over the puncture. Press down firmly.
5. Check the inside rim of the tyre to make sure there are no thorns or glass that will puncture it again. Fit the inner tube when it is slightly inflated.

PRODUCTIVITY

Growing your own food is probably one of the most satisfying things you can do. We love the idea of not only being productive but also of growing in harmony with nature. When you get the balance right, you'll discover just how awesome nature can be and how much it is on your side. This chapter provides you with all the information you need to make the most of your plot, large or small. We put a particular focus on how to save time in the garden and extend your growing season – so you can sit out in the summer sun and still be harvesting fresh crops in the depths of winter.

APPROACHES TO GROWING

Growing your own food is a big step for first-time gardeners and the range of techniques can be daunting. However, the fact that there are so many points to one simple truth: there is no right way to do it. Experiment with whatever appeals – some methods take so little effort, you won't even need to lift a spade – and you'll find an approach to suit you and your plot.

KEEP AN OPEN MIND

You don't have to live on a five-acre smallholding to follow organic principles or design a permaculture plot. These are simply the "tools" that can be used to cultivate your space, large or small.

Techniques have changed drastically over time, from traditional methods to intensive pesticide-based systems, and back again to natural approaches. As with many aspects of sustainability, the traditional methods are often more valuable.

We have found it worthwhile to keep an open mind about growing. We try anything once: in our opinion, once can be an accident, twice a mistake, and three times is enemy action. If a method doesn't work for you, try an alternative. Remember, sustainability is about getting the most out of your space without damaging its future productivity.

GROWING ORGANICALLY

Organic cultivation is based on the belief that the environment and the way we grow our food are interdependent. Its principles take into account soil and climate, while avoiding the use of toxic chemicals.

How it works

Enhancing soil health by composting is one of the foundation stones of organic practice, mainly because there is lots of life in the soil, when compared to artificial fertilizers. These lack the humus content – the decayed plant material and soil organisms – of compost, which brings vitality and holds moisture.

Crop rotation (see page 120) reduces the build-up of soil-based diseases and pests.

Weeds are controlled mechanically by hoeing or hand weeding, or by mulching (see page 112).

Using green manure – growing crops that are dug back into the soil to improve fertility – is another key organic method (see page 118).

Pest control is based on minimizing the opportunities for pests to flourish, using techniques such as companion planting (see box opposite) and encouraging natural predators (see pages 132–135).

PERMACULTURE

Permaculture means different things to different people. We understand it as working with nature rather than against it. Permaculture uses natural ecosystems as a model for growing food, linking plants, animals, and microbes. If a plot is designed around what is already growing, permaculture can fit any situation.

How it works

A no-dig policy and low-energy approach are often followed. However, it's not just lazy gardening! In permaculture, disturbing the soil is

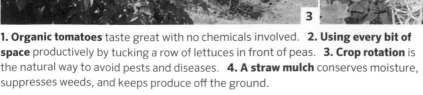

1. **Organic tomatoes** taste great with no chemicals involved. **2. Using every bit of space** productively by tucking a row of lettuces in front of peas. **3. Crop rotation** is the natural way to avoid pests and diseases. **4. A straw mulch** conserves moisture, suppresses weeds, and keeps produce off the ground.

COMPANION PLANTING

Companion planting is based on crops that grow together in a beneficial partnership, reducing the time needed for pest control and maintenance. Here are a few tried-and-tested combinations – experiment with your own and plant your beds with a bit more companionship.

Sweetcorn, beans, and squash
The sweetcorn serves as a trellis for the beans to climb up; the beans fix nitrogen for the sweetcorn; and the squash grows around the base, keeping in the moisture and suppressing weeds. Try substituting sunflowers for sweetcorn.

Onions and lettuce
Rows of onions or garlic next to lettuce and other salads create a "smell" barrier that stops slugs and snails getting too close.

Poached-egg plants and peas
Poached-egg plants are a perennial you can move each spring to your legume bed. The flowers attract lots of hoverflies, which eat the aphids that would otherwise spoil your peas.

Basil and tomatoes
Aphids hate basil, so plant it next to your tomatoes. It's a great combination on the plate too.

Nasturtiums and cabbages
Nasturtiums are a favourite food of cabbage-white caterpillars. Grow them as a trap crop to lure butterflies and caterpillars away from brassicas. Add aphid-free flowers and leaves to salads.

Radishes and cucumber
Planting a ring of radishes around cucumber plants reduces the risk of cucumber beetle.

Carrots and onions
One of nature's matches made in heaven: the onions drive away carrot fly and the carrots keep away the unwanted onion fly.

Keep aphids under control by sowing poached-egg plants (*Limnanthes douglasii*) at the same time as your peas.

thought to destroy its natural fertility and lead eventually to erosion.

Growing perennial crops is an important feature of permaculture. When they are well established they use up less energy than sowing a new crop each year, and weeds are less of a problem as there is less bare soil for them to colonize.

Zoning is another fundamental. Whatever needs most attention – vegetable beds, salads, and herbs – is grown close to the house. Orchards and livestock zones are further away; further still, you come to field crops.

Mulching is one of our favourite permaculture principles (see page 112).

NO-DIG GARDENING

The normal reasons for digging are to incorporate manure and compost, remove or bury weeds, and create a good tilth for sowing. However, in no-dig gardening all manure, compost, and other organic material is applied to the surface two or three months before sowing, and is gradually incorporated by

worms and other organisms. No-dig gardening feeds the soil not the plant.

How it works

Worms and their allies do the digging for you. Earthworms work on a permanent and ongoing basis that is arguably far more efficient than using a spade or fork!

Weeds are dealt with by taking preventive action. Shallow hoeing or mulching in winter removes weeds before they have time to seed. A decent layer of light-excluding mulch left in place for at least a growing season can be used to clear the ground of tougher weeds. You can plant through the mulch (see page 112).

Conserving your energy is one of the key reasons for adopting a no-dig approach. Some people love digging but others are put off the whole idea of gardening by this strenuous aspect. For those who are elderly, less able, or less inclined to get muddy, no-dig is an accessible entry into the world of growing your own.

NO WATERING

We first encountered this technique when we visited the The Lost Gardens of Heligan, the award-winning garden restoration project in the far south-west of England. It saves water in the garden and saves a lot of effort too. We've had great success with this method on our courgettes.

How it works

Once a young plant has been transplanted and thoroughly watered in, it is left to grow without being watered again. At first the plant will look a little unhealthy and may struggle, but this is only while the roots go down deeper in search of the natural water table.

A strong taproot develops as a result, and the plant thrives in dry periods due to its resilient root formation, without extra watering.

SQUARE-FOOT GARDENING

Square-foot gardening is a new approach that evolved from the need to optimize space and reduce maintenance. It's ideal in an urban environment with limited space but works just as well for an intensively planted plot or on an experimental smallholding or city allotment.

How it works

The basic concept is to grow crops in a series of squares instead of rows. Each is about 30cm (12in) square and has a different vegetable, herb, or flowering plant in it.

Obvious benefits include companion planting (see page 111): pest-deterrent plants like marigolds, onions, garlic, and chives are so close to the surrounding crops that their effectiveness is magnified.

Succession planting (see page 146) is much easier with this method: once a square has produced its harvest, it is easily replanted with a new mini crop.

BIODYNAMICS

First propounded by the philosopher and scientist Rudolf Steiner at the beginning of the 20th century, biodynamics treats a plot of land as an interdependent and self-nourishing system. It has a lot in common with organic principles, but biodynamics is still regarded as being at the wacky end of gardening.

How it works

Biodynamics goes a step further than the usual ecological gardening practices such as companion planting, green manures, and composting, by using a series of herbal remedies to treat the soil, plants, and even the compost heap.

An astronomical calendar is used to plan sowing and harvesting times.

The dynamic aspect takes note of biological processes. A typical biodynamic technique is to take sap rising in the morning as a sign that leafy vegetables should be harvested first thing, when they are fresh.

PROJECT Easy no-dig mulching and planting

This is the easy way to plant a vegetable bed, with no digging or weeding. It relies on mulching – blocking the light so weeds can't grow. Cardboard is an ideal mulch as it's readily available and biodegrades over time.

YOU WILL NEED

Cardboard	Topsoil
Large stones	Spade
Watering can	Sharp knife

1. Lay the cardboard mulch direct onto uncultivated ground, overlapping the pieces so that there are no gaps for weeds to grow through. **2. Weigh down the edges** with stones to stop the cardboard blowing around. **3. Water thoroughly.**

4. Cover with a layer of topsoil around 7.5cm (3in) deep. **5. Leave for two weeks** and then make a planting hole by digging into the soil and cutting away a circle of cardboard. **6. Plant up** (a squash is shown) and firm the soil around. Water well and let it grow!

Permaculture forest garden

A forest garden is made up of mainly perennial plants, growing as they would in the wild. A well-managed garden will yield fruits, nuts, and herbs, and annual crops such as salads can be sown in between the perennials. Growing a mix of diverse plants side by side reduces disease and pest attack: it's companion planting on a large scale.

Making a forest garden

A forest garden is ideal for well-established plots as it can incorporate mature trees. But don't let that limit your imagination. You can create a forest garden in pots on a patio by planting a dwarf fruit tree, a grapevine, herbs, and salads. Forest garden plants benefit each other: vines use trees for support, and wild garlic grows in the shade of fruit bushes, leaving no space for weeds.

A forest garden is planted in layers of edible perennials: here gooseberry bushes are planted alongside fruit and nut trees, with lavender as low-growing ground cover.

The forest garden plot

Perennial plants need far less attention than annual crops. Once they are planted, there is very little digging to do in a forest garden, apart from preparing small salad beds.

Fruit bushes like redcurrants, blueberries, and gooseberries can cope with shade from the trees – mushroom logs will also thrive.

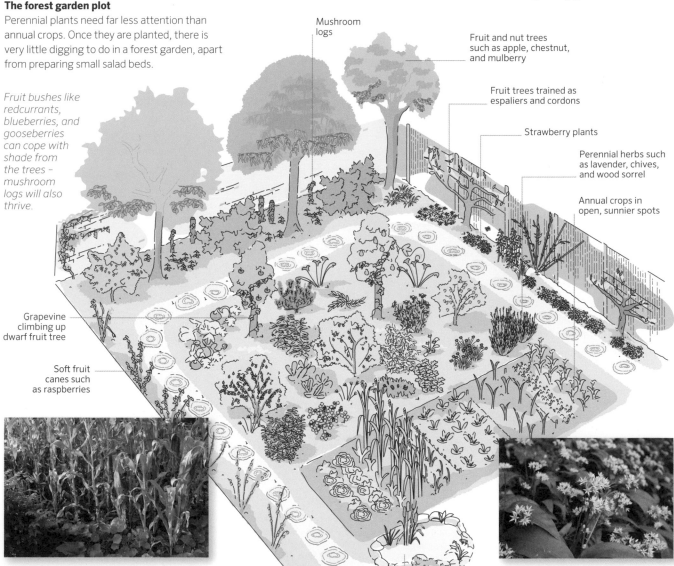

Mushroom logs

Fruit and nut trees such as apple, chestnut, and mulberry

Fruit trees trained as espaliers and cordons

Strawberry plants

Perennial herbs such as lavender, chives, and wood sorrel

Annual crops in open, sunnier spots

Grapevine climbing up dwarf fruit tree

Soft fruit canes such as raspberries

Sweetcorn and nasturtiums are annuals. The corn supports the nasturtiums, which shade the soil and keep it moist.

Pond attracts frogs for pest control

Perennial edible ground cover plants, such as wild garlic and alpine strawberries, help suppress weeds.

GROWING WITH HYDROPONICS

Imagine a method of growing plants that doesn't need soil, uses a fraction of the water, and gives yields up to four times greater than traditional soil-based growing. It almost sounds too good to be true – and in the past hydroponics has suffered from being associated with white-coated scientists. Today it's a technique that anyone can try, and we have been researching how to use it at home.

WHAT IS HYDROPONICS?

Hydroponics is a method of growing plants without the need for soil, by directly feeding the roots with a mineral nutrient solution. Setting up a system at home does involve an initial outlay of time and resources but, in the long run, it saves labour and conserves water.

We first became interested in hydroponics because of the challenge facing many people living in densely populated urban centres, with little space to grow their own food. Hydroponics makes efficient use of space and labour, and has a low impact on the environment.

At Newhouse Farm we tried and tested the Nutrient-Film Technique (NFT), where a very shallow stream or "film" of nutrients, 1–2mm (less than ⅛in) deep, provides plants with all the nourishment they require for healthy growth. Because the nutrient flow is so shallow, the roots are still able to access oxygen, without which they would rot.

The advantages:

- **Hydroponics can be used** on stony or otherwise difficult sites.
- **There is no risk** of developing soil-borne diseases.
- **There are no weeds** to compete for space, nutrients, and water.
- **Nutrients are delivered directly** to the plants' roots, so up to four times more plants can be grown than on the same area of land.
- **Tailoring the nutrient mix** to suit each particular crop increases yields.
- **Because hydroponics uses water efficiently,** consumption can be reduced by up to 90 per cent.
- **Combining hydroponics** with a greenhouse or polytunnel stretches the growing season, enabling you to eat out-of-season produce.

WHAT TO GROW

Plants grown hydroponically require a constant stream of diluted nutrients washing around their roots. Crops that we have had success with are herbs, summer squashes, salads, tomatoes, chillies, and aubergines. It's important to support tall-growing vegetables with wire or string as they grow, as you would normally.

The only vegetables we would not grow hydroponically using NFT are root vegetables: there simply isn't space in the growing channels for large roots. The bottom line with hydroponics is to experiment and learn as you go along.

CARING FOR CROPS

With hydroponics, successful growing depends on monitoring the nutrients and pH level of the solution.

Starting plants off

The seed-germination medium we use is coconut fibre, a by-product from the husking industry, which comes ready shaped into cubes or plugs. Germination is up to five times faster than with traditional methods as the fibre retains water more effectively than soil. Place seeds in the middle of the plugs and feed them on a start-up nutrient (you can order this online) for a couple of weeks.

Transferring plants

Once plants reach a decent size and the tips of their roots are poking through the bottom of the cube, move the cube and plant into the growing channels.

Feeding with nutrients

Organic nutrients can be bought online from hydroponics suppliers: worm casts, krill, alfalfa, and kelp are all suitable. If you want to make your own, use the liquid fertilizer from the bottom of a wormery (see page 139) or make comfrey concentrate (see page 119).

A 45-litre (10-gallon) tank will be large enough for a small-scale domestic growing system. It will need a top-up every couple of weeks to keep the water level high and to replenish the nutrients absorbed by the plants as they grow.

Monitoring pH levels

Check the pH of the nutrient solution in the tank regularly using a waterproof pH monitor. Adjust the pH by adding "pH down" powder (handle it with care), or by altering the amount of nutrients you add.

Tomato plants thrive when grown hydroponically. With a constant supply of water and nutrients you need never worry about the fruits splitting and it saves lots of time. Set up your system under cover and you can extend the growing season and eat tomatoes for most of the year.

Setting up a system

A hydroponics system involves growing plants in channels fed by a stream of nutrient-rich water pumped round in a continuous cycle. Ready-made kits are available but to really understand the process we recommend putting together a system yourself. You can grow plants hydroponically outside, but growing under cover gives maximum yields.

The irrigation system

To pump the nutrient solution to the growing plants, we used a simple plant irrigation system available from garden centres. This uses electricity, so there can be a small running cost. We ran a 12V pump from a battery powered by a solar panel and/or wind turbine.

Growing channels

You can buy special growing channels, but we've found flat-bottomed guttering works just as well. Make lids from cedar planks or plastic sheeting to keep the inside dark; this reduces algal growth and simulates the normal environment for roots. A thin layer of capillary matting down the length of the channel gives an even flow rate and healthier roots.

Growing channels need to slope gently down towards a collection channel; we adjusted their height with slivers of wood until the solution was trickling slowly down. If the gradient is too shallow, the solution collects in puddles around the roots and increases the risk of disease.

The length of channel is up to you. Ideally they will have closed ends with small drainage holes, but open-ended channels do exactly the same job, just with more splashing!

The shape of a geodesic dome avoids warm air pockets forming, making it ideal for growing plants under cover.

Grow in winter with a large energy-efficient light bulb to simulate summer light, a fish tank heater to warm the nutrient solution, and sheeps wool or recycled textiles for insulation.

A typical set-up

A simple trestle table is ideal for the growing channels, leaving plenty of space for the nutrient tank underneath.

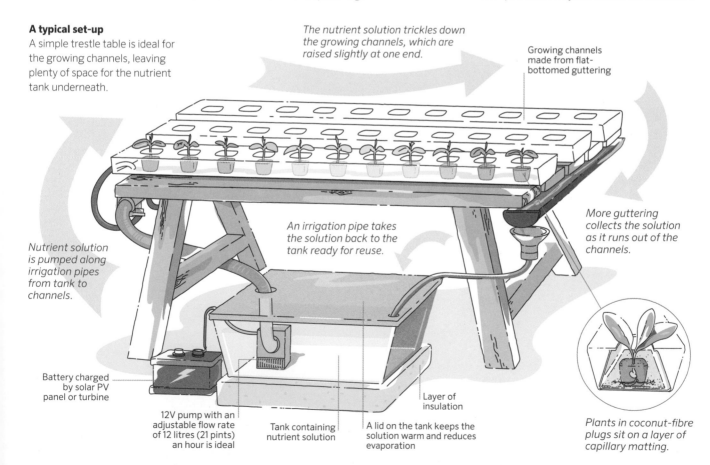

The nutrient solution trickles down the growing channels, which are raised slightly at one end.

Growing channels made from flat-bottomed guttering

Nutrient solution is pumped along irrigation pipes from tank to channels.

An irrigation pipe takes the solution back to the tank ready for reuse.

More guttering collects the solution as it runs out of the channels.

Battery charged by solar PV panel or turbine

12V pump with an adjustable flow rate of 12 litres (21 pints) an hour is ideal

Tank containing nutrient solution

A lid on the tank keeps the solution warm and reduces evaporation

Layer of insulation

Plants in coconut-fibre plugs sit on a layer of capillary matting.

ACHIEVING A PRODUCTIVE PLOT

Growing your own food is relatively easy; plants require little more than nutrients and water to survive. But to improve productivity and get impressive harvests, a little more effort is needed.

Regular watering, feeding, and thorough maintenance make your plants stronger and more resilient to weeds and garden pests. Providing your crops with optimum soil conditions is the key.

BROWN AND GREEN COMPOST

Combine equal parts "brown" and "green" materials, such as those suggested below, to create good, nutritious compost:

Browns (high in carbon)
• Straw
• Dead leaves
• Cardboard
• Shredded paper
• Loo roll tubes
• Egg cartons and egg shells
• Twigs and plant stems
• Sawdust
• Vacuum dust

Greens (high in nitrogen)
• Grass cuttings
• Uncooked fruit and vegetable scraps and peelings
• Loose tea leaves and coffee grounds
• Manure
• Urine
• Weeds and fresh plants

Layer your compost with "brown" kitchen waste followed by "green" grass cuttings. Mix every few weeks, then leave to compost.

LOOKING AFTER THE SOIL

It's good gardening practice to think of the soil structure as the foundation of a productive plot. Whether you have a raised bed in an urban area or a large smallholding, you'll get better results if you add plenty of rich organic materials to the soil. We both make huge amounts of our own compost year on year to replenish our raised beds. We enrich our soil with green manure crops and animal manure, and make our own liquid fertilizers to boost certain crops in the growing season. Nurturing the soil is a bit like mixing together the right ingredients in a cake: it directly determines how well your fruit and vegetables "rise". We find it's a very hands-on aspect of growing and really gets you back to the roots of a productive plot.

COMPOST COOKED FOOD

Above all, you should never waste food and always try to be inventive with leftovers. If you do have cooked food, meat, fish, and dairy products to dispose of, however, it's possible to compost them using the Bokashi system. This turns cooked food waste into compost in an odourless, compact unit, using a type of bran that contains microorganisms. The process is anaerobic, so the container must be airtight. The microorganisms ferment the waste, then you can add it to your conventional compost bin.

MAKE YOUR OWN COMPOST

Composting is the process whereby the natural decomposition of organic matter is accelerated to create a rich growing material. Bacteria, fungi, and microorganisms thrive in the right conditions and will break down waste quickly to provide a great source of quality compost to use around your plot. Successfully making your own compost depends on getting the mixture just right.

Any compost heap will naturally break down eventually and turn into something useful, but to avoid a smelly wet mix or a pile of dry matter, you must get the balance right between materials high in carbon and those that are high in nitrogen. We always try to mix one part "greens" with at least one part "browns" (see box, left). We also

Add cooked food scraps and a handful of bran. Bokashi also makes strong liquid fertilizer: dilute with water to feed plants or use neat to unblock drains!

1. **Green plant material** is rich in nitrogen, so add it to your compost heap. **2. Mix it** 50:50 with dry "brown" material such as straw, cardboard, or dead leaves. **3. Cone-shaped bins** are ideal for cool composting. **4. Cool compost in progress.** The bottom layer is ready to use but the top goes back in the bin for a bit longer. When it's ready, spread home-made compost on a vegetable bed, then fork it in.

add activators to speed it all up (see box, below right); most work by providing a nitrogen boost to kick-start the process.

Cool composting

The cool composter is probably the most common style of compost bin. It doesn't retain much heat but instead encourages a large worm population to break down materials, along with a huge collection of microorganisms. Just lift the lid and add small amounts of waste regularly. Don't add weeds with seed heads: these must be hot-composted to destroy the seeds or you will end up sowing weeds on your vegetable beds when you dig in the compost.

Cool composters are easy to use. We recommend emptying them out every couple of months and turning the compost before putting it back in. Use a fork to mix it up. If it looks too wet, add more "browns"; if it's too dry and fibrous, add more "greens". We have several cool

composting bins that we use on rotation for a ready supply throughout the year.

Hot composting

Composting at higher temperatures kills weed seeds and reduces the spread of plant diseases, but you won't find many worms in the mix – it's too hot for them. Heat-loving anaerobic microorganisms do all the work in a hot compost heap.

When we first arrived at Newhouse Farm we cleared a lot of uncultivated land for vegetable beds. This in turn presented the challenge of what to do with all the unwanted weeds. We didn't want to waste the plant material, so we built a few big hot-composting systems out of pallets.

Wooden pallets are available all around the world – free of charge if you scavenge them. Fix them together using some big nails or string to create two adjoining wooden boxes that can be used in rotation, each about one cubic

metre (1.3 cubic yards) in volume. The wood helps to retain the heat created by the composting process. For extra insulation, line the insides of the boxes

TRY THIS

Compost just happens, but if you want it in a hurry, add one of these activators to speed up the process:

Urine – men's works better than women's (it's the hormones).

Grass cuttings should be added sparingly in thin layers.

Comfrey leaves can be added fresh, or use the old ones left over from making concentrate (see page 119).

Seaweed must be collected fresh, from below the tide line.

Manure should be added a little at a time.

Nettles – don't forget to wear gloves when picking.

Topsoil contains lots of microorganisms which will help to get the heap going.

Compost from an old heap is also full of microorganisms.

you have built with big bits of cardboard and stuff natural fibre insulation into the gaps for an efficient hot composting bay.

It may also be worth using a shredder to cut up some of the materials if you are processing large quantities. To maintain the temperature in a hot compost heap, you need to turn the compost regularly to aerate it. Use a fork to shift the outside to the inside and vice versa; it's hard work but well worth it. Covering the heap with fleece, a layer of straw, or some bits of old carpet will help to retain the heat too.

Using a wormery

One of the main advantages of composting with a wormery is that when it's designed in vertically stacked tiers, it takes up so little space. In an urban environment we'd recommend one as your first choice. Build your own (see page 139) or buy one ready-made. The main attraction is the liquid fertilizer or worm tea that collects in the base. It's high in nitrogen and phosphates – a rocket fuel for growing plants.

Making leaf mould

Leaf mould is an excellent way of making your own compost to use as a soil conditioner, to improve both drainage and water retention, or for potting up seedlings. Simply collect fallen leaves and place them in old sacks, animal feed bags, or even a separate compost bin. Leave them for about a year and then enjoy – a simple compost recipe but worthwhile. If you are making leaf mould on a larger scale, we recommend building a big cube-shaped cage using four wooden posts and some chicken wire. Shred the leaves before adding or collect them using a lawn mower: it chops the leaves up and mixes them with grass, which is high in nitrogen, creating a magic compost mix full of microorganisms and water.

USE GREEN MANURES

Green manures are crops grown specifically to help build a good soil structure and maintain fertility. They are not harvested; instead they are dug straight back into the soil. Once considered an agricultural technique in farming, growing green manures is increasingly popular with small-scale organic gardeners.

At Newhouse Farm we used them for a few years as part of our crop rotation and were pleased with the results. Not only did green manures look attractive, they also helped to protect bare soil from erosion, and provided a good covering blanket to keep the beds free of weeds. Green manuring is simple. Choose a suitable crop from those listed on the crop rotation plan on page 121 and allow it to grow for a short time. Then dig it back into the soil with a sharp spade to replenish the nutrients.

You must dig green manure back in while it's young and sappy. If you leave it too long, then it defeats the point and looks more as if you have planted a bed full of weeds! Several green manures – including alfalfa, clover, trefoil, and fenugreek – belong to the legume family and can actually take up nitrogen from the air, adding

PROJECT **Dig a compost trench**

Compost trenches are a really good alternative to a conventional compost heap, particularly during winter when normal compost bins have cooled down and are slower to deal with waste. They are completely smellproof, quick and easy, and a great source of food for young plants; the decomposed material provides a rich supply of nutrients right where it is needed, at the roots of the growing plants.

1. Choose a space in a vegetable plot that you intend to plant into next season. In autumn, dig a trench one spade deep and a spade's width across. **2. Fill it gradually** by adding vegetable scraps and kitchen waste, covering them with the original soil from the trench as you go. **3. Once the trench is full,** leave it for a couple of months, then sow seeds or plant directly into it. Runner beans and peas do particularly well on top of compost trenches; the following year, pumpkins and courgettes can be planted in the same spot.

Make comfrey fertilizer

Comfrey is an excellent plant to grow in the garden. Not only does it attract a whole host of bees and pollinating insects but, by following a few simple steps, you can turn the leaves into a liquid fertilizer rich in nitrogen, potash, and phosphorus – essential for healthy plants. Comfrey is a deep-rooted herbaceous perennial that absorbs nutrients from way down in the soil. It grows best in full sun and can be cut down four or five times in a season, but remember to stop by early autumn to allow enough regrowth before winter. In our opinion, the only negative issue with composting comfrey is the smell – but that just adds more character.

TRY THIS

Cut the bottom off a large single-use plastic bottle and pack it with comfrey leaves. Stand the bottle, without its top, upside down in another container. In two weeks you'll have comfrey concentrate on a smaller scale.
Substitute nettles for comfrey in either method if you haven't got any comfrey on your plot.

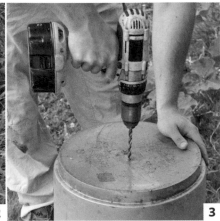

1. **Pick handfuls** of comfrey leaves. It grows prolifically so don't worry about stripping the plant. 2. **Build a stand** for your comfrey distillery – we used old breeze blocks. Place a container such as a watering can underneath to collect the concentrate. 3. **Drill a hole** in the base of a butt or large container with a lid, capable of holding approximately 45 litres (10 gallons).

4. **Pack the comfrey leaves** tightly into the container. 5. **Weigh down the leaves** with some bricks to speed up the process. As more leaves grow back on the comfrey plant, pick them and add to the container. After about 10 days the leaves will start to decay into a black liquid. You should collect about 2–3 litres (3.5–6 pints). 6. **Decant the comfrey concentrate** into a jar or bottle and store in a cool dark place. Don't seal the bottle too tightly as it may ferment in warm weather. Once the concentrate has stopped dripping into your collector, add the leftover comfrey leaves to the compost heap as an activator. Dilute the concentrate 1:15 with water and feed your plants for amazing results.

extra goodness to the soil. We allow a month or so after digging them in before we plant up the beds.

A few favourite green manures:

Alfalfa Sow from spring to midsummer and grow for a few months.

Crimson clover flowers attract insects, especially bees. Sow from early spring to late summer.

Fenugreek is one of the fastest-growing green manures; useful if you have a tight crop rotation. Sow from early spring to late summer.

Trefoil does well in shade; use it to under-sow sweetcorn or sprouts. Sow from early spring to late summer.

Mustard can be sown from spring to summer to grow for about eight weeks. Goes to seed quickly in hot weather, so dig it in nice and early.

Grazing rye is ideal over winter, when it suppresses weeds. Sow late summer to late autumn.

PRACTISE CROP ROTATION

Before we head out into the garden at the beginning of each year, we plan an organized system of crop rotation. This sounds complicated but it makes sense. Simply put, some families of plants are more prone to particular diseases and have specific pests that prey on them. In addition, different plant families also need different nutrients from the soil. By moving the crops from plot to plot each year you can give them a better chance to get strongly established before problems start. At the same time the soil gets the chance to replenish itself.

Working out a rotation plan

Trying to work out a definitive rotation plan proved nearly impossible when we first started, as different experts give different advice. So we tried to understand exactly which elements of each system we liked and what logic we agreed with. After many headaches, we came up with a plan that suited us and appeared to be in line with most expert advice. We divided our crops into groups, set out in the chart opposite. Green manures are included in our crop rotation too.

1. Crop rotation stops soil depleting and avoids a build up of pests. **2. Pollinating insects** are key to a productive plot. **3. Sow clover** as a weed-suppressing green manure. **4. Labelling is crucial:** you've got to be organized for crop rotation to succeed.

CROP ROTATION PLAN

We divided our crops into five groups but followed a four-year, four-bed rotation by letting alliums and umbellifers share a bed. We also included a sixth group, salads, which we succession planted when and where we could, both within beds and under cover – that's why they're not in the rotation plan. We didn't include potatoes in our crop rotation because they were prone to blight where we lived and needed to be grown separately.

	YEAR 1	YEAR 2	YEAR 3	YEAR 4
BED 1				
BED 2				
BED 3				
BED 4				

	GROUP	CROPS INCLUDE		SOIL REQUIREMENTS
	Alliums	Garlic Leeks	Onions Shallots	High organic matter; may need lime added to the soil to reduce acidity.
	Umbellifers	Beetroot family (quinoa, spinach, Swiss chard, spinach beet) Carrots Celeriac Celery	Fennel Parsley Parsnips Potatoes – unless blight is a problem where you live	Root crops need stone-free soil; a fine tilth. Soil should not be freshly manured. Some root crops also help to break up soil structure, especially potatoes.
	Legumes	Alfalfa (green manure) Broad beans Clover (green manure) Fenugreek (green manure) French beans	Lupins (green manure) Peas Runner beans Tares (green manure) Trefoil (green manure)	Well-drained and moisture-retentive, but not nitrogen-rich. Legumes fix nitrogen in their roots for future crops. When harvesting, leave the roots.
	Brassicas	Broccoli Brussels sprouts Cabbage Calabrese Cauliflower Kale	Kohl rabi Mustard (green manure) Oriental brassicas Radish Swede Turnip	Leafy crops need nitrogen-rich soil; may need lime added to reduce diseases like club root.
	Miscellaneous	Buckwheat (green manure) Courgettes Cucumbers Grazing rye (green manure) Peppers Phacelia (green manure) Pumpkins	Squashes Sweetcorn Tomatoes	Some plants have so few soil-dwelling pests or diseases that they can be fitted in anywhere in the rotation.

SET UP A GREENHOUSE IRRIGATION SYSTEM

When you grow plants outside you don't have to worry too much about watering, as you expect nature to do it for you. But once you start growing plants in polytunnels, greenhouses, cold frames, or cloches, they depend on you for water. Installing an automatic irrigation system will make life an awful lot easier at the height of the growing season.

GETTING WATER WHERE IT'S NEEDED

To set up an irrigation system you first need a water source. Rainwater butts are an ideal simple solution: we have one inside our greenhouse that's filled from guttering attached to the greenhouse roof. Bear in mind that rainwater butts have a limited capacity; in a dry spell you may find yourself topping up the butt with water from the mains or elsewhere.

Water is heavy and it takes a lot of energy to move it. If you can collect and store water on land that's higher than your greenhouse, then you can use gravity to help with the watering. If you can't use gravity to reduce your workload, then technology can come to your rescue.

A simple little demand pump is all it takes to move water to exactly where you want it. A demand pump can use mains electricity or be battery operated, and will only pump – and therefore use energy – if a tap is turned on, in this case by an electronic timer. The pump is controlled by a pressure switch: when the tap is turned off again, the water pressure builds up and switches the pump off.

DELIVERING THE RIGHT AMOUNT

The key to irrigation is to deliver the right amount of water. You'll know when you get it wrong. Splitting tomatoes are a tell-tale sign: if a tomato plant receives too little water, then is deluged with it, the fruits can't handle the sudden surfeit and the skins will split. Don't forget plants only need water during the day when they are photosynthesizing. Water in the morning and harvest in the evening when flavours are at their most concentrated. Don't take our word for it – try it with your tomatoes!

PROJECT Install an automatic watering system

A few hours arranging leaky pipes, drippers, and sprayers and connecting them to a water supply will save you many more hours of watering. Your system will pay for itself in terms of increased yields and the plants get watered very gently, without disturbing the compost.

YOU WILL NEED

Small block of wood
Hose pipe
Water source
12V demand pump
Battery-powered pump timer

Rechargeable, deep cycle, or leisure battery
Leaky pipe system, dripper system, spray system, or any combination of these

CONNECTING UP TO THE WATER SUPPLY

1. Make a float from a block of wood for the hose drawing water from the butt, so it takes water from the top, rather than the bottom where sediment collects. **2. Mount the pump and timer** on a wooden box unit using the supplied fixings. Our unit has a sloping roof to keep the battery dry and you could mount a solar panel on the roof to charge the battery. Connect the pump to the battery; the timer has its own internal battery. Connect the timer to the pump with a length of hose, and connect a supply hose to the timer.
3. Connect the pump to the water butt with another length of hose, keeping the distance between them as short as possible.
4. Bring all the supply hoses to a single junction close to the timer beside the pump. You can run several systems at once but the water pressure will drop. Connect the junction hose to the timer.

INSTALLING A LEAKY PIPE

1. Install the system before planting up. Lay the perforated leaky pipe on the compost in a way that will distribute moisture to the maximum growing area. Use small sticks to hold it in position and serve as markers.   **2. Bring both ends together** and join them to the water supply hose using a T-junction. Connect the supply hose to your automated system (see opposite).
3. Add compost to cover the pipe.   **4. Plant up the bed,** avoiding the sticks marking the leaky pipe.

INSTALLING DRIPPERS AND SPRAYERS

1. Arrange sprayers so the nozzles deliver water to each pot.   **2. Continue building up** the system; drippers are ideal for hanging baskets.   **3. Add dual-purpose** attachments to act as misters or drippers.   **4. Add a small end stop** when you get to the end of a run of piping.

TRY THIS

If you have only a few plants, these small-scale systems will keep them watered for a couple of days.

Cut the bottom off a single-use plastic bottle, turn it upside down, and push the neck into the soil next to a plant. Fill it with water, which will seep into the soil.

Take a container with a lid. Make a hole in the lid, drape the ends of longish strips of fabric through the hole and insert a piece of pipe. Spread out the fabric strips and bury the container in compost, with the pipe sticking out so you can use it to fill with water. The fabric "wicks" transfer moisture from the container to the compost.

Lay a piece of fabric on a tray and stand plant pots on top. Drape the fabric into a container of water so that the plants can draw up moisture by capillary action.

GROWING UNDER COVER

Growing under cover is the process of extending the growing season by using a greenhouse, polytunnel, or smaller-scale options such as cloches and cold frames, to create warmer conditions for your plants. This allows you to produce fresh vegetables for longer, grow some foods out of season, and grow exotic crops from warmer climates.

WHY DO IT?

To become self-sufficient you need to maximize the growing potential of your plot, starting crops early to ensure successive harvests and avoiding lean periods later on.

A greenhouse or polytunnel may be expensive, but the benefits soon pay off – from growing salads in winter to giving tender plants a head start in spring. Some foods travel across the world to reach us, leaving a trail of environmental issues in their wake. A greenhouse is a way to grow them at home, rather than giving them up or feeling guilty every time we eat a strawberry out of season.

Even a small greenhouse can be surprisingly productive, and can allow you to grow all sorts of different fruits and vegetables throughout the year.

TYPE	PROS	CONS
Cloche	• Cheap to construct – free if you make it out of scrap. • Ideal for gradually hardening off seedlings (see page 146). • Protects vulnerable young plants from pests.	• Can get easily knocked over or blown away. • Only useful when plants are small. • Time-consuming technique.
Windowsill	• Almost everyone has one. • Big plants like tomatoes will be extremely productive.	• Tall plants may reduce daylight entering the room. • Pots must stand in a tray to protect paintwork when watering. • You must rotate the pots.
Cold frame	• Easy access for weeding, watering, harvesting. • Simple and cheap to make.	• Not enough room for tall plants such as tomatoes.
Greenhouse	• Productive growing space with the right layout. • Relatively easy to assemble.	• Can be expensive.
Polytunnel	• Cheaper than a greenhouse. • Large space available for growing.	• Less durable than a greenhouse. • Not suitable for a windy site.
Geodesic dome	• No warm air pockets. • More space for tall plants. • Admits maximum amount of light. • Aerodynamic shape – great on windy sites. • Fits into the landscape because of its completely curved nature.	• Challenging to assemble from a kit or to build from scratch. • Expensive.
Conservatory	• Great for exotics. • Part of the house so plants get maximum care.	• Very expensive. • Temperature control is essential.

1. **A traditional glass cloche** can protect a single plant. 2. **The shape of a geodesic** dome prevents frost pockets forming. 3. **A DIY cloche** can be made from a single-use plastic bottle. 4. **Harden off** plants in a cold frame before planting out. 5. **Open vents** and the door in a greenhouse to keep the air circulating.

SMALL-SCALE OPTIONS

From mini cloches for individual plants to home-made cold frames, you can still give plants on smaller plots extra protection without putting up a polytunnel or greenhouse. Even city dwellers growing herbs and salads on windowsills can reap the benefits of growing under glass.

Cloches

Cloches are essentially protective covers for plants. We have tried tent-style cloches covered with polythene, low plastic tunnels with inverted U-shaped wires, glass frames, and hard corrugated plastic sheets bent into a curve. We also cut the bottom off big single-use plastic bottles and place the top half over the plant, unscrewing the cap for ventilation in warmer weather. One of our favourite plastic-free methods is simply to place a clean jam jar over a newly transplanted seedling – a modern take on a Victorian glass bell cloche.

Windowsills

Growing plants on a windowsill may not sound like the most aspirational approach to growing food under cover but it can be successful. Young plants thrive in warm, slug-free conditions and, because you see them regularly, they get more attention and grow well, provided you rotate them so that all sides receive equal light.

Cold frames

Traditionally used for hardening off seedlings before planting out, we've found cold frames are ideal for raising seedlings from scratch and for extending the season of salads and herbs. A cold frame's key design feature is that it can be accessed easily for watering and harvesting. Using an old window for the top saves lots of effort and is a great way to recycle (see pages 126–127).

6 STEPS TO SUCCESS

Follow these guidelines to get the best from plants grown under cover.
Ventilate often, even in winter, to prevent problems like red spider mite developing. A heat sink (see pages 130–131) helps to keep air circulating, or open the door and roof vents to draw air through.
Don't overplant: too many plants too close together are more susceptible to pests and diseases.
Nourish the soil in permanent beds with a good supply of liquid fertilizer or new compost. If you use soil intensively it can become depleted of nutrients.
Monitor the temperature: frost is still an issue under cover. Even a water butt or small pond will help to warm the air, but a fan-assisted heat sink is ideal (see pages 130–131).
Clear away detritus from dead plants and fallen fruits etc. to reduce the chances of pests and diseases building up.
Water automatically by installing an irrigation system and plants need never go thirsty (see pages 122–123).

PROJECT **Build a cold frame**

In its simplest form a cold frame is a box with a glass lid that helps seedlings to grow by enhancing the environment they are in and protecting them from extremes of weather. Rather than buying new materials to build one, we always recommend looking around to see what bits and bobs you have available. Reusing old windows or panes of glass is a much greener option – and it's cheaper.

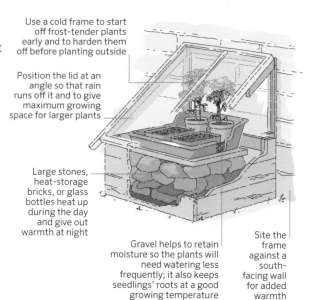

Use a cold frame to start off frost-tender plants early and to harden them off before planting outside

Position the lid at an angle so that rain runs off it and to give maximum growing space for larger plants

Large stones, heat-storage bricks, or glass bottles heat up during the day and give out warmth at night

Gravel helps to retain moisture so the plants will need watering less frequently; it also keeps seedlings' roots at a good growing temperature

Site the frame against a south-facing wall for added warmth

YOU WILL NEED

Screwdriver	Wood: planks, uprights, and battens
Spirit level	Old window plus extra glass
Drill	Sliding hinges
Tape measure	Rawlplugs
Hand saw	Nails
Hammer	Stones or glass bottles
Punch	Old paving slab
Screws	

MAKING THE BASE

1. Make a simple box by screwing horizontal planks to a framework. The size of your box will depend on what you use for a lid. We based ours on an old window. Make the uprights from standard 50 x 25mm (2 x 1in) lengths of wood. We used some leftover planks for the sides – 150 x 25mm (6 x 1in) are ideal – and we saved on wood by positioning the cold frame against a raised bed and wall.
2. Check the sides of the box are level.   **3. Add a brace** across the back of the box to keep it square.

MAKING THE LID

4. Use an old window for the lid. Screw two sliding hinges to opposite ends of the top edge of the window frame. Slide on the other half of each hinge.   **5. Screw the other hinge halves** to a length of wooden batten the same width as the lid.   **6. Flip the hinges shut** and hold the lid in its final position, resting it against the wall and the front edge of the cold-frame base. Mark the position of the batten on the wall, then lift the lid down.   **7. Slide the bolts** out of each hinge to separate lid and batten.   **8. Drill the batten** to the wall using rawlplugs and screws, following the guide marks you made earlier.

CONSTRUCTING THE SIDES

9. Measure and cut timber for the triangular side-wall frames, with the lid in place. Remove lid. **10. Mark the angle** of the joint for the side-wall frame. Do this at both ends. Cut the angles. **11. Drill the side frame** in place using long screws at an angle. Measure the triangular space and cut a pane of glass to this size. **12. Nail narrow battens** to the inside edges to stop the glass falling inwards.
13. Fit glass against inner battens and nail additional narrow battens on to the outside to hold it in place. Take care not to smash the glass. Use a nail punch to keep your hammer that extra safe distance away! If you don't feel confident about cutting glass to fit, use perspex or reuse bubble wrap packaging and staple it to the frame.

USING THE COLD FRAME

14. Fill the cold frame to within 10cm (4in) of the top with large stones, heat-storage bricks, or glass bottles filled with water, allowing space for warm air to circulate and be absorbed. **15. Place an old paving slab** on top of the stones, again leaving room around the sides for air circulation. Adjust it until it's level. **16. Fill a plant tray** with gravel and level the surface. Place it in the cold frame. **17. Slide the lid** back on to the hinges. Stand trays and pots of seedlings on the gravel – taller plants at the back, small ones at the front. Keep a length of wood handy to prop the lid open to acclimatize plants gradually and harden them off (see page 146).

ALTERNATIVE USES
Make maximum use of your cold frame with these ideas.

Make a hotbed
Line the base with biodegradable mulch or thick cardboard then half fill it with fresh horse manure. Mix in some straw and leave to ferment. Wait for the strong smell of ammonia to go, then add about 30cm (1ft) of topsoil. Plant or sow when the soil temperature is about 27°C (81°F). Great for initial germination of plants.

Drying seeds
Use your cold frame to speed up the drying process when preparing seeds to store. Keep the lid slightly open to allow moisture to escape.

This cold frame doubles up as a solar dryer. Spread seeds thinly to dry. Use trays to keep different varieties separate.

LARGE-SCALE STRUCTURES

With the warmth and protection of a polytunnel or greenhouse, it's more than viable to become self-sufficient in a range of vegetables. What you lose in ground space on your plot, you more than make up for with the extra weeks of growing time gained.

Greenhouses

A greenhouse has to be one of the nicest places to work in winter. The warmth and smells make it a bit like going on holiday.

As with all methods of growing under cover, choosing where to site it is crucial. Our greenhouse was orientated east-west on a south-facing wall, which gave lots of light, and the wall acted as a heat sink.

Greenhouses are expensive to buy new but are often available second-hand if you are prepared to dismantle one. They can be surprisingly productive for their size if you get the layout right. Optimize vertical space by installing tiers of shelves for pots and trays, and by hanging baskets inside. Try growing tomatoes, herbs, aubergines, chillies, and salads, and starting off all sorts of seedlings.

In winter the temperature should not fall below 4°C (40°F) at night and it can get surprisingly warm in sunny weather. Installing a heat sink (see pages 130–131) will keep the air warm in winter, and will also circulate air in summer to keep the temperature down and reduce the risk of disease. To stop a greenhouse overheating, hang a damp towel in the open doorway and keep the roof and side vents wide open.

Polytunnels

Commercial polytunnels are a familiar part of our landscape but they're not just for farmers and market gardeners. We built our first one at the bottom of a small suburban garden.

Polytunnels are often made from polythene or PVC stretched over metal frames and are simple to put together. Dig down around the frame and lay some turf over the edges of the plastic; this keeps it stretched tight and avoids any embarrassing kite fiascos in high winds.

At Newhouse Farm we built raised beds down either side and installed benches for potting up seedlings. We added a pond to help retain heat and encourage frogs to keep down pests. In summer the pathway down the middle was full of pots holding the taller plants like tomatoes and cucumbers, which grew incredibly well tied up to the centre of the frame and were easy to harvest by walking along the rows.

Geodesic domes

Geodesic domes are not new; they have been around since the beginning of the 20th century. Championed by inventor Buckminster Fuller, they are designed to have a minimum external surface area with far fewer internal building supports, giving maximum growing space.

Geodesic domes also make the most of solar energy. Sunlight shining into a conventional structure like a greenhouse creates warm air pockets that can damage plants. The dome shape promotes good air circulation and prevents pockets developing.

We ordered our geodesic dome in kit form. Putting together a building that gains its strength only when the last piece is put in place is a bit of a nightmare, but a lot of fun. We think it looks like a spaceship. For more on geodesic growing, see pages 114–115.

Conservatories

Most conservatory owners know that they are excellent places to grow plants. You can start off an array of seedlings in a conservatory and make it a permanent home for exotics. Citrus and olive trees, grapevines, and physalis plants do particularly well.

UNDERCOVER JOBS

Exotics such as chillies have to be grown under cover in temperate climates. You'll get an earlier harvest from many crops if you start off seedlings in a greenhouse or polytunnel, and a protected environment can fool other vegetables into cropping out of season.

Take cuttings from chillies. Remove a few leaves to increase the chances of success, and pot up in a rehydrated coir plug or cutting compost (see page 147).

Start off sweetcorn seedlings under cover for an earlier harvest. Harden them off (see page 146) before planting outside when there's no longer any risk of frost.

Using a polytunnel

Polytunnels are great places to grow everything from spinach to salads, but their usefulness doesn't end there. Use the space to dry garlic, onions, and herbs before storing, and to bring tender plants such as citrus trees under cover in winter – and make a workstation where you can repot plants or take cuttings. Our polytunnel was a carefully planned mix of permanent plantings such as vines and olives, crops such as tomatoes and cucumbers that would spend their entire life under cover, and others that benefitted from a warm start but were destined for the vegetable plot. We even had a propagator in the polytunnel, which acted as a double skin to retain the sun's heat and was ideal for starting off tomato and melon seedlings.

Hang up garlic to dry in the airy warmth of a polytunnel before storing it.

Maximize productivity
Use your undercover space to start plants off early, keep salads and herbs producing over winter, and grow more exotic crops such as melons and aubergines that won't survive outside in a temperate climate.

Pest control is simple if you keep chickens. Let them loose at the end of the season to root out and eat any overwintering pests.

Growing tomatoes in a polytunnel gives big yields and they are easy to harvest. Support plants using strings attached to the frame.

Beetroot seedlings for planting out and to cut as salad leaves.

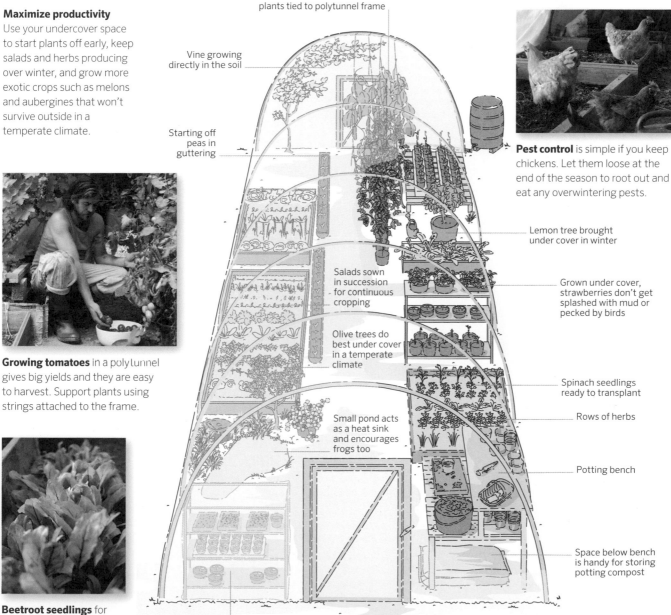

Tall tomato and cucumber plants tied to polytunnel frame

Vine growing directly in the soil

Starting off peas in guttering

Salads sown in succession for continuous cropping

Olive trees do best under cover in a temperate climate

Small pond acts as a heat sink and encourages frogs too

Large propagator with shelves and its own cover

Lemon tree brought under cover in winter

Grown under cover, strawberries don't get splashed with mud or pecked by birds

Spinach seedlings ready to transplant

Rows of herbs

Potting bench

Space below bench is handy for storing potting compost

MAKING A HEAT SINK

Even growing under cover, plants still need extra protection from frost. Greenhouses and polytunnels (see pages 128–29) heat up quickly but, sadly, cool down just as fast. The challenge is to find a way of keeping the temperature above the danger zone where plants can be damaged – we're not saying warm, just warm enough. That's where the heat sink comes in.

WHAT IS A HEAT SINK?

A heat sink is something that can capture the sun's heat and store it during the day, then give it off when required – like a night-storage heater.

Rather than trying to generate heat to keep our greenhouse frost-free, we like the idea of capturing it. Even in the depths of winter the sun warms greenhouses and polytunnels, but the lack of insulation means the heat disappears quickly as the sun sets.

So we decided to stop some of that free heat escaping and store it. Our system uses a small fan that runs continuously. During the day it blows warm air through a box buried in the ground and filled with lumps of glass – the heat sink. At night it blows cooler air through the warm glass, which warms the air. This warm air stops the temperature in the greenhouse falling below freezing. For added frost protection we also built our lean-to greenhouse against a south-facing stone wall, which itself acts as a heat sink.

CONSTRUCTION

Our heat sink is a hole roughly 1cu m (1.3cu yd), dug in the floor of the greenhouse. We reckoned this would be big enough for our 2.4 x 4m (8 x 13ft) greenhouse. We insulated the hole with discarded polystyrene to stop heat being lost to the earth, before filling it with imploded glass to absorb the heat. Imploded glass has amazing thermal properties but we realize it's not always that easy to get hold of this specialist type of glass: rocks, stones, glass bottles or jars – filled with water – will work just as well. Making the pipework is a great opportunity to recycle odd bits of leftover plumbing. You'll need a variety of connectors and lengths of standard plastic piping. Drill holes to perforate the pipes before assembling. Start with a T-junction connector where the inlet pipe enters the ground and connect a series of pipes to form a big loop that returns to the T-junction. There's no need to apply sealant between joints.

THE BENEFITS OF A HEAT SINK

Very little power is needed to run the system, which relies on an inexpensive computer fan designed to run for years on very low power. You can generate the power yourself if you use a solar PV panel to charge up the battery.

No nasty fumes from conventional paraffin greenhouse heaters.

Unexpected frosts won't catch you out. The heat sink runs regardless, so no need to constantly check the weather forecast.

Prevent diseases such as botrytis by keeping the system running all year round to circulate air.

THE RESULTS

It's very hard to say exactly how efficient our system is, but combined with the south-facing rear wall, it has managed to keep our greenhouse frost-free. Salads continue to grow deep into the winter – no mean feat considering we regularly have frosts with temperatures sometimes dipping as low as -6°C (21°F). A winter of extra productivity and the system paid for itself: not a bad investment.

How the heat sink system works

During the day a fan sucks hot air into a pipe opening at the apex of the roof **(1)** and down into a network of perforated pipes surrounded by heat-absorbant material **(2)**. At night, the fan draws down cool air into the sink **(3)**, where it is warmed by the absorbed heat **(4)**. This newly warmed air then escapes via a vent and in turn warms the greenhouse **(5)**.

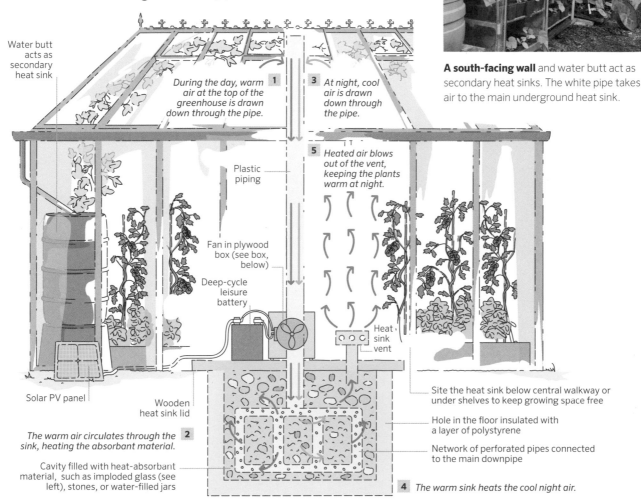

A south-facing wall and water butt act as secondary heat sinks. The white pipe takes air to the main underground heat sink.

Water butt acts as secondary heat sink

1 *During the day, warm air at the top of the greenhouse is drawn down through the pipe.*

3 *At night, cool air is drawn down through the pipe.*

5 *Heated air blows out of the vent, keeping the plants warm at night.*

Plastic piping

Fan in plywood box (see box, below)

Deep-cycle leisure battery

Heat sink vent

Solar PV panel

Wooden heat sink lid

Site the heat sink below central walkway or under shelves to keep growing space free

Hole in the floor insulated with a layer of polystyrene

Network of perforated pipes connected to the main downpipe

The warm air circulates through the sink, heating the absorbant material. **2**

Cavity filled with heat-absorbant material, such as imploded glass (see left), stones, or water-filled jars

4 *The warm sink heats the cool night air.*

INSTALLING THE FAN

1. Wire up the fan. **2. Make a small plywood box** to house the fan. Make a lid to seal the box. Cut out holes in the sides to fit the air pipes, and drill holes for the wiring. **3. Connect the fan** to a deep-cycle leisure battery using crocodile clips. Screw the lid onto the fan box. Keep battery and fan box in a waterproof crate to protect from damp. Cut holes in the crate to match the holes in the fan box for the air pipes. **4. Push the pipes** through the holes in the crate and into the fan box. **5. Connect the battery** to a small solar panel outside the greenhouse and the system is up and running.

ENCOURAGING WILDLIFE

Simply by gardening organically, without chemicals, you've already made your plot wildlife-friendly, even if you do nothing else. By adding a few extras such as nectar-rich flowers and a small pond or bird table to attract the right wildlife, we've noticed a positive effect on the productivity of our cultivated areas for little effort and outlay.

MAKING HABITATS

Different creatures like different environments. A shady log pile will shelter frogs and toads, while a rock pile in a sunnier spot will encourage slow worms. You need them on your plot because they all eat slugs.

A nettle patch attracts ladybirds. Adults and larvae eat huge amounts of aphids (up to a 100 a day!) and polish off pests such as potato beetles, mealy bugs, and whitefly.

Set up a hedgehog house – a wooden box with a narrow entrance tunnel – to encourage hedgehogs to stay. They eat slugs and other pests such as cut worms and wire worms.

Create wildlife corridors

Use these to link habitats and help insects and small animals to move around the garden. All you need to do is leave areas along boundaries, under hedges, and against fences to grow unchecked. You could even ask your neighbours to do the same and the result would be a natural wildlife highway across your neighbourhood.

Planting for wildlife

Native plants are the best choice for encouraging biodiversity as they support a wider range of wildlife. Bees like flowers that are rich in nectar such as borage, lavender, lemon balm, mint, chives, and other herbs. Useful insects such as hoverflies and parasitic wasps, which prey on aphids, prefer flat flowerheads, especially those packed with lots of tiny flowers, so plant dill, fennel, yarrow, daisies, and poached-egg plants.

Don't deadhead flowers – seedheads make good snacks for small seed-eaters. Cornflowers, teasel, and sunflowers are ideal.

FEEDING

Put seeds and nuts out when natural food is in short supply, then birds will be used to visiting your garden by the time insect pest populations build up. Feed early bees and butterflies when flowers are scarce by investing in a nectar station. It looks a bit like a bird feeder but it holds a sponge soaked with sugar water and is bright yellow to attract insects.

NEST BOXES

We see birds as both friends and foes. It's great to have their company in the garden, but our fruit bushes need netting to prevent them from harvesting before us. On the other hand, birds like to eat insect pests, and so we encourage them to visit. Nest boxes help: set them up out of reach of cats, and facing between south-east and north to avoid them overheating in strong sunlight.

INSECT HOUSES

As well as a bee B&B (see opposite), leave clay pots stuffed with moss and dry grass at ground level for bumble bees, which hibernate underground.

Cater for overwintering ladybirds by opening a ladybird hotel. You can make your own by drilling a log full of holes (see the bee B&B instruction for tips) and setting it vertically on a pole. Or cut short lengths of hollow bamboo and twigs, tie them into bundles, and hang them around your plot.

These small changes can make a big difference to a useful insect's survival rate.

1. Bird boxes quickly attract feathered residents who will help manage the insect pest population. **2. Ox-eye daisies** have flat open flowers that attract useful insects. **3. Poppies** are another favourite of bees and other beneficial insects.

Construct homes for wildlife

There are occasions when you can improve on nature. A home-made bee B&B makes an ideal shelter and nesting site for solitary pollinating bee species such as mason bees. And we'd say that digging a pond is one of the most wildlife-friendly steps you can take in your garden: it's a water source for small birds and mammals, and a vital breeding habitat for frogs.

YOU WILL NEED
Saw
Drill with various size bits
Hammer
Wood offcuts and a fairly thick log
Bracket, nails, screws

BUILD A BEE B&B

1. Use a flat piece of wood for the base. Cut a log to size to fit the base. **2. Drill a range of holes** in one end of the log using drill bits of various sizes. Drill at a slight upward angle so that rain can't drip in. **3. Add a bracket** to the base and screw the base to the log. **4. Make a roof** with some spare wood. **5. Attach the B&B** to a south-facing wall or fence, near bee-attracting plants, and wait for the bees to move in.

MAKE A WILDLIFE POND

Dig your pond at least 60cm (2ft) at its deepest. It must have a shallow area too, as varied depths will allow your pond to suit different plants and pondlife. Rake the soil to remove sharp stones, then spread out the underlay and liner. Use one or two rocks to hold them in place while you fill with water. Edge with stones or rocks.

YOU WILL NEED
Spade, rake
Underlay, such as carpet
Pond liner
Edging stones or rocks

A bucketful of water from another pond will encourage creatures such as pond snails and water beetles to visit your pond.

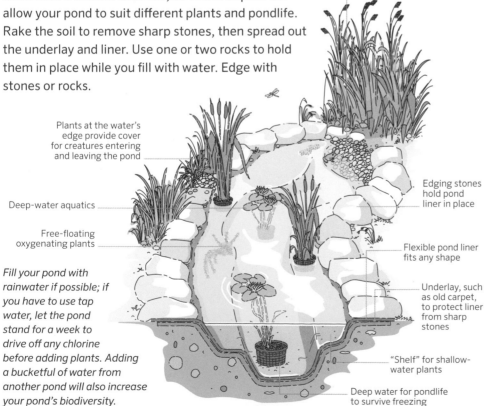

Shallow "beach" area for creatures to climb out easily

Plants at the water's edge provide cover for creatures entering and leaving the pond

Deep-water aquatics

Free-floating oxygenating plants

Fill your pond with rainwater if possible; if you have to use tap water, let the pond stand for a week to drive off any chlorine before adding plants. Adding a bucketful of water from another pond will also increase your pond's biodiversity.

Edging stones hold pond liner in place

Flexible pond liner fits any shape

Underlay, such as old carpet, to protect liner from sharp stones

"Shelf" for shallow-water plants

Deep water for pondlife to survive freezing

Dragonflies spend part of their lifecycle under water. They eat midges and other insect pests.

Gardening in harmony with nature

We encourage wildlife on our plot because improving biodiversity is an effective method of pest control and boosting productivity. This dedicated wildlife garden illustrates different ideas and the benefits they bring, so that you can choose suitable elements to add to your own plot. And don't forget, urban gardens can be wildlife havens too.

KEY TO GARDEN PLAN

1	Nettle patch
2	Bird nesting box
3	Bat box
4	Small pond
5	Rock-pile habitat for slow worms
6	Bee B&B
7	Flowers for useful insects
8	Ladybird hotel
9	Nectar station for bees and butterflies
10	Bumble bee house
11	Bird table
12	Hedgehog house
13	Log-pile habitat
14	Mini wildflower meadow

A hedgehog house (12) placed in a quiet corner of your garden can be a home for a natural predator of unwanted slugs and snails.

Flowers such as Michaelmas daisies (7) will attract butterflies that will then move on to pollinate your crops.

A mini wildflower meadow (14) benefits all wildlife – tall grasses provide safe cover for little creatures and birds like to eat the seeds, while wildflowers attract butterflies and bees.

Make a log pile (13) by stacking up logs, vines, and other twiggy trimmings to create a habitat for frogs, newts, and toads, as well as solitary bees and insects which shelter in the gaps.

Birds can act as pest control – some snatch insects on the wing, while other species, such as sparrows, love to munch aphids and caterpillars at ground level.

Feed the birds (11) by hanging up fat balls packed with nuts and seeds.

A small pond (4) provides a place for frogs, newts, and toads to breed – and the good news is they all eat slugs and other garden pests.

A bee B&B (6) made from bamboo in a plant pot will encourage solitary bees to take up residence.

Spiders are an important part of a garden's ecosystem, playing their part in keeping down insect pests and being in turn a valuable food source for small garden birds like wrens.

GROWING IN URBAN SPACES

Space may be restricted in towns and cities, but it's still possible to grow some of your own food in a small garden or even on a balcony or large windowsill. By using simple and intensive gardening methods, you'll soon be picking salads and herbs, or even fruit and vegetables – and they can all be grown conveniently just outside your back door.

THINK OUTSIDE THE BOX

Balconies, flat roofs, and even sloping roofs can all help to optimize your area for growing food, while using vertical walls makes sense in an urban environment where space is often severely limited.

But you don't have to sacrifice everything for food production. Balancing productivity with a place to relax after work is easy if you rethink what is considered ornamental. Chard with colourful stems, architectural artichokes, and peas and beans grown for flowers as well as crops will all make your space more beautiful. And include flowers that work harder; some act as natural pest controls (see page 111) while others can be eaten (see page 181).

GROWING IN CONTAINERS

Almost any fruit or vegetable will grow in a container, and this form of gardening offers a manageable introduction to growing your own. Container growing is cheap, easy, suitable for any size of plot, and moveable. We spent years moving from house to house with our collection of kitchen-garden plants in pots, ranging from herbs to fig trees.

Use any type of container for growing plants, from terracotta pots and plastic bottles cut in half to old paint tins and buckets.

Drill drainage holes in the bottom of your containers. Cover the holes with a layer of broken pots or small stones to prevent them clogging up.

Plants in pots are not self-reliant: water them regularly, feed them, and replace the compost each spring.

GROWING IN RAISED BEDS

Raised beds are ideal for areas where there's no soil or it's poor quality.

1. Urban balconies can accommodate lots of produce if you interplant and grow trailing plants or climbers that optimize the space. **2. A raised bed** can be handy for growing salad close to the house and will look great in summer if you plant a selection of edible flowers. They can also be cheap to make if you can get your hands on scrap materials. **3. Wall mounted planters** are excellent for squashes, which will hang down and make the most of south-facing walls.

Building a raised bed (see page 138) makes planting and harvesting so convenient; you can even make them waist height to avoid back strain.

A wide range of materials can be used to build a raised bed.

Raised beds on hard surfaces are like containers, and you'll have to water and feed plants regularly.

GROWING IN HANGING BASKETS

Hanging baskets aren't just for flowers; you can grow fruit and vegetables in them too. Strawberries, tumbling tomatoes, salads, and herbs will thrive in them. Use a bracket to fix baskets to a wall, or hang edible hampers from trellis or fencing.

Pests such as slugs and snails find scaling hanging baskets a mission impossible, especially if you hang them with copper wire, which they will be loath to cross.

Baskets dry out quickly, which means they need regular watering. Consider an automatic irrigation system (see pages 122–123).

GROWING VERTICALLY

Vertical growing is all about making the most of the space you have available by growing upwards instead of outwards – and you don't even have to have a garden to do so.

Wall-mounted fabric plantholders can maximize growing space and work just as well inside as they do outdoors. You can even make your own from worn-out clothes. Add small pebbles at the bottom of the pockets to help drainage.

Grow climbing plants – anything from runner beans to grapevines – to transform walls, fences, pergolas, and trellis into productive areas.

PROJECT **Plant up a hanging basket**

We used a basket we made from willow withies, but you've probably got one lying around that will do just as well. You also need a tough plastic liner, such as a section of an old compost bag. We planted our basket with alpine strawberries. They're easy to grow from seed and produce fruit the same year. Or try salad leaf mixes, herbs such as thyme and basil, and tumbling cherry tomato plants.

YOU WILL NEED
Basket
Coco liner
Compost, straw
Copper wire
Strawberry plants

1. Line the basket and pierce a few holes in the liner for drainage. Fill with compost. Use copper wire to make a handle to hang the basket.   **2. Make holes** in the compost and insert the plants. Firm compost around plants. Water well.   **3. Mulch with straw** to reduce evaporation and keep soil moist.   **4. Hang the basket up** in a sunny spot. Water daily.

PROJECT **Build a raised bed**

Building a raised bed is a great way to turn a small space, even one where there is no soil, into a productive plot. Of all the different gardening methods we've experimented with, our raised beds take the least effort to maintain. They are easy to weed and water, there's less bending involved – ideal for older gardeners – and the soil doesn't get compacted as it is never walked on.

YOU WILL NEED

Saw	Screws, nails
Electric drill	Pond liner or tarpaulin
Hammer	Broken pottery or stones
Wood – salvaged floorboards are ideal and cheap	Compost

Decide where to build your bed based on the available heat and light. We made ours against a south-facing wall to absorb more heat during the day and retain it overnight. It's shown planted with squash, chamomile, lettuce, and runner beans.

BUILDING THE FRAME

1. Cut wood to length. Vary the lengths of the pieces, so that in the first layer the front overlaps the sides, in the second layer the sides overlap the front, and so on. Staggering the end joints in this way strengthens the frame. The average size for a raised bed that is easy to reach from all sides is 60cm x 90cm (2ft x 3ft). **2. Screw each layer** together. We used long Posidrive screws. **3. Build up the frame** in layers. **4. Strengthen the finished frame** by adding corner posts.

LINING AND FILLING THE BED

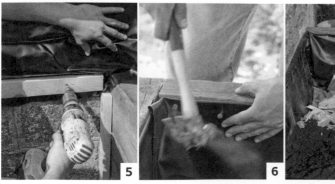

5. Line the bed. We used a piece of old pond liner and attached it to the wall using a wooden batten. **6. Fold the liner** to fit, then nail the edges to the frame. The liner stops moisture soaking into the wood and retains moisture in the compost.
7. Fill the bottom 10cm (4in) with broken pottery and stones for drainage. Add good-quality compost, water it well, and start planting.

TRY THIS

Lift some slabs on your patio and use them to build a raised planter. Stand them on end and bury them up to half their depth for stability.
Dismantle an old rockery and reuse the stone – it stores heat and keeps your plants' roots warm.
Experiment: build walls from sheets of tin, old car tyres (see page 48), or glass bottles cemented together. Or stack a few tyres on top of each other for an instant raised bed or potato tower.
Create an E-shaped bed. The recesses allow you to reach plants at the back easily – particularly useful for larger raised beds.

PROJECT **Make a wormery**

Conventional compost bins take up a lot of space and the process can be slow. A wormery turns kitchen waste into superb compost quickly, thanks to the action of brandling worms. These aren't earthworms but worms that live in leaf litter on top of the soil. A wormery re-creates their ideal habitat, which is warm, dark, and damp. The easiest thing to do is to buy a kit that consists of a series of stackable units, and comes with a tub of worms. As you fill each unit with waste you simply add another on top. The worms also produce a liquid nutrient that filters down through the wormery – we call it worm tea. Dilute it about 1:10 with water before feeding it to your plants.

A wormery in action takes up minimal space and even has room for a plant. Copper tape keeps slugs away.

DO FEED YOUR WORMS
Fruit and vegetable peelings
Loose leaf tea and coffee grounds
Paper and cardboard (shredded)
Old socks (not synthetic fabrics)
Flowers
Pet and human hair
Fallen leaves
Small amounts of grass cuttings
Dust from the vacuum cleaner
Raw eggshells

DON'T GIVE THEM
Meat, fish, and dairy products
Dog and cat faeces
Too much citrus or too many onions
Cooked food

ASSEMBLING THE WORMERY

1. Stand the base unit with the tap on a level piece of ground close to your kitchen. Push one of the corrugated rings firmly into place. These stop the worms escaping. **2. Stack the unit** with the fine mesh bottom on top of the base unit and add the tub of worms to it. **3. Feed the worms** with some fresh kitchen scraps. Chop up any big pieces first. Feed them small amounts until they settle in. **4. Cover them** with shredded paper. Lift the paper up to add food. You can tell when the worms are coping with the waste you give them: they will be active and you'll see they are composting what you put in.

5. Add another corrugated ring and composter unit. Add a final ring to the top unit and put the planter on top to act as a lid. **6. Fill the planter** with potting compost. **7. Plant the planter** with vegetables or herbs. We added a courgette. The top planter doesn't benefit from the worms below – it's just a way of making every bit of space in your plot productive and a neat finishing touch to your wormery.

CARE TIPS

Start using the unit above when the bottom composter is full of scraps. Take a couple of handfuls of worms and waste from the bottom layer to start up the next one. When the worms are in full production they will compost scraps within a couple of weeks.
Drain off the worm tea weekly.
Inspect your worms regularly. If the compost looks dry over the summer, add a little water. If it looks too damp, then add more paper, cardboard, or a handful of straw.
Don't add more scraps if there are no worms on the surface of your food waste. Too much food will starve your worms of air, leading to a soggy slimy mess and eventually even worm fatalities.

TAKING ON AN ALLOTMENT

If you have a small garden, an allotment provides valuable extra space in which to grow your own food. Most are also very social places, where you can swap seeds, exchange produce, and pick up tips from other keen plot-holders. An allotment allows you to practise planting on a larger scale, too, and you can grow enough to preserve and store your produce for year-round use.

WHAT TO GROW

Try our recommended annual crops for first-time allotment holders.

Garlic is very easy to grow and can be stored for months.

Leeks are high-yielding crops for a winter harvest.

Beetroot is another high-yielding crop. You can use the young leaves in salads, as well as the roots.

Squashes store well for winter but they require a lot of space to grow.

Rainbow chard is one of our favourites. It has incredibly colourful stems, and it grows quickly.

Radishes grow rapidly – sow them between slower-growing crops.

Broad beans can be sown in early spring and autumn for a steady seasonal supply, but watch out for pest damage and disease.

Climbing and runner beans make the most of vertical space. Grow them at the northern end of your plot so they don't shade other crops.

Courgettes and summer squashes need little work for a high yield. Try a no-watering policy (see page 111).

Potatoes are a no-brainer on an allotment. If you want to optimize your space, try growing them in stacks of tyres, barrels, or old compost bags, as well as in the ground.

Courgettes taste best when small, before they turn into marrows.

ASSESS YOUR TIME

The problem with allotments is their sheer popularity. Waiting lists can be long, and by taking on an allotment you will inevitably be delaying someone else's chance to grow their own food, so if your garden at home isn't used for crops, cultivate that first. Also think about how much time you will be able to give; if you don't keep your allotment looking ship-shape, you might be asked to give it up. The obligation to keep your plot tidy and productive is a great incentive to get the most out of it, but the pressure can be daunting. We would recommend applying for your first allotment with a friend or neighbour if you are in any doubt about how you will cope.

READ THE RULES

Always check your local allotment bye-laws, as they can vary across the country. Depending on where you live, you may be allowed to keep small animals or bees, though you'll likely find restrictions on planting trees, building structures, and digging ponds, among other things.

FIRST STEPS

You rarely get to choose your plot, and will probably be forced to work with what you're given. This presents a great opportunity to do a mini site survey (see pages 204–205). Look at the slope of the land and type of soil. Note which areas get the most sunshine and which are shady.

Before you start planting, condition the soil with plenty of well-rotted manure and fertilizer (see pages 116–121), and perhaps some sharp sand, mushroom compost, or wood ash if it is heavy clay.

When deciding what to grow, think about what will be most suitable for each of your productive sites. For example, grow herbs and salads that you use most days at home – perhaps on a balcony or windowsill – and plant crops that require less attention on your allotment. You can also start seeds off at home, then transplant them to the allotment.

GROW PERENNIALS

One of the great things about perennials is that they need little maintenance – ideal if you don't have much time. They are also relatively hardy. You could go for soft fruit such as strawberries, raspberries, and gooseberries, and a variety of vegetables, including perennial rocket, perpetual spinach, globe and

TIME-SAVING TIPS

Mulch areas to reduce weed growth (see page 112) until you are ready to work them.

Try solarization, which also helps to suppress weed growth. This novel technique involves laying clear plastic sheets over large freshly dug areas and keeping them in place with stones or wood. Once the ground has heated up to about 26°C (80°F), plant seedlings that you've grown on your windowsills at home, directly into the ground. Cut two slits in the plastic in a cross shape, peel it back and plant. Keep plants well watered. Any weeds that start to sprout under the plastic won't be able to withstand the heat and will wither and die. This is also a great way to reuse plastic.

A no-watering policy for summer squashes and courgettes reduces the workload (see page 111).

Jerusalem artichokes, and, of course, rhubarb. Although it's not edible, comfrey is also worth growing as a useful fertilizer (see page 119).

If your site has restrictions on planting trees, try dwarf varieties or training trees as cordons or step-overs – you may be able to fit them in without breaking the rules.

ACCESS TO WATER

Keeping plants well watered can be a challenge. Sites usually have taps, but they may not be close to your plot. To preserve water, plant your crops further apart than the recommended spacing, giving the roots more room to search for moisture. Also mulch between rows to reduce evaporation and help retain soil moisture. If you have a shed on your allotment, attach a water butt to capture rainwater from the roof.

SHEDS AND STRUCTURES

If you are allowed to put up a shed, it will save you carrying tools to and fro – especially useful if you live some distance away. Add a decent lock to store equipment safely. If you have a garden at home too, you'll need to double up on basics, such as a spade and fork. For the ultimate well-organized shed, see pages 142–143.

Other permanent structures to consider are a cold frame to protect small plants (see pages 124–127), and a double compost bin made out of old pallets to recycle all your allotment's organic waste.

1. All the features of an ideal allotment, including shed, compost heap, and organized beds. **2. Water butts** collect a significant amount of rainwater from a shed roof. **3. Plant out seedlings** quickly and gently. **4. Fruit cages** and fleece tunnels protect plants from pests.

SETTING UP A WELL-EQUIPPED SHED

The shed is the hub of a sustainable plot as it keeps it functioning efficiently, while the right tool for the job makes all the difference. We love our tools and tend to favour manual over mechanized. While machines are sometimes best for getting a job done, in general, we believe more traditional tools help you become more connected to the land and more productive overall. Hand tools are nearly always cheaper and easier to maintain.

KEY TO SHED

1. Spades
2. Garden fork
3. Rake
4. Post driver
5. Scythe
6. Long-handled shovel
7. Axe
8. Broom
9. Pitchfork
10. Sledgehammer
11. Hoe
12. Mattock
13. Pickaxe
14. Tool rack
15. Fishing rod (p.267)
16. Shelving
17. Squashes storing over winter (pp.166–167)
18. Egg incubator, infra-red lamp (p.242)
19. Bee-keeping equipment (pp.270–271)
20. Demijohns for brewing (pp.196–201)
21. Hand trowel
22. Hand fork
23. Watering can
24. Petrol can
25. Plant pots
26. Dibber
27. Bin for animal feed
28. Wheelbarrow
29. Bucket
30. Spare wellies
31. Riddle
32. Lawnmower
33. Bee suits (pp.270–271)
34. Screw press (pp.196–201)
35. Apple chopping box (pp.200–201)
36. Straw for animal bedding
37. Sack trolley
38. Chitting potatoes (p.156)
39. Industrial wood chipper (pp.92–93)
40. Rotovator

Hanging tools up saves space in a small shed.

You never know when bits and bobs will come in handy for an eco-project.

A dibber (26) is useful for repotting small plants and for planting onion sets. Look out for different sizes.

Spades (1) receive plenty of use so it's worth investing in really solid ones that won't bend.

A riddle (31) is ideal for sieving soil or home-made compost to get rid of stones and bits of twig.

Keep a good selection of hand tools for the garden.

Store squashes (17) on the shelves if your shed is frost-free. Turn them regularly so they don't rot.

Shelves and hooks keep a big shed tidy and well organized. Use a floor-standing rack to stop tall tools falling over. Put trays and boxes on the shelves to store small items neatly.

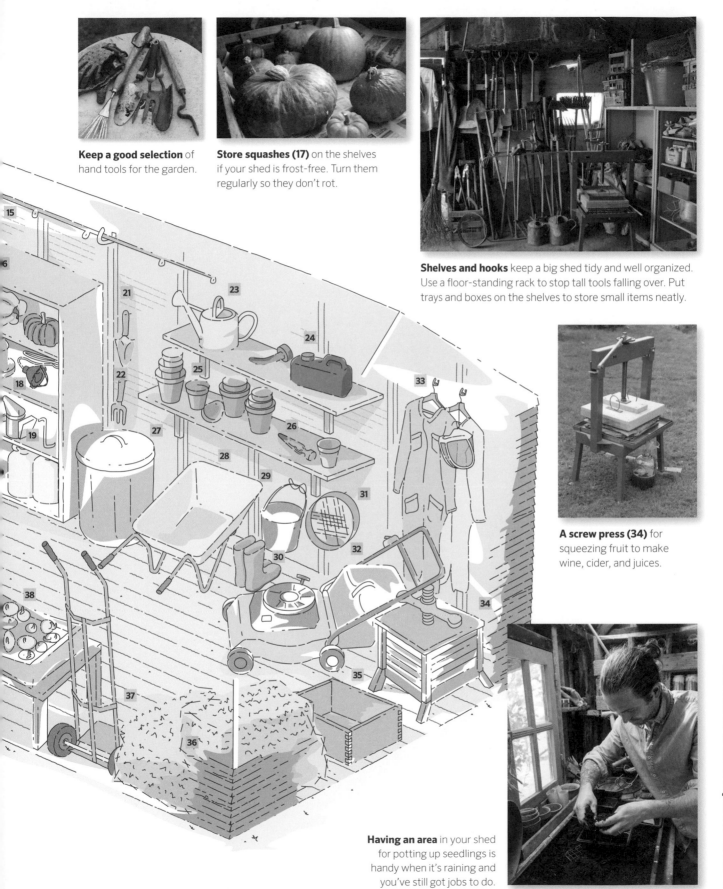

A screw press (34) for squeezing fruit to make wine, cider, and juices.

Having an area in your shed for potting up seedlings is handy when it's raining and you've still got jobs to do.

WHAT TO GROW

The variety of plants that you can grow for food is absolutely amazing, from quick-growing salads to slow-maturing walnut trees. If you are able to grow under cover too, you can extend the growing season and try out some more exotic crops, for the sheer fun and challenge of it, but also to reduce your food miles and still be able to enjoy a few tropical treats. Our philosophy is: try growing anything once and if it grows well and you enjoy eating it, grow it again! There is something very special about planting fruit and nut trees that you know will provide produce for generations to come. And don't ignore nature's larder of wild foods to forage...

SOWING AND PLANTING

Successful sowing and planting is all about getting the timing right – from starting seeds off under cover so that the seedlings are ready to plant out when there's no risk of frost, to waiting for the soil to warm up before sowing outside. By using succession sowing – where you sow more seeds every month – you can also guarantee a harvest throughout the growing season.

PREPARING THE SOIL

Choose a warm area for a seed bed and add plenty of well-rotted organic matter, such as home-made compost, to improve it. Spread over the soil in autumn and fork in or leave the worms to do it for you. Compost not only adds nutrients, it also improves structure, making clay soils easier to work and sandy soils more moisture retentive.

Dig heavy clay soils in autumn and leave the frost to break up large clods. Any left in spring can be easily broken up with the back of a rake. Another option if your soil is clay or full of rubble, is to build raised beds on top of it (see page 138).

Before sowing seeds, rake the soil to achieve a crumbly, stone-free structure, known as a "fine tilth".

FROM SEED TO BED

Buy seeds from a catalogue or garden centre; you can also save seeds from vegetables you have grown (see page 166), or even organize a local seed-swapping event. Having space to start some seeds off under cover is invaluable. Polytunnels, cloches, and cold frames protect seedlings from frost and pests (see pages 124–125 and 128–129), and in temperate regions, they allow you to start your growing season earlier and make the most of short summers.

We sow seeds in a range of trays, pots, and modules, using our own home-made compost, sieved to remove lumps and twigs. Pots are ideal for big seeds such as courgettes and beans, while module trays or egg-boxes allow you to sow small seeds in individual units, which minimizes root disturbance when transplanting.

THINNING OUT

Seedlings grown in trays need thinning out so that they have more room to grow. Water them first, then hold a seedling between your thumb and forefinger and pull gently. Leave seedlings spaced at 2.5–4cm (1–1½in). Transplant thinnings into seed trays to grow on – or eat them as microgreens.

HARDENING OFF

Plants grown under cover need to be acclimatized before transplanting outside. Set trays outside and bring them in at night, or open the lid of a cold frame during the day and close it at night, until there is no risk of frost.

SUCCESSION SOWING

Sowing at intervals throughout the year maximizes the growing season and avoids gluts. For example, start peas off under cover in February to harvest in May; then sow a few rows each month from March to July, and you'll be picking until October.

1. **Raking** to create a fine tilth ensures good contact between seed and soil.
2. **Add compost** before planting out seedlings to give them a nutrient boost.
3. **Use a drill marker** – string tied to two stakes – to sow seed in a straight line.
4. **Thinning and transplanting** basil seedlings, to give them space to grow.

Basic techniques

When you can see a pot or seed tray is full of roots, it's time to transplant seedlings. Sow seed outside when the soil has warmed up and is not too wet; follow packet instructions, but as a general rule, sow thinly and not too deep. Plant fruit trees and bushes in winter when they are dormant.

TRANSPLANTING SEEDLINGS

1. Gently ease the seedling (an alpine strawberry) out of the module. Hold it at the base of the stem and push up from below. **2. Use a dibber** to make a hole – either in a pot of compost or outside in the soil – and set the plant at the same depth it was growing at in the tray. **3. Firm the soil** round the roots and water well.

SOWING OUTDOORS

Use a spade to make a ridge in the seed bed. **1. Sow seeds** thinly into a shallow drill made by drawing your hand or the side of a rake through the top of the ridge. These are radish seeds. **2. Cover** with finely sieved soil tilth. **3. Tamp down** the sides of the ridge with a spade. Water and leave to grow. Thin out the seedlings after a week or two, if overcrowded.

PLANTING UP CUTTINGS

1. Measure out equal volumes of home-made compost and sharp sand. Sieve the compost to remove lumps. **2. Mix the compost and sand** well. Sand improves drainage and reduces the risk of rotting. **3. Use the mixture** to fill pots and push one cutting – these are gooseberries – into each pot. Water well and leave them to grow in a sheltered spot. Do not allow the pots to dry out.

PLANTING TREES

Dig a hole twice as wide and as deep as the rootball. Drive a stake into the hole before planting. **1. Plant the tree** at the same depth it was growing at – you'll see a soil line on the stem. Mix the excavated soil with compost and use it to fill in around the roots. Firm in gently with your foot. **2. Tie** an old bicycle inner tube round the trunk and nail it to the stake. **3. Water well.** Add a mulch, avoiding the stem.

PROBLEM SOLVING

Our way of dealing with pests, diseases, and weeds is to use an integrated approach that takes into account the soil and the crops themselves. This means getting to know your pests and their natural predators. It's a method that places great store in the old expression "prevention is better than cure". It's not perfect, but it's cheap – and there's no risk to people or the environment.

DEALING WITH WEEDS

Keeping on top of weeds before they become a problem is the best approach. Mulching is extremely efficient at suppressing weeds (see page 112). Where it isn't possible to mulch, hoe off weeds regularly when they are small, especially on dry, sunny days when they will shrivel and die. Bigger weeds may need removing with a small fork or trowel. Always deal with weeds before they set seed. Remember: "one year's seeds means seven years' weeds." Digging is a satisfying way of clearing overgrown areas but sadly the benefits can be short-lived. It exposes dormant seeds that will germinate and create as many weed problems as you've solved. Try covering the ground with a cardboard mulch instead, let the weeds die down, then plant through. Many weeds are edible, so check pages 186–187 before choosing to compost them.

Try these two ways to tackle weeds. Mulching (top) suppresses weeds by blocking out light. Hand weeding is the best way to get rid of long-rooted dandelions.

START WITH THE PLANTS

To get the best results from our crops, we start by buying seed of pest-resistant varieties – usually traditional, local, proven varieties. Then we grow strong healthy plants that will be more resistant to attack. We boost our plants by feeding the soil. Worm tea (see page 139) is useful as it contains all sorts of beneficial fungi and bacteria, not just nutrients.

We also focus on the environment: crop rotation (see pages 120–121) reduces the risk of disease building up in the soil; good spacing between plants allows air to circulate, which helps reduce disease; winter digging exposes pests to bird predators; and companion planting in spring and summer (see page 111) attracts beneficial insects to deal with pests. Interplanting – growing two crops together – confuses crop-specific pests like carrot fly. These preventive methods are highly effective, but you should still keep a watchful eye.

ANIMAL PESTS

Moles are difficult to deal with. They eat huge amounts of worms, damage plant roots, and build mole hills in undesirable spots. Get a specialist trap and check it daily.

Birds can be deterred with a traditional scarecrow, which works best when moved around over the season. Recycle plastic bottles to make good bird scarers, or string up old CDs to create moving reflections. We use netting to stop birds getting at our soft fruit or young brassicas. Fine netting will also stop cabbage white butterflies and turnip moths laying eggs on leaves.

Rabbits can devastate a garden. The best option is to fence them out with 2.5cm (1in) wire netting. Bury it at least 15cm (6in) below the surface and make it at least 60cm (2ft) high.

SMALL PESTS

Aphids can decimate a salad or bean crop. We recommend interplanting with marigolds (*Calendula officinalis*), which works by attracting hoverflies that eat the aphids. Other predators include ladybirds, which you can encourage by building a ladybird hotel and placing it near crops (see page 132). When aphids appear on broad beans or peas, we pinch out the tips where they cluster, and throw these to the pigs.

Leatherjackets are big, fat, white, ugly larvae of the crane fly. They eat all sorts of plant roots. Whenever we dig one up, we feed it to the ducks.

Caterpillars are easy to spot and pick off by hand – or use barriers such as fine netting or fleece, which allow light to reach plants but keep out butterflies and other insect pests, and even birds and rabbits.

Cabbage root fly can be foiled by laying a 15cm (6in) square of rubber carpet underlay around the base of plants. This stops the fly laying eggs close to the roots.

Slugs and snails are deterred by a barrier of copper tape fitted around individual plants or pots. It works wonders, but is expensive. Try this if you don't like squashing or cutting them in half with a pair of scissors (a deed best done at night while wearing a head torch). We have tried other deterrents, such as egg shells, coffee granules, and soot, but find beer traps work best for us. The smell attracts slugs and snails, who fall in and drown. Empty the traps regularly to avoid an unpleasant concoction developing.

DISEASES

Cleanliness helps to avoid problems in the first place. Scrub pots and trays before using, pick up fallen fruit and leaves, and burn diseased plants.

Blight affects tomatoes and potatoes, causing brown and black marks on leaves and fruit. Remove and burn infected leaves quickly. It is a wet-weather disease, so provide plants with good drainage. We've dealt with the problem by growing our outside tomatoes upside-down!

Club root is a soil-borne disease affecting brassicas. If it becomes a problem, grow plants in a cardboard box filled with fresh compost and sunk into the soil.

Grey mould, or botrytis, affects plants grown under cover. It starts with brown spotting on the foliage, followed by furry mould. Improve air circulation, avoid overwatering, and burn infected shoots.

Rust appears as yellow, red, and brown pustules on leaves. It can kill an entire onion crop. Burn affected leaves, don't wet the foliage when watering, and avoid high humidity.

Silver leaf can develop on fruit trees, especially plums. A silvery hue appears on leaves. Cut back branches 15cm (6in) beyond the affected area.

1. String up CDs and bottle tops to deter birds. **2. Make a bird scarer** from a recycled plastic bottle. Cut side "wings" so it spins in the breeze. **3. Pour beer** into jars set level with the soil to trap slugs and snails. **4. Eject aphids** with a high-pressure blast of water. **5. Pick off caterpillars** as you spot them and feed to your chickens. **6. Netting** keeps birds and butterflies from brassicas. **7. Jagged edges** on cut-down plastic bottles, plus copper tape, discourage slugs and snails from reaching lettuces.

VEGETABLES TO GROW

Nothing beats the satisfaction of picking and cooking your own home-grown produce, from the first runner beans to late winter leeks. The biggest problem is deciding what to grow. We see it as a matter of balancing the "must haves" with the "should haves" and adding a few of the vegetables we think are a bit decadent, such as globe artichokes. You also need to take into account the suitability of your soil: there's more information on this on pages 116–121.

THE MUST HAVES

These are the basics that give you crops to eat all year. With a little planning and succession planting you'll be able to pop out to the garden to harvest something useful at any time of year. Sow seed outdoors, unless otherwise stated.

Beetroot

Beta vulgaris

HOW TO GROW Beetroot has a long season: succession sow monthly throughout spring and summer, starting in April. Sow thinly, a couple of seeds every 10cm (4in), about 2.5cm (1in) deep and in rows about 30cm (1ft) apart. Thin them out when the seedlings are about 2.5cm (1in) tall. Try growing a Golden Beetroot variety for another bright colour.

PROBLEM SOLVING Beetroot needs regular watering or it will be tough and woody to eat.

HARVEST, COOK, STORE When harvesting, twist off the leaves 5cm (2in) above the roots to stop them bleeding. Store in clamps or in a box of sand in a cool place (see pages 166–167). Young leaves can be used in salads and the small beetroots from thinning rows are tasty too. Naturally high in sugar, beetroot also makes a great base for cakes and brownies.

Broad beans

Vicia faba

HOW TO GROW Depending on your climate, sowing is usually February to May. Sow about 10cm (8in) apart and 5cm (2in) deep. We tend to plant in blocks so we can use trellis laid horizontally and staked at 30cm (1ft) tall to support the growing beans, though staggered pairs of rows with a 60cm (2ft) gap between them works well.

PROBLEM SOLVING The biggest pest is blackfly: pinch out all the growing tips where they gather. If you do this before they really take hold, you should have lots of succulent, untouched bean tops to eat.

HARVEST, COOK, STORE You should be able to pick beans from late spring all the way through to early autumn. They can also be dried for winter use, when they'll need soaking for 8 hours and boiling for 40 minutes. In spring, steam bean tops or blitz them with olive oil and Parmesan cheese for a yummy pesto.

Broccoli

Brassica oleracea Cymosa Group

HOW TO GROW Purple-sprouting is the hardiest and most prolific variety – that's why we grow it. It's great for late autumn and winter. Sow the seed in individual plugs or very thin drills in April under cover. When robust enough, transplant seedlings to their final growing position outdoors: leave about 45cm (18in) between them as the mature plants are huge.

PROBLEM SOLVING Birds love broccoli, so hang up a bird scarer or protect plants by stringing cotton between twigs to stop birds landing. They can get top-heavy, so stake plants as they grow or try the trellis support system, as with broad beans.

HARVEST, COOK, STORE Cut stems before flower buds open. Cut the central spear first. Keep cutting and the plants will carry on producing for most of the winter, provided you don't strip all the stems at once. Purple-sprouting is great stir-fried. Broccoli can be frozen, but the result is such a poor imitation of the fresh vegetable that we don't bother. Sprouted broccoli seeds are delicious in salads (see page 58).

Brussels sprouts

Brassica oleracea Gemmifera Group

HOW TO GROW It wouldn't be Christmas without Brussels sprouts. For the festive season sow your Brussels around Easter in individual plugs or thinly in drills under cover; transplant seedlings at the end of May, leaving at least 60cm (2ft) space between them, and then wait for winter. Kalette, a new hybrid of Brussels sprout

Celeriac is a valuable addition to your vegetable plot. You can use everything from the leaves to the roots.

and kale, is a delicious alternative for sprout-haters to try.

PROBLEM SOLVING Brussels are a low-maintenance veg, but watch out for birds, which will eat seedlings as well as helping themselves to the final crop. Protect seedlings by stringing cotton between twigs; hang bird scarers near mature plants.

HARVEST, COOK, STORE Start picking sprouts from the base of the stem – use a sharp knife. Sprout tops are also good for eating. Sprouts can be frozen (blanch them first) but we prefer to eat them in season.

Carrots
Daucus carota
HOW TO GROW Sow carrots thinly in rows 15cm (6in) apart and set the seeds about 1cm (½in) deep. Sow in succession from late winter through to early summer. Cardboard tubes from the middle of loo rolls make great biodegradable pots for carrot seeds. Heritage carrots come in all shapes, sizes, and colours, and taste great. Seeds for these unusual varieties are now readily available.
PROBLEM SOLVING Companion plant with onions to deter carrot fly, which will help protect the onions against onion fly.

HARVEST, COOK, STORE You should be able to harvest from June to December if you have succession planted. Lift maincrop carrots in October before the frosts and store in a clamp (see page 167) or root cellar (page 215). Pulled straight from the ground, wiped on a sleeve, and then crunched, carrots are a good way to get kids to eat vegetables. We also like roasting them or grating into carrot cake.

Celeriac
Apium graveolens var. *rapaceum*
HOW TO GROW There is no doubt that every kitchen garden needs celery or celeriac – the flavour is invaluable in so many dishes. We chose celeriac as it's easier to grow. Raise seedlings under cover in early spring. Plant them out at the end of May, about 30cm (12in) apart and 45cm (18in) between rows.
PROBLEM SOLVING Generally maintenance-free, though slugs can be a problem.
HARVEST, COOK, STORE Harvest the roots from autumn and all through winter – leave them in the ground rather than lifting and storing. Eat them raw or cooked: they are a pain to prepare because it is difficult to scrub all the dirt

off, but practice makes perfect. The leafy tops can be used in warming winter broths.

Courgettes, squashes, and pumpkins
Cucurbita family
HOW TO GROW We grow a lot of squashes and courgettes. Sow seeds under cover in April, in individual pots. Plant them out at the end of May. We always follow a no-watering policy (see page 111), which saves us a lot of hard work. Squashes and pumpkins take up a lot of space, but are good trailing down over raised beds, or grow them up netting and trellis.
PROBLEM SOLVING Slip a square of old carpet or slate under larger fruits to keep them clean and stop them rotting.
HARVEST, COOK, STORE It's essential to keep picking courgettes and summer squashes while they are young: miss a couple of days and all of a sudden you have marrows in your vegetable rack. Frying garlic, summer squashes, and a handful of tomatoes in olive oil makes a very quick and easy, colourful meal, or use a spiralizer and try low-carb courgetti instead of pasta. Store squashes and pumpkins on open shelves in a frost-free outbuilding. They will keep for months and are a useful standby for winter soups.

Kale
Brassica oleracea Acephala Group
HOW TO GROW Kale is an old-fashioned vegetable that has returned to popularity in recent years. It is very hardy and tasty, plus it's dead easy to grow and will flourish nearly everywhere. Sow seed in May and transplant the seedlings in July to their final position. Some of our favourite varieties include Cavolo Nero, curly kale, and red Russian.
PROBLEM SOLVING More or less trouble-free.
HARVEST, COOK, STORE Pick kale from November all the way through to the following May. It's a great standby – we nip out and whip off some nice-looking leaves when we need greens in a hurry. Like most brassicas it can be frozen, but we think it is only truly appetizing when freshly picked.

We think it's hard to beat parsnips roasted with some sea salt and rosemary but, if we have a glut, we like to ferment some too.

PROBLEM SOLVING Onions grown from sets have a tendency to "bolt" or flower too early. If this happens, cut off the flower stalk and use the onion straight away.

HARVEST, COOK, STORE The delicate flavour of shallots makes them popular with chefs but we mainly use them very early in the season as a substitute for spring onions – mature shallots are small and fiddly to prepare. Spring-onion-size shallots are usually ready long before the first true spring onions. Main-crop onions are ready to harvest around July.

Parsnips
Pastinaca sativa

HOW TO GROW We know parsnips are not as popular as they were but they are nonetheless on our "must have" list. Our efforts at growing them have been quite hit and miss. The little seeds can blow away easily, so take care when planting: spread them very thinly, a couple of seeds every 15cm (6in), in rows 30cm (12in) apart, at a depth of 1cm (½in). Sow them around Easter. They are slow to germinate. Thin out when the parsnips are the size of a golf ball.

PROBLEM SOLVING Dig over the soil and pick out stones to avoid growing misshapen roots. Improving the soil with well-rotted compost helps to produce long, straight parsnips.

HARVEST, COOK, STORE We dig our parsnips from October to the following Easter. The parsnips are happy sitting in the ground waiting for you to come and get them, but the tops wither away so you have to remember where you planted them. We really enjoy eating them, either roasted or curried in soup. They are a family favourite, and let's not forget they were one of our staple foods before the potato turned up.

Peas
Pisum sativum

HOW TO GROW Sow peas under cover in late winter and plant out when the risk of frost has passed. Continue sowing every couple of weeks (outside, once the weather warms up) until early summer.

Leeks
Allium porrum

HOW TO GROW Growing prize leeks is a mystical art, but we're just interested in growing them to cook with. Sow them in drills about 15cm (6in) apart and 1cm (½ in) deep. When they're big enough to transplant – about 20cm (8in) tall – get a handful of the plants, twist off all but 5cm (2in) of the green, and separate them. Using a dibber, make holes in the ground, about 5cm (2in) deep, and then place one leek in each hole. A good watering will fill the hole sufficiently; there's no need to firm the soil in place. They grow well, even in cold, wet areas.

PROBLEM SOLVING More or less trouble-free.

HARVEST, COOK, STORE Leeks may look like they're growing in the plot most of the year, but that's because they can be left in the ground and lifted as needed, from September all the way through to May. It's comforting to know they are out there

to be harvested all winter. Always ease the plant out of the ground with a fork or you will end up tearing off the outside layers or even pulling the leek apart. And please don't forget exactly how good the green part of the leek is. For some reason supermarkets have a tendency to chop off all the greenery to make handling and display more convenient – what they are really doing is throwing away the goodness. By all means chop off the tips, but cook the rest!

Onions and shallots
Allium family

HOW TO GROW Onions are usually pretty cheap to buy but there is something special about having a plait of your own hanging up ready to use. We grow onions from sets – miniature onion bulbs – rather than from seed, because the plants have better disease resistance. Plant them out in late winter, spacing them 10cm (4in) apart, in rows about 25cm (10in) apart.

PROBLEM SOLVING Use bundles of pea sticks – or any twigs and small branches – to help keep pea plants off the ground and growing upwards. This helps to reduce slug damage. Taller varieties may need further support, such as netting stretched between posts.

HARVEST, COOK, STORE Start picking in June and continue until September. The first peas of the year are great to eat straight from the plants; when tiny we munch the whole pods as raw mangetout. The growing tips are tasty sautéed in butter and the empty pods can be turned into a lovely country wine (see page 198).

Radishes
Raphanus sativus
HOW TO GROW It is so quick and easy to grow radishes that many gardeners think it beneath them. By summer you may have had a surfeit, but we always keep a row or two on the go. Sow seed in drills about 15cm (6in) apart and 1cm (½in) deep, from February to August. Radishes are also great sown as a quick crop in between rows of slower-growing veg.
PROBLEM SOLVING Flea beetles may attack the leaves but the radishes themselves aren't usually affected.
HARVEST, EAT, STORE Pull radishes while they are still small: the larger they are, the tougher they can be, and

the hotter. Enjoy them fresh in salads – you can't store them for any length of time.

Rhubarb
Rheum x hybridum
HOW TO GROW Yes, we know you eat it like a fruit, but rhubarb is a stalwart perennial of the veg plot. It's easiest grown from crowns – rooted pieces of

---

FORCING RHUBARB

We "force" rhubarb by covering plants with a heap of well-rotted manure in spring and putting an old half barrel on top. We take the barrel off when the pale shoots emerge from the top of the rich manure. Tradition says that once a plant has been forced, it needs to rest for two years before you do it again, but we force our rhubarb plants every year and they are fine!

---

plant. Set them 90cm (3ft) apart. Create more plants by dividing every five years.
PROBLEM SOLVING Generally trouble-free.
HARVEST, COOK, STORE Leave new plants to establish for a year, then harvest any time from February until July. Pull the stalks, don't cut them. As well as crumbles and fools, rhubarb can be used in chutney and jam. It freezes well. Don't forget the leaves are toxic.

Runner beans
Phaseolus coccineus
HOW TO GROW Sowing runner beans in pots in the greenhouse and transplanting them out as early as the frost allows is one of our yearly rituals. You can also sow straight into the ground later on. Grow on wigwams; leave 23cm (9in) between plants, stick in bamboo canes and tie them together at the top.
PROBLEM SOLVING Lots of people are put off runner beans because they think they are tough and stringy. There is absolutely no reason to suffer tough beans. You can buy "stringless" varieties, but by far the easiest answer is to pick them young and often.
HARVEST, COOK, STORE Keep picking so the plants keep producing: surpluses can be blanched – plunged in boiling water briefly – and frozen.

Salad leaves
Lactuca and others
HOW TO GROW Growing salad, especially lettuce (*Lactuca*) and rocket (*Eruca*), is one of the easiest things to do well. Sow thinly in drills 30cm (12in) apart and 1cm (½in) deep. This ensures that the seeds are well spread out and not as much thinning is needed. Choose "cut and come again" varieties and sow in succession – you'll be cutting salad all summer long. Also consider sowing over-wintering salad varieties.
PROBLEM SOLVING To protect salad leaves from slugs, use beer traps and if you have any single-use plastic bottles, use them to make protective collars. You can also encourage natural predators (see pages 148–149).
HARVEST, EAT, STORE Cut the leaves in the morning when they are freshest.

Salad leaves can grow for most of the season if you start them early under cover and continue to succession sow.

Vine-ripened tomatoes are a gardener's delight. Try growing heritage varieties for extra colour and flavour.

THE SHOULD HAVES

If time and space are tight then you may have to sacrifice some of these "should haves", but if you can squeeze them in, they are all worth growing.

Asparagus
Asparagus officinalis
HOW TO GROW It is worth having an asparagus bed for the sheer seasonal indulgence, even though it takes time for the plants to establish. Plant one-year-old "crowns" in April (see box, below).
PROBLEM SOLVING Go out after dark with a torch and pick off any slugs – feed them to your hens or ducks.
HARVEST, COOK, STORE Cut your first

PLANTING ASPARAGUS CROWNS

Dig a trench and make a shallow mound at the bottom of it. Set the crowns on the shallow mound, placing them at least 45cm (18in) apart. Spread out the roots and cover them with soil. Keep mounding up the soil as the plants grow, so that you end up with them growing on a ridge. This will get higher over the years as you add mulches and well-rotted manure. Asparagus plants are perennial, so site the bed carefully – you can't start moving them around. For the same reason, you must pull out every bit of couch grass, ground elder, dock, and any other pernicious weeds before you plant your asparagus, because they will be impossible to dig out later.

Spinach
Spinacia oleracea
HOW TO GROW We reckon Popeye had the right idea; spinach is good for you – and it's very easy to grow. To pick spinach all year round, keep succession planting in rows about 30cm (12in) apart and at a depth of 2.5cm (1in). Don't be tempted to plant too much at once, as the leaves are best eaten young. There are varieties of perpetual spinach but in our opinion they don't taste anywhere near as good as annual spinach.
PROBLEM SOLVING Can "bolt" or run to seed if it doesn't get enough water.
HARVEST, COOK, STORE We cut off handfuls with a pair of scissors: it will grow again quite quickly, allowing for a second harvest. Sadly, after all the effort of picking and washing, spinach cooks down to very little. The young leaves are great raw in salad. Blanch the leaves before freezing them.

Tomatoes
Lycopersicon esculentum
HOW TO GROW Tomatoes are an easy crop to grow on windowsills, balconies, under cover and outside. We grow the majority of our tomatoes in the greenhouse, starting them off in small pots and trays in March, and transplanting them to bigger pots as they mature.

Keeping them in pots does require quite a lot of compost and watering, but they do extremely well. Tomatoes need feeding regularly. We use our worm tea (see page 139) and comfrey fertilizer (see page 119); they help the tomato plants to remain healthy for significantly longer each season. Support plants with canes or tie them in to strings running down from the greenhouse roof. There are some incredible heirloom varieties available to plant, all with different flavours.
PROBLEM SOLVING We have problems with blight, which is a disease that damages tomatoes and potatoes: we've tried to counter the problem by growing the plants upside down, which is meant to prevent blight from establishing. Use a hanging container and plant young tomato plants through a hole in the base; the plants' roots grow upwards. From our experience it's reasonably effective – and a good talking point.
HARVEST, COOK, STORE Pick tomatoes when they are completely ripe: cut them just above the flower stalk. Try growing different varieties for cooking, salads, etc. If you have a glut, freeze tomatoes – skin them first, then cook them for 10 minutes or so and sieve to make your own home-made passata. At the end of the season green tomatoes make great chutney, or ripen in a bowl with a banana.

spears two years after planting and then annually in late spring. Start cutting when the spears are 10cm (4in) long. Use a knife to cut them off just below the surface of the soil. Stop cutting in June and allow the plant to grow. We love the spears simply steamed and served with butter. You could freeze some if you have spare.

Cabbage
Brassica oleracea Capitata Group
HOW TO GROW There is nothing nicer than a neat row of cabbages if you have the space. Sow spring cabbages in summer and Savoys in spring for winter eating.
PROBLEM SOLVING Birds will peck at the seedlings, so set up a bird scarer (see page 149). Caterpillars can be a problem later on; pick them off by hand and feed them to your hens.
HARVEST Cut spring cabbages from April onwards and Savoys in winter. We don't think brassicas freeze that well from a taste point of view.

Cucumbers
Cucumis sativus
HOW TO GROW Cucumbers are relatively easy to grow and every year we marvel at how many we get from each plant. In our first year at Newhouse Farm we picked well over a hundred from one plant! Okay, it's confession time – our cucumbers look nothing like the ones you buy in the shop. They tend to be a lot shorter and fatter and, dare we say, tastier. Sow seed under cover in April or outside in June.
PROBLEM SOLVING Spray greenhouse cucumbers with water to increase humidity and discourage red spider mite.

HARVEST, COOK, STORE Carry on cutting until the first frosts. Try pickling them if you have a glut, or make tzatziki.

Fennel
Foeniculum vulgare
HOW TO GROW Fennel hates having its roots disturbed or being transplanted, so sow in early spring in small pots or trays under cover, or in late spring directly where it is to grow. Choose a sunny spot in light, free-draining soil. Bronze fennel is highly ornamental and makes a great addition to an edible flower border or herb garden.
PROBLEM SOLVING Fennel requires little maintenance, but look for out for greenfly on the soft shoot tips or on leaves. They suck sap and excrete sticky honeydew, encouraging the growth of black moulds. Squash aphids with your finger and thumb, or use biological control in the greenhouse by introducing ladybirds.
HARVEST, COOK, STORE Harvest leaves as required from spring to autumn. Seeds can be used fresh over the summer months or dried for winter use, and the flowers and pollen are delicious as a garnish. Grill the thinly sliced bulb on the barbecue, or braise it with lemon and herbs.

French beans
Phaseolus vulgaris
HOW TO GROW French beans are less prolific than either broad or runner beans. From a 3m (10ft) row you can expect 9kg (20lb) broad beans, 27kg (60lb) runner beans, and a mere 4.5kg (10lb) French beans – however, they are very fine eating. Sow seeds outside in May and carry on succession sowing until late June.

PROBLEM SOLVING Young seedlings are vulnerable to slugs so set plenty of beer traps. Support your beans with trellis or willow wigwams and avoid them being easy targets for ground pests.
HARVEST, COOK, STORE Pick beans from July to October. Like runner beans, you can leave the pods on the plant to fatten up and then dry the beans for winter use – or blanch and freeze the tender young beans.

Globe artichokes
Cynara cardunculus Scolymus Group
HOW TO GROW Globe artichokes are magnificent, thistle-like perennial plants that take up a lot of space for very little produce. That said, we love them and they taste great. For best results plant rooted offsets (like suckers taken from the base of a mature plant): it's easier than growing from seed.
PROBLEM SOLVING Set beer traps for slugs while plants are small. Once full size they can cope with a few tattered leaves.
HARVEST, COOK, STORE Wait a year before harvesting the heads, then cut them every summer. Pick them before they flower, when they become a popular hideout for insects. You can pickle the bases or "hearts" but preparing them is a bit fiddly and time-consuming. Use lemon juice to stop them browning when "turning" them.

Jerusalem artichokes
Helianthus tuberosus
HOW TO GROW Jerusalem artichokes are prolific, but we know very few people who consume the whole harvest. Our pigs usually benefit from part of ours. Although they are tasty, we have nicknamed them "fartichokes" because of their propensity to cause wind. Think carefully about where you site them: these 2.2m (7ft) high daisies are invasive and once you've planted them they are nearly impossible to get rid of. But on an exposed plot they make a great, productive wind break with pretty flowers.
PROBLEM SOLVING More or less trouble-free.
HARVEST, COOK, STORE Dig up tubers from November onwards. Roast them along with other vegetables or make them into soup.

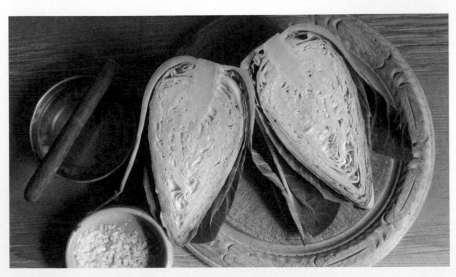

Ferment cabbage with a small percentage of salt to preserve it after harvesting.

Sprinkle salt over slices of aubergine to draw out some of the moisture and intensify the flavour. Grill, then serve on bruschetta.

Kohl rabi

Brassica oleracea Gongylodes Group

HOW TO GROW Kohl rabi is not very popular in Britain but has a following in Europe. Sow seed from April to July.

PROBLEM SOLVING Use a bird scarer to keep birds away from the young plants.

HARVEST, COOK, STORE Leave in the ground and dig up as needed from mid-summer onwards. Simply trim off the leaves, scrub, and slice. They make a cracking gratin or mash them with butter. Also excellent in slaw.

Potatoes

Solanum tuberosum

HOW TO GROW You can buy local potatoes cheaply in season. But they never taste as good as those dug from the garden and put on the plate within the hour! Plant seed potatoes from late March to April depending on whether they are earlies or maincrop.

PROBLEM SOLVING Protect against frost by drawing soil over leaves with a hoe.

HARVEST, COOK, STORE Dig earlies from June, maincrop from September. Harvest promptly before slugs move in. Home-grown potatoes are sweet and tasty with a little butter. Store maincrop varieties in a clamp (see page 167).

CHITTING POTATOES

Seed potatoes should be encouraged to begin sprouting (or to have "chitted") before you plant them outside. Several months before planting, lay the potatoes out in a dry, cool, frost-free place. It should be light, but out of direct sunlight – a shed with a window is ideal.

Swedes and turnips

Brassica napobrassica and *B. rapa*

HOW TO GROW We plant more than we will ever need of these great winter staples and use the less perfect ones to feed our pigs. Sow seed in May.

PROBLEM SOLVING Keep birds away from the young plants.

HARVEST, COOK, STORE Dig up roots from October onwards. Store in a clamp (see page 167). Some of us love mashed swede; others claim it is punishment, not nutrition.

Sweetcorn

Zea mays

HOW TO GROW Sweetcorn needs plenty of sun to ripen the cobs successfully. Sow under cover in April and plant out after the danger of frost has passed.

PROBLEM SOLVING Stake plants on windy sites.

HARVEST, COOK, STORE Pick the cobs in late August/September. Eat as soon as possible after picking before the sugar content turns to starch.

WHY NOT TRY

We wouldn't give these top priority, but that doesn't mean you shouldn't experiment with the following veg.

Aubergine

Solanum melongena

HOW TO GROW In certain regions, you might get away with growing aubergines outside, but it's safest to have some in the greenhouse too. Start seed off in pots in March under cover. Stake and feed plants as for tomatoes.

PROBLEM SOLVING Red spider mite can be a problem. Reduce the risk by spraying with water to keep the humidity up.

HARVEST, COOK, STORE Pick aubergines once the skin is shiny and they are at least 10cm (4in) long. Watch out for the prickles on the stalks and green caps. Some of our favourite ways to cook aubergine are slow-roasted and served as baba ganoush, or thinly sliced and layered in a moussaka or lasagne.

Cauliflower

Brassica oleracea Botrytis Group
HOW TO GROW Cauliflowers are challenging to grow, needing well-manured soil that hasn't been recently dug and, above all, regular watering. Grow different varieties and you can cut cauliflowers from spring to autumn. Sow seed in spring and thin seedlings as they grow. Transplant so that mature plants will be 60cm (2ft) apart.
PROBLEM SOLVING Keep birds away from the young plants with home-made scarers (see page 149).
HARVEST, COOK, STORE To eat them at their best, cut them while the heads are still tightly packed – loose florets mean they are past their prime. Fold in the leaves and cover with a piece of slate for a week first. We've got a great recipe for piccallili and also love making cauliflower rice to serve with curries.

Chard

Beta vulgaris var. *flavescens*
HOW TO GROW Chard, or leaf beet, is easy to grow – we often find it self seeds and pops up in the veg bed unannounced. Sow seed in April and thin the seedlings to 30cm (1ft) apart. Rainbow chard has ruby red, orange, yellow, and pink stems.
PROBLEM SOLVING Trouble-free, apart from the odd bit of slug damage to leaves.
HARVEST, COOK, STORE Pick leaves all summer long. Chard gives you two veg for the price of one: steam the stalks and cook the leaves separately like spinach. Freeze them in the same way too. Chard is also excellent fermented for kimchi.

Chillies and sweet peppers

Capsicum annuum
HOW TO GROW Chilli peppers are really easy to grow. We have tried everything from small Scotch bonnets that blow your socks off to more refined banana peppers. All are fast-growing annuals that do best in warm summers and like a bit of humidity. In temperate climates they are definitely best grown under cover and sown in spring. Sweeter capsicum peppers also come in all shapes and sizes. Our favourites are classic bell peppers and Romano. Grow in the greenhouse, just as you would tomatoes (see page 154).

Cauliflower's robust flavour suits cooking over coals and heavy spicing. We love the combination of turmeric and charred florets.

PROBLEM SOLVING Look out for red spider mite – misting with water can help prevent attacks and chillies like the humidity. If pepper stems flop over, stake them up.
HARVEST, COOK, STORE Pick peppers in late summer. We use chillies in fresh salsas and curries and with roasted seeds and nuts. To dry them, string them up and hang in a dry, airy spot. You can freeze chopped chillies – remove the seeds first.

Ginger

Zingiber officinale
HOW TO GROW If you like fresh ginger, try growing your own from a root from the grocer or supermarket. Look for a portion of root that has a bud on it. Start it off in a small pot in spring and repot as it grows. Grow ginger under cover in temperate climates. It's a tall plant with grass-like leaves.
PROBLEM SOLVING As ever in the greenhouse, red spider mite is a problem. Spray with water.

HARVEST, COOK, STORE Tip out the pot in autumn and harvest your roots. Replant a bit and use the rest. We use it to make ginger beer (see pages 272–273). You can freeze ginger.

Horseradish

Armoracia rusticana
HOW TO GROW Horseradish grows in moist, rich soil and can tolerate a shady position.
PROBLEM SOLVING It is highly invasive: plant it in a secluded part of the garden where you can easily restrain it.
HARVEST, COOK, STORE Instead of buying jars of horseradish sauce we now make our own. You can dig up and harvest horseradish about nine months after planting: save the smaller roots for replanting. It tastes extremely good with smoked mackerel. Also try grating into sea salt with some black pepper for a tasty horseradish salt. Freeze the roots or store in a clamp (see page 167).

PRUNING, TRAINING, AND GRAFTING

Pruning will make your fruit trees more productive, increasing yields and fruit quality, as well as making harvesting a whole lot simpler. Training is a more specialized form of pruning and is used to restrict a tree's growth to create formal, space-saving shapes. Grafting is a method of making more fruit trees, which mature quickly and bear fruit sooner, for very little outlay.

PRUNING FRUIT TREES

The point of pruning is to increase yields and create an open tree shape that's more convenient for harvesting. It's also needed when a tree is suffering from any of the three Ds: dead, dying, or diseased branches. Pruning has many more complex quirks, depending on different trees, and it's worth consulting a book that's entirely dedicated to the subject.

Pruning generally falls into one of two categories: either summer or winter pruning. Winter is a great time to tidy up the tree and remove any dead or diseased bits.

Summer pruning

This stimulates growth of fruit buds ready for the next year, opens the tree up to sunlight so that the fruit will ripen better, and is a good time to prune trees that you are training into espaliers, fans, and cordons, because the shoots are more flexible then. Do it once the leaves are dark green and the bark has started to turn brown and woody at the base of the shoots – towards the end of August. In cold areas it may be autumn before shoots are mature enough to prune. See the directory on pages 160–165 for specific advice on different trees.

Tools to use

For pruning thin shoots up to around 2cm (¾in) in diameter you need pruning shears. To deal with bigger branches, invest in a pruning saw and a pair of long-handled loppers. You'll need them for coppicing woodland too (see pages 220–221).

TRAINING

Training fruit trees is an ideal way to grow fruit on a small plot. It can take advantage of warm south-facing walls and fences that help fruit ripen quicker, and makes harvesting easier. All trained shapes need support of some sort. Run horizontal wires across a wall or fence; use posts and wires for free-standing trees (see below). Once established they are easy to maintain.

Popular styles of training

By training fruit trees into neat space-saving shapes, you can fit them into small spaces and improve productivity.

GRAFTING FRUIT TREES

We tend to leave grafting to the experts but it's worth learning. Grafting combines the properties of two different varieties: strong roots and trunk (the rootstock) with top-quality fruit (the scion). All trees you buy have been grafted.

Budding is useful for grafting a favourite apple, for example, onto a less productive tree. Whip-and-tongue grafting creates a new tree.

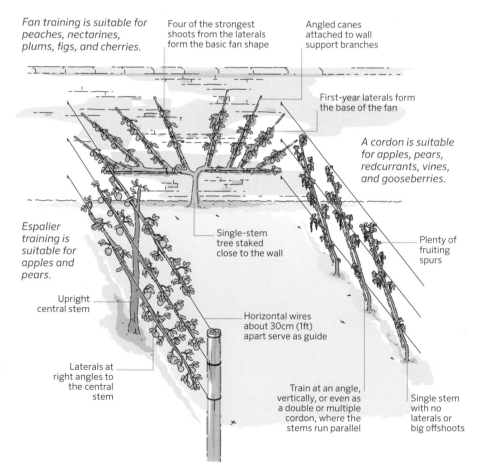

Fan training is suitable for peaches, nectarines, plums, figs, and cherries.

Four of the strongest shoots from the laterals form the basic fan shape

Angled canes attached to wall support branches

First-year laterals form the base of the fan

A cordon is suitable for apples, pears, redcurrants, vines, and gooseberries.

Espalier training is suitable for apples and pears.

Single-stem tree staked close to the wall

Plenty of fruiting spurs

Upright central stem

Horizontal wires about 30cm (1ft) apart serve as guide

Laterals at right angles to the central stem

Train at an angle, vertically, or even as a double or multiple cordon, where the stems run parallel

Single stem with no laterals or big offshoots

PROJECT Prune an espalier apple tree

Summer pruning is crucial to maintain the shape of a formally trained espalier tree. The basic techniques shown here apply to other styles of training, as well as to pruning straightforward bush-type fruit trees. Use sharp shears and always cut above an outward-growing bud.

1. Cut cleanly with the shears at a shallow angle facing away from the leaf. Leave a small gap between the cut and the last leaf bud. **2. Cut back leafy shoots** that have grown up directly from the main horizontal branches to three leaves above the base clump of leaves. **3. Tie in sideshoots** growing out of the vertical stem by training them along a wire and incorporating them into the shape. If the tree is becoming overcrowded, remove them completely.

4. Trim back any shoots that are growing out at an angle from the espalier instead of along its length. **5. Take off the top** of the tree. **6. The finished tree** has a precise tiered shape that makes harvesting easy and saves space in the garden.

PROJECT Grafting techniques

BUDDING

To be successful, budding needs to be done in midsummer. Choose a healthy cutting about 30cm (12in) long for the scion and graft it to a strong healthy branch of your rootstock tree.

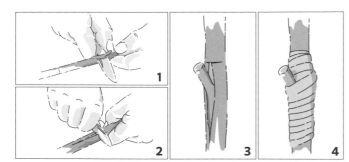

1. Cut a piece of bark with a leaf bud from the scion **2. Cut a T-shaped slit** in the bark of a branch of the rootstock tree. **3. Peel back** the flaps of bark and insert the bud. Fold the flaps of bark back down around the bud. **4. Bind the stem** with grafting tape. Your chosen variety will grow from the newly grafted bud. As the bud grows, cut back the old stem of the rootstock above the grafting point so that all the energy goes into the new variety.

WHIP-AND-TONGUE GRAFTING

For this technique, best carried out in February, you will need a year-old rootstock. Start by taking a healthy cutting about 23cm (9in) from the scion tree in December. Heel it into the ground to keep it dormant. Rootstock and scion should be roughly the same diameter – ideally about 2.5cm (1in).

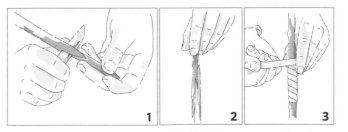

Cut off the rootstock at 15–30cm (6–12in) above ground level.
1. Make a sloping cut across the rootstock 5cm (2in) long and slice into it thinly to form an upward-pointing tongue. Make a similar cut in the scion, just behind a bud, to form a downward-pointing tongue. **2. Interlock the tongues** on scion and rootstock. **3. Bind together** with grafting tape.

FRUIT AND NUTS TO GROW

Growing a wider range of fruit and nuts is one of the advantages of a smallholding. In an average-sized garden it's easy to add climbers such as grapevines and kiwis, while blueberries and dwarf fruit trees thrive in pots on a patio.

SOFT FRUIT

The great thing about soft fruit is that you don't have to wait long for results. Unless otherwise stated, plant as bare-root plants from October to March. They need little maintenance other than weeding and mulching, and yields can be phenomenal.

Blackberries
Rubus fruticosus
HOW TO GROW At Newhouse Farm, we never planted blackberries – but that's because we were lucky enough to have lots in the hedgerows. Once a year they cease to be thorny invaders and become prickly baskets of tasty fruit. If you need to buy plants look out for thornless varieties – ideal for smaller spaces – and juicy hybrids like loganberries and tayberries. Grow on wires and posts or train against a wall.
PROBLEM SOLVING On a smallholding and in many gardens, brambles can be a nightmare. If you inherit some wild brambles on your plot, try to tame them by shaping them into easily accessible blocks.
PRUNE If growing blackberries on wires, cut out all dead wood in autumn and tie in new canes.
HARVEST, COOK, STORE Pick fruit from July to August. As well as crumbles, we always make jams and jellies, and open freeze surplus berries on trays before bagging them up.

Blackcurrants and redcurrants
Ribes nigrum, R. rubrum
HOW TO GROW Blackcurrants are easy to grow and thrive on cool, heavy soil – even on clay. Mulch them generously each spring using well-rotted manure. Bushes are heavy cropping and the fruit has

Gooseberries are deliciously tart and are tasty in sweet pastry desserts or served as a spiced jam or relish with smoked mackerel.

a high vitamin C content. Redcurrants are yummy and particularly useful to grow against awkward north-facing walls and in shady areas – ideal for an urban garden.
PROBLEM SOLVING Put up netting to stop birds stripping your crop. Watch out for redcurrant blister aphid on the leaves in summer. Cut stems back as far as the first fruits if symptoms appear.
PRUNE For blackcurrants, prune between November and March. Remove one-third of the stems: they can become overcrowded in the middle of the bush. For redcurrants, follow the same instructions as gooseberries.
HARVEST, COOK, STORE Pick in July and August. Separating the small berries from their trusses is tedious. Some people prefer to cut off entire fruiting branches and strip the berries in the kitchen, but we pick ours by hand so that the stored energy in the branches can return to the plant. We make our blackcurrants into pies, jams, and jellies. Redcurrants are great in summer pudding and in jellies to serve with roast lamb. Both berries freeze well.

Blueberries
Vaccinium corymbosum
HOW TO GROW Blueberries are a mountain fruit that does really well in a temperate climate. They grow best in groups so that they can cross-fertilize each other; you'll need two varieties that flower at the same time for pollination. Unless your soil is acidic, you'll need to grow them in raised beds or big tubs filled with ericaceous compost, which has a low pH. Plant bushes in April or May.
PROBLEM SOLVING Always water bushes with rainwater to keep the soil pH low.
PRUNE Not necessary.
HARVEST, COOK, STORE Pick berries from July to September when they have turned from light blue to a pewtery deep blue-black and are soft to touch. We like fresh blueberries in fruit salads or pack them into muffins. We freeze any surplus for making smoothies.

Cape gooseberries and tomatillos
Physalis peruviana, P. ixocarpa
HOW TO GROW These are annuals with a difference – each cape gooseberry is

Find just the right spot for your kiwi plant and, when it's established, you'll be amazed at how productive it can be.

HARVEST, COOK, STORE Pick grapes from August to October. Grapes are one of the fruits we grow for their fermenting potential – they taste great too.

Kiwi fruit
Actinidia deliciosa
HOW TO GROW Kiwi fruits are exotic but do well in a temperate climate. Choose a self-fertile variety so you only need one plant. For best results grow against a south-facing wall or in a polytunnel.
PROBLEM SOLVING Generally trouble-free.
PRUNE You don't have to prune them, but if you don't, kiwis take up lots of space; pruning also encourages more fruit. Do it between November and February. Train side shoots on wires and, after a couple of years, nip out growing shoots about 15cm (6in) beyond the last fruiting buds.
HARVEST, EAT, STORE Pick fruits from October to December. They can be eaten straight from the vine as they don't keep very long.

enclosed inside a papery lantern. The tomatillo is a close cousin. Plants must be grown in a greenhouse or inside. Sow seed February to March. Look after them like tomatoes: use canes or string to support the growing plants.
PROBLEM SOLVING Spray regularly with water to reduce red spider mite.
PRUNE Not necessary.
HARVEST, COOK, STORE Pick from August to October when the berries are bright orange. Tomatillos are excellent in a Mexican salsa. Cape gooseberries taste sweet but smell rather odd.

Cranberries
Vaccinium oxycoccos
HOW TO GROW We grow cranberries to go with our home-reared Christmas turkey. Plant bushes in March. They grow well in a sunny, sheltered position and need regular watering. If growing them in a pot, stand it in a wide saucer topped up with water, and use ericaceous compost as cranberries need acidic soil.
PROBLEM SOLVING Water with rainwater rather than tap water to keep soil pH low.
PRUNE Not necessary.
HARVEST, COOK, STORE Pick berries in September and October. Cranberries are easy to freeze, ready for Christmas.

Gooseberries
Ribes uva-crispa
HOW TO GROW Gooseberries need rich, moist soil and regular watering. Plant bushes in autumn: mulch with leaf mould.

PROBLEM SOLVING Pick off sawfly caterpillars by hand and buy plants resistant to American gooseberry mildew.
PRUNE Cut back old wood in winter; prune strong lead shoots by half to an upward-pointing bud. Cut out cross shoots and shorten laterals. Summer prune in July so that fruits have maximum exposure to sunshine to ripen. Aim for an open cup shape, so you can reach in to pick the berries. Take cuttings in early autumn. They will start producing fruit after two years.
HARVEST, COOK, STORE Pick from May to July. The smaller the berries, the more sour they are. We also grow red dessert varieties, which are much sweeter and don't need cooking. Like other berries, gooseberries freeze well.

Grapes
Vitis vinifera
HOW TO GROW We have planted a small vineyard on a south-facing slope and have had a few vines growing under cover for years. We water indoor vines regularly and give them liquid tomato feed once a week in spring. Outside we feed our vines by mulching in spring and autumn. They grow on thin topsoil on a clay-based subsoil.
PROBLEM SOLVING We grow mulberry trees at the end of the rows of vines. They have an anti-fungal property that helps protect the vines from disease.
PRUNE Shape the frame in the first few years. Then prune annually in December (see box, right).

TRAINING GRAPES

Train the main stem upwards with a cane. Train some lateral stems horizontally on wires. In winter, prune the laterals to two or three buds from the main stem. In summer, stop them from getting too long by "nipping" out the growing tip three or four leaves beyond each developing bunch of grapes.

Melons
Cucumis melo

HOW TO GROW We pride ourselves on our melons. We sow seed in pots in April and add a cane for each plant to grow up and some netting to support the fruit, but you can grow them on the ground. Melons are very sensitive to cold so do best under cover.

PROBLEM SOLVING Plant melons on mounds to avoid neck rot – damage between the base of the stem and the top of the root brought on by wet conditions at soil level.

PRUNE Pinch out growing tips two leaves after a developing melon. This prevents the plant from wasting energy on growing and instead puts it back into the fruit.

HARVEST, COOK, STORE Melons ripen from August to October. They don't store well.

Raspberries
Rubus idaeus

HOW TO GROW Raspberries are always expensive in the shops, but luckily they grow well in wet and cold areas and have a long picking season. Raspberry canes like well-drained soil in a sunny position. Support them with a couple of stakes at each end of the row with some wire running between them.

PROBLEM SOLVING Cover with netting to prevent bird theft.

PROTECTING YOUR FRUIT CROP

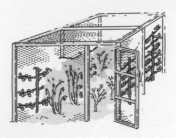

A fruit cage is extremely effective as a defence against birds. We leave the door to our fruit cage open in early summer so that birds can eat any pests, but close it when the currants and raspberries are ripening. We keep it closed in winter too, when birds like to snack on the buds of the currant bushes.

Netting fruit bushes or rows of strawberries is another option. Make a tunnel by pushing bent hoops of plastic piping into the ground and stretching netting over the top. Peg netting into the soil to stop birds sneaking underneath. Don't lay it directly on top of fruit – birds will peck straight through it!

PRUNE In winter, prune the majority of canes all the way back to just 3cm (1¼in) above the ground. Leave a few unpruned. As time goes on, leave more and more canes unpruned. Try pruning the canes at different levels so that you have a nice spread of fruit.

HARVEST, COOK, STORE Raspberries ripen from July onwards, and can carry on until October depending on the variety. They are great fun to pick, though they don't all make it safely to the kitchen – gardener's perks! They are delicious as they are, made into jam, or preserved in liqueurs for a winter treat.

Strawberries
Fragaria x *ananassa*

HOW TO GROW Strawberries are one of our must-haves. Plant them from March to April on ridges for drainage. Plants have a relatively short life of cropping (three to four years) so try to propagate some of your own from runners – mini plants on the end of a long stem.

PROBLEM SOLVING Avoid slug attack by growing plants in hanging baskets – but they will need heavy watering. When growing conventionally we mulch plants with straw in summer to keep the fruit from rotting by lifting it off the ground. Net the plants to stop birds getting the fruits before you do.

PRUNE Not necessary.

Try smoking strawberries over oak for a deep, charred flavour. Serve with balsamic vinegar or black pepper meringues.

HARVEST, COOK, STORE Pick fruits in July. If you have a glut, make some jam or open freeze the berries on trays before bagging them up.

FRUIT TREES
Fruit trees are an excellent long-term way to cut down on your food miles – especially if you store some apples for winter eating. Plant varieties that suit your local environment: look around and see what is growing well in nearby gardens. Many dwarf varieties of fruit tree will even grow well in a container on a patio. All are best planted when the trees are dormant, between October and March. Fruit trees are an investment for the rest of your life, so take your time and do it properly. Training fruit trees not only saves space; it increases yields too. We explain some common styles on page 158.

Apples and pears
Malus domestica and *Pyrus communis*

HOW TO GROW Both trees like annual heavy mulching with rich compost. Pears like to grow in a more sheltered spot than apples and are more fragile. Pears normally need another tree within 100m (110 yd) to serve as a pollinator, or you can buy a family pear tree that has three different varieties that will self-pollinate grafted onto one trunk. Apples and pears are available in dwarf varieties.

PROBLEM SOLVING Codling moth larvae burrow into apples. Use pheromone traps to catch adult moths before they mate.

Plums can be preserved as a jam or spiced sauce, but when they are perfectly ripe they rarely make it out of the garden.

PRUNE Pears don't need specialized pruning. Trim long shoots in summer and again lightly in winter. For apples, see information on page 159.

HARVEST, COOK, STORE Pick from July to October. Depending on variety, some apples are best eaten early, others will keep well into next year. Store apples and pears in trays in a cool, dry place. As well as tarts and pies, apples make a good base for herb jellies.

Apricots, peaches, and nectarines
Prunus armeniaca, P. persica, and *P. persica* var. *nectarina*

HOW TO GROW Apricots, peaches, and nectarines like a sunny, sheltered position. We fan-trained our nectarines on a south-facing wall.

PROBLEM SOLVING Flowers open early when pollinating insects are rare, so hand pollination helps. If short of space, grow patio apricots in pots – they produce full-sized fruit even though they are barely 1m (3ft) tall.

PRUNE Remove dead wood as necessary.

HARVEST, COOK, STORE Pick from July to September. Apricot jam is delicious. Eat the best peaches and nectarines; preserve the rest in brandy or dry them (see pages 168–171).

Cherries and plums
Prunus avium and *P. domestica*

HOW TO GROW Cherries and plums are easy to grow. They work best fan-trained against a south-facing wall or in a sunny, sheltered position.

PROBLEM SOLVING Protect fruit from birds with netting. Plums are a wasp's favourite food, so pick ripe fruit daily.

PRUNE Pruning increases the chance of silver-leaf disease, so do it as little as possible.

HARVEST, COOK, STORE Pick cherries in July, plums from August to September. We make plum jam and chutney, and try to save some cherries for bottling in brandy.

Figs
Ficus carica

HOW TO GROW Figs do remarkably well in a temperate climate. They grow well in

containers and do even better if you move them under cover in winter and water and feed regularly. They grow best when their roots are restricted: it improves fruiting.

PROBLEM SOLVING Generally trouble-free.

PRUNE Between December and January. Remove all unripened fruit and fruitlets or they will reduce next year's crop.

HARVEST, COOK, STORE Harvest figs from August to September, when they feel soft and when the "eye" on the bottom of the fruit has opened up a bit. They should be moist, sweet, and fragrant. Pick them just before they would fall off the tree naturally. Eat immediately.

Lemons
Citrus limon

HOW TO GROW Lemons are the easiest citrus trees to grow in a temperate climate. It's best to grow them in large pots and bring them inside over winter. When growing lemons in pots, you need to "top dress" them every year or so by removing the top 2.5 cm (1in) of compost and replacing it with fresh. This is best done in spring.

PROBLEM SOLVING Mist with water to reduce red spider mite.

PRUNE Shear them right back to a

compact shape in winter. Try rooting the prunings.

HARVEST, COOK, STORE Pick lemons from July to October. Watch out for their spiky bits when harvesting.

WHY NOT TRY?
We wouldn't give these more unusual fruits top priority, but that doesn't mean you shouldn't have a go.

Medlar
Mespilus germanica

HOW TO GROW Medlars are medium-sized trees with highly unusual fruits that look like little wooden rose hips. They are tough, hardy trees and like a sunny open spot. You can even grow them in a large container, such as a half barrel.

PROBLEM SOLVING Generally trouble-free.

PRUNE Avoid pruning as fruits are produced at branch tips.

HARVEST, COOK, STORE Pick in October when the fruits have almost rotted, or "bletted", to a rich golden yellow. Eat the flesh with a teaspoon – enjoy it with cheese and a nice glass of port – but don't eat the seeds. Medlars don't keep: make jelly with any surplus.

Mulberry
Morus nigra

HOW TO GROW Mulberries don't travel well, so really the only way to taste them is to grow your own. We grow the black variety rather than white mulberry, whose leaves silkworms eat. We've planted them alongside our grapevines. They come into flower quite late so tend to avoid any frost damage.

PROBLEM SOLVING Generally trouble-free, though birds help themselves: you'll know about it when you find mulberry-stained bird poo on the paths or, worse, the washing.

PRUNE Thin branches in winter if necessary to maintain the tree's shape.

HARVEST, COOK, STORE Pick from August to September. Watch out for juice stains: wear old clothes when harvesting. The fragrant fruit tastes of raspberries dipped in red wine. The berries soon deteriorate once picked, so freeze any surplus immediately or make them into jam, jelly, or wine.

Olives
Olea europaea

HOW TO GROW We have yet to get a decent crop from our olive trees, but patience is a virtue and so we will wait. In the UK they do best under cover. Olives grow particularly well in sandy soil interspersed with clay layers. You may need to water small trees in particularly hot summers.

PROBLEM SOLVING Grow olives on a slope to avoid their roots sitting in too much water.

PRUNE Heavy pruning needed. Consult a specialist book.

HARVEST, COOK, STORE Pick olives in late summer. You can't eat them straight from the tree: try pickling them in brine first (see page 174).

Quince
Cydonia oblonga

HOW TO GROW Quince trees grow to about 5m (16ft) high and produce fruit like enormous golden pears. They like damp soil and prefer a sheltered spot.

They don't flower until late spring so are usually safe from frost damage.

PROBLEM SOLVING Thin the fruits as they develop to get bigger, better quinces.

PRUNE None required.

HARVEST, COOK, STORE Quinces ripen in October. They are too hard to eat raw so have to be cooked first. They won't last much longer than a few weeks, so use them quickly: try adding a few slices to an apple pie. We also enjoy quince poached in rosewater syrup with elderflower cream.

NUT TREES

Nuts are a great source of protein and relatively easy to grow. Plant trees any time from October to March. Harvesting normally takes place between September and November. Most nuts can be stored for about six months in a dark, dry environment. Here are some of our favourite nuts to grow.

Almond
Prunus dulcis

HOW TO GROW Best grown as fans on south-facing walls. Almonds are just like peach trees – you can hardly tell them apart – and need similar conditions. They have beautiful blossom that appears before the leaves in spring.

PROBLEM SOLVING Throw a sheet of plastic over trees in winter to keep buds dry and prevent peach leaf curl disease.

PRUNE Try to avoid pruning: it increases the risk of diseases such as peach leaf curl. As a last resort, prune in spring when new growth starts.

HARVEST, COOK, STORE Pick almonds in autumn and rub off the leathery skin (wear gloves to do this); then spread on a tray in a thin layer to dry before storing.

Hazelnuts and cobnuts
Corylus avellana

HOW TO GROW Hazelnuts are very tasty and the trees can be used for hedging and

Roasted, salted, or smoked almonds are a family favourite and always a great nibble when entertaining.

Preparing and storing nuts can be laborious, but it's well worth the effort. They also pack a punch of protein – ideal for meat-free diets.

coppiced for firewood, plant supports, and making your own hurdle fencing. Cobnuts are a cultivated variety of hazelnut.

PROBLEM SOLVING Birds and squirrels love the nuts and often strip the crop before you get a chance to pick. You can't get ahead of the game and pick unripe nuts – they need to ripen on the tree. Visit trees daily and start picking as soon as the husks start to turn brown.

PRUNE Keep trees to a manageable height by pruning in winter – or follow a coppicing routine (see page 220).

HARVEST, COOK, STORE Pick nuts when ripe in autumn. Eat them straight away or, if you are going to store them, take off the husks first.

Sweet chestnut
Castanea sativa

HOW TO GROW A low-maintenance tree that erupts with loads of yummy nuts in spiny shells. They are huge trees, so not for a small plot.

PROBLEM SOLVING Generally trouble-free.

PRUNE No pruning necessary.

HARVEST, COOK, STORE Knock down the nuts in autumn and take them out of their shells – wear gloves! You will find up to three nuts per shell. Leave them to dry, then store. If you roast chestnuts on a fire or in the oven, pierce the shells first with a skewer or they'll explode – messily.

Walnut
Juglans regia

HOW TO GROW Walnut trees are definitely a long-term investment. They are very slow-growing and can reach 33m (110ft) high, so consider the location carefully before planting. Grafted varieties can crop within three to four years but with traditional types, it's more like 10 to 15 years.

PROBLEM SOLVING Grow in a sheltered spot to avoid frost damage to flowers.

PRUNE Shouldn't be necessary, but if you need to take out a branch or two, do it in autumn to avoid the sap bleeding.

HARVEST, COOK, STORE If you want to pickle walnuts, harvest them before the shell forms. Otherwise pick them later, rub off the sticky skin and dry them – they store well.

WHY NOT TRY?

Here are a couple of unusual trees that we plan to try. They are a long-term investment but look to be worth experimenting with, especially using permaculture principles in a forest garden (see page 113).

Honey locust
Gleditsia triacanthos

HOW TO GROW These multi-purpose trees are native to North America. The seedpods are edible, the flowers attract bees, the trees can be coppiced, and the leaves used as fodder.

PROBLEM SOLVING Protect young trees from severe frost until established.

PRUNE Shouldn't be necessary. Watch out for thorns on mature trees.

HARVEST, COOK, STORE Collect pods as they fall. The pulp inside is edible – use it to make beer! – but the pods are more commonly used to feed sheep, cattle, or pigs.

Pecan
Carya illinoinensis

HOW TO GROW We tend to associate pecans with the southern USA, but in fact they are hardy. The main drawback is how long they take to mature and produce nuts – anything from 10 to 20 years. A mature tree can be 30m (100ft) tall and 20m (65ft) wide.

PROBLEM SOLVING Water well while establishing to avoid powdery mildew.

PRUNE Rarely needed, but if you do have to take out a branch or two, do it in autumn to avoid bleeding.

HARVEST, COOK, STORE Pick in October when husks open. The nuts keep best if frozen. Dry them at room temperature for a couple of weeks first.

ORCHARD BIRDS

Letting geese, ducks, and hens have free run of an orchard or allowing them to forage under fruit and nut trees benefits both the birds and the trees. The birds supplement their diet with all manner of insects and stop pest populations building up. Geese are traditional orchard birds and will also keep grass under control by grazing. Use tree protectors, or don't let them in until trees are mature.

STORING THE HARVEST

Growing your own produce is one of the most exciting aspects of living a more self-sufficient lifestyle, especially with an effective storage regime to back it up. A well-thought-out system will mean your excess harvest won't end up as fodder for livestock or cause the compost bin to overflow. More importantly, it will help you to maintain a supply of fresh food all year round.

HARVESTING FOR STORAGE

As a general rule, harvest produce for storage when it is at its peak. It should look healthy, ripe, and ready to eat, but there are a few exceptions. With vegetables such as squashes, wait until the stem has dried out and shrivelled a bit, and the skin doesn't dent when you press it with your fingernail. Equally, wait a bit longer before picking apples and pears, which are best harvested for storing near the end of the growing season. For best results store only late-ripening varieties and leave them on the tree for as long as possible. You'll know they are ready when they can be gently lifted off the branches.

Try not to damage crops as you pick them or dig them up, and don't waste time trying to store damaged or diseased produce – either send it to the compost bin, or give it to pigs or chickens as fodder.

CLEANING AND TRIMMING

Brush off most of the excess soil from root crops but don't scrub or wash them – you can do that just before you cook them. Similarly, don't trim off any unshapely knobbly bits: the exposed cut could start to rot in storage. But do remove the leafy tops from roots before you store those, as they will take some of the moisture from the root and can turn slimy later, ruining anything they are touching.

IN THE STORE ROOM

For thousands of years people lived without fridges and plastic-wrapped food. They used natural larders, pantries, root cellars, and store rooms to keep food edible. Today we keep our produce fresh for months in our store room. A store room needs to be at a cool and constant temperature, and should be frost-free. It doesn't have to be dark like a root cellar (see pages 214–215), but it shouldn't have direct sunlight, which will make the room too warm and can discolour produce. The humidity should be ambient: too wet and produce will develop mould and fungal problems; too dry and your vegetables will shrivel.

Ventilation is also important: fly-proof mesh across windows and vents lets air in while keeping pests out.

SEED-SAVING TIPS

Leave seeds to ripen on the plant, but cut them before they scatter naturally. Cut the stalks and place the seed heads upside down in a paper bag to collect seeds as they dry.
Briefly soak seeds from fruits like tomatoes in water to wash off the "jelly" before drying on a hard surface – on kitchen paper you'll never get the paper off the seeds!
Shell peas and beans from the pod before drying and storing.
Dry seeds in a cold frame to speed up the process (see page 127).
Store them in labelled paper envelopes, in an airtight container in a cool, dry room.
Save bags of desiccant that come with electronic equipment and use to keep the moisture out of your store.

Runner bean pods dry out and split to reveal ripe beans ready to store.

1. Cut squashes for storage when the stem attaching them to the plant has dried out. **2. Beetroots** will store for longer without their leaves. **3. Cold-smoked garlic** stores well and has a lovely aroma.

Hanging produce

This is a great way of making stored food inaccessible to vermin while allowing good air flow. We hang vegetables like garlic, onions, and sweetcorn from strings, while suspending squashes in netting avoids having to turn them regularly when they are sitting on shelves.

Using trays

Onions and apples store particularly well on trays with slats or chicken-wire bases to allow air to circulate. Apples are best wrapped individually in paper to stop one rotten apple spoiling the rest. Deeper trays can be used to store beetroot, Jerusalem artichokes, carrots, and parsnips in sand or sawdust. Put a layer of sand in the bottom of the tray and arrange the roots so that they don't touch. Layer sand and roots, finishing with a final covering of sand. Fill trays in situ to save the effort of moving them. You can also store roots in a barrel (see page 214). For other methods of preserving food, see pages 172–173.

1. Use deep trays to store layers of root vegetables in sand. **2. Hang onions** and garlic from the roof. Herbs and chillies store well this way too. **3. Keep home-made drinks,** preserves, cured hams, and eggs, as well as fruit and veg, in a well-organized store room. **4. Wrap apples** individually in a single sheet of newspaper.

PROJECT Make a potato clamp

Clamping is a traditional way of storing potatoes if you haven't got a store room or root cellar. Site a clamp in a sheltered spot as it isn't guaranteed frost protection. Store only best-quality potatoes and watch out for slugs. You can use the technique to store any root vegetables and fodder beets for livestock too.

YOU WILL NEED
Fork
Colander
Straw
Spade

1. Dig up your potatoes and leave to dry – ideally in a colander – for a couple of hours, before laying them on a base of straw or bracken. **2. Arrange them in a pyramid** and cover with more straw. Then leave them to breathe for an hour or two. **3. Cover the pyramid** with a layer of earth 15cm (6in) thick. Allow some straw to poke through the soil, to let air get to the crop. Pat the sides flat with a spade; make them steep so that rain can run off.

DRYING FRUIT, HERBS, AND VEGETABLES

Sun-dried tomatoes and dried mushrooms and strawberries are sold at a premium as luxury foods, but drying food has long been a traditional way of preserving produce for winter use – and it's easy to do at home with little effort or cost. The drying process also often intensifies the flavour of the food, and apple slices and vegetable crisps make delicious healthy snacks.

HOW IT WORKS

Drying is a really effective way to preserve food. It prevents the development of unwanted enzymes, bacteria, yeasts, and fungi by removing the moisture they thrive on. There are various ways of doing this, from using the sun's heat or an oven on a low setting, to mixing food with desiccants such as salt and sugar.

AIR DRYING

Air drying is one of the oldest and simplest ways to preserve produce. All you need is a dry, dark, well-ventilated place to hang herbs, fruit, and vegetables – an airing cupboard is ideal – and food will dry quickly and efficiently. Avoid hanging food where humidity levels are high as it will go mouldy and rot.

Leave herbs to hang for a week or so, then crumble the leaves in your fingers to remove them from the stems. Hang the stems upside down in a paper bag to collect herb seeds.

SUN DRYING AND SOLAR DRYERS

Laying out food on racks to dry in the sun is an effective preserving method, but it is more difficult in temperate regions where a few days of back-to-back sunshine without rain are rarely guaranteed. You also need to bring racks under cover every evening to prevent the produce from rehydrating. A solar dryer can simplify the process (see opposite).

OVEN DRYING

You can dry sliced vegetables and fruit in a fan-assisted oven set to the lowest temperature – about 50°C (158°F) – with the door open. (To make sure air is circulating in a conventional oven, prop the door open fractionally by slipping a skewer between the door and frame.) To

DRYING TIPS

The freshest, top-quality fruit and vegetables are best for drying.
Use sharp tools when harvesting and slicing produce.
Always harvest produce in the afternoon. Produce covered in morning dew takes longer to dry and so could go mouldy.
Test food frequently. Bend herbs or chillies in your hand: if they snap they're ready; if they are bendy, leave them for a little longer. Squeeze fruit and veg – no juice should come out.
Store dried foods in airtight containers in a cool, dark place. We take eco-pride in saving every glass jar that comes into the house and reusing them for storing our dried produce.

1. Salt fish to draw out the moisture and firm up flesh before cooking. **2. Hang herbs** in an airy place to dry. Suspend chillies by their stems, tying them to a length of string at 5cm (2in) intervals.
3. Dry borlotti beans in their pods until brittle, then shell and dry the beans.

dry mushrooms, brush off any dirt first, but do not wash them. They take from 4 to 6 hours to dry and are ready when they shrink to half their size but are still pliable. Tomatoes take longer as they are fleshier – between 8 and 12 hours. Apples and pears may take up to 24 hours. If you want to save fuel, use an earth oven (see pages 60–61). Leave all oven-dried foods to cool before storing in airtight jars.

Using a microwave

You can also use a microwave to dry produce. Slice fruit and vegetables, wrap them in a paper towel, and cook for a minute at a time on a high temperature setting. Test between each session. Use the microwave to dry herbs too: wrap them in paper, and place a small cup of water in the microwave at the same time – herbs (and some vegetables) don't contain much moisture and could damage your microwave.

ELECTRIC DEHYDRATOR

Electric food dehydrators are table-top units with a heating element, a fan, and a series of trays or racks that stack on top. They are efficient but expensive – the perfect present for an enthusiast who likes gadgets.

DESICCANTS

These draw the moisture out of produce while preserving colour and flavour. Salt and sugar are the most commonly used. This is a more expensive method, as you need enough salt or sugar to fill a container and completely cover the food.

We use this technique to make aromatic herb "rubs" for cooking joints of meat. Layer rosemary, marjoram, oregano, and salt in a container, and leave the herbs to dry. Then crumble them up with the salt and store in an airtight jar for a very tasty herb mix.

DRY THIS

Dry apple and pear rings using a solar dryer or thread them on a length of string and hang them above the stove.

Air dry ripe pea or bean pods until brittle, then shell them. Finish drying in a solar dryer. Rehydrate before cooking, or save to sow in spring.

Slice beetroot, carrot, and parsnip thinly and dry in the oven or in a solar dryer to make vegetable crisps.

Air dry bunches of herbs, such as rosemary and sage, before they flower, when aromatic oil levels are highest.

Mash ripe strawberries and add 2 tbsp sugar. Roll the mixture out to a depth of 2–3mm (1/16–1/8in) on a piece of muslin. Air dry on a rack for at least a few hours and up to a couple of days, then slice into tasty snack-sized pieces.

Hang up chillies to air dry.

Slice large mushrooms, but leave small ones whole. Dry in the oven or solar dryer.

Blanch then dry chopped kale in either the oven set to a low temperature or a solar dryer for lovely, healthy crisps.

PROJECT Use a solar dryer

A solar dryer needs no fuel and costs nothing to run. Place it in direct sunlight, facing south, and move it every couple of hours to track the sun. If you know you're not going to be able to do this, position it to face the sun when it's at its highest point. See pages 170–171 for how to make a dryer.

YOU WILL NEED

Apple peeler, corer, and slicer
Sharp knife
Chopping board

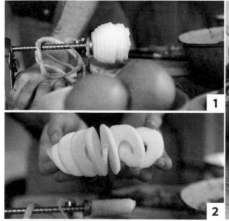

1. An apple peeler, corer, and slicer is a handy gadget that does all the functions at the turn of a handle. It works on pears too. **2. The apple** comes off in a continuous spiral. Cut into rings and dip in a solution of water and lemon juice (one tablespoon each) mixed with a teaspoon of sugar. Pat slices dry before putting in the dryer. **3. Cherry tomatoes** are simply cut in half. **4. Arrange food** on wire shelves, which allows air to circulate. Poppy seedheads don't need any preparation. Close the door, position the dryer, and leave for a few days. Test produce regularly (see box, opposite).

Make a solar dryer

Solar drying is an ancient cooking method. Sunlight is converted to heat in the solar collector – an insulating box with a glass front – and the hot air rises into the dryer or cooking chamber. A solar dryer is not nearly as powerful as a conventional oven. We use ours for drying fruit and vegetables (see pages 168–169), rather than cooking a full meal – you would need a very sunny day for that.

YOU WILL NEED

Drill and various size bits
Hole saw
Tape measure
Saw and hacksaw
Spirit level
Glass-fronted cabinet
Plywood, spare timber, battens

Nails and screws
Black paint, white paint
Mesh – metal or muslin
Corrugated iron
Door knob, hinges, catch
Wire shelves

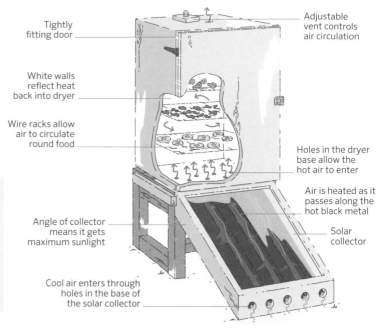

Tightly fitting door

Adjustable vent controls air circulation

White walls reflect heat back into dryer

Wire racks allow air to circulate round food

Holes in the dryer base allow the hot air to enter

Air is heated as it passes along the hot black metal

Angle of collector means it gets maximum sunlight

Solar collector

Cool air enters through holes in the base of the solar collector

MAKING THE SOLAR COLLECTOR AND STAND

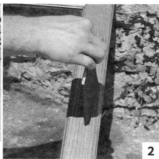

1. Reuse an old cabinet. The glass door will form the top of the solar collector and the cabinet will form the dryer. Drill a hole at one end of the cabinet using a 25mm (1in) hole saw to form the air vent (see step 16).   **2. To find the best angle** for the solar collector prop it against a wall at midday. Adjust the angle of the glass door until a small notebook held upright on the frame casts virtually no shadow. At this point the collector will be at the optimum angle to receive the maximum amount of sunlight all day.   **3. Measure the height** of the door against the wall; this will be the height of the dryer base.   **4. Cut a piece** of 50 x 25mm (2 x 1in) batten to the same width as the cabinet and drill a row of holes using a 25mm (1in) spade drill bit.

5. Measure the glass door and make a shallow 3-sided box from plywood to the same width but longer, to fit beneath the dryer, and just deep enough to hold the corrugated iron. Use the drilled wood for the end and 50 x 25mm (2 x 1in) for the sides.   **6. Paint** the inside white.
7. Tack mesh across the holes to keep insects out.   **8. Add wooden spacers** for the iron to rest on to let air circulate.   **9. With a hacksaw** cut the iron to fit inside the box. Paint it black.

10. Screw the glass door to the top of the box. **11. Hold the box** at the correct height for the optimum angle, as calculated in steps 2 and 3. Use a spirit level to mark a horizontal edge along each side of the box at the open end. Cut along the lines with a saw. These form the edges where the dryer cabinet is fixed on top, and determine the size of the stand. **12. Slide** the black iron inside the collector. **13. Build a stand.** Use two square frames for the ends, and screw them to two side pieces of 50 x 25mm (2 x 1in) wood. The stand should be the same width as the dryer, and the height as calculated in steps 2 and 3 at the back, and the height less the depth of the collector at the front. Attach the collector using timber side braces.

MAKING THE DRYER

14. Mark two staggered rows of holes in the base of the dryer to allow air to enter from the solar collector. **15. Drill** the holes using an 8mm (⅛in) drill bit. **16. Stand the dryer** on top of the collector, with the air vent at the top, to check the fit, but don't join them together yet. **17. Cut a door** for the dryer from a piece of plywood and attach it with hinges. Nail equally spaced runners made from thin wooden battens to the sides of the dryer to hold wire shelves for drying the food.

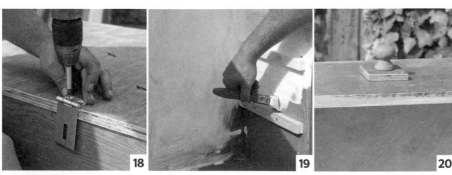

18. Add a catch to keep the door securely shut. **19. Paint the inside** of the dryer white. **20. Attach a door knob** to a square piece of wood big enough to cover the vent hole, then screw the wood in one corner so that it can be swivelled open to let air out or closed to stop rain getting in. Screw collector and dryer together tightly, with no gaps. **21. Add wire shelves.** For food preparation tips, see pages 168–169.

PRESERVING THE HARVEST

Certain times of the year bring with them a bountiful supply of produce, but the harvesting season is often short and sweet. However, you can make the most of these seasonal gluts by storing and preserving your produce, allowing you to enjoy it all year round. Some preserving methods also intensify flavours and make tasty treats to cheer you up on cold winter days.

FILLING THE LARDER

If you are living and eating in tune with the seasons, you will notice a change in the foods available locally at different times of the year. Yet, by eating purely seasonal produce you could find yourself with a vitamin deficiency in the cold dark days of winter, which a supply of stored summer crops can help to remedy. Besides, there comes a time in the depths of winter when all you really want is a tasty fruit pudding or tomato sauce with pasta. To achieve this, you need to look ahead in summer and early autumn, and store your produce when it is ripe. Get the technique right, too, so that you don't end up with a jar of inedible mould.

BEST PRESERVING METHODS

Pickling and making chutney

(see pages 174–175) can actually improve the taste of your produce, as well as preserving it. The processes rely on preserving food in vinegar and flavouring it with spices and other ingredients. Both are delicious with cold meats, cheese, and curries.

Making jams and jellies (see pages 176–177) is a great way to preserve all those fruits that grow in excess during the summer and early autumn.

Drying (see pages 168–171) is the oldest method of preserving produce, and also one of the cheapest and simplest. We frequently dry herbs, tomatoes, seeds, chillies, salamis, and hams, and have built a solar dryer (see pages 169–171).

Curing (see pages 252–253) is a preserving technique that is normally reserved for meat and fish. It usually involves using large amounts of salt to draw the moisture out of meat or fish and so make it inhospitable to harmful microorganisms.

Hot and cold smoking (see pages 224–225) are methods to preserve food, but are more reliable if you cure or dry it first. Today, smoking is primarily used for flavouring, rather than as a preserving technique. Having said that, if you smoke something for long enough, it will also dry out enough for storage.

Clamping and using a sandbox (see pages 166–167) are other traditional storage techniques that are very easy to do at home.

Freezing allows you to store a wide selection of meats, fish, fruits, and vegetables. We also often freeze soups and leftover meals once they have cooled down. The benefit of freezing fruit and vegetables is that most of the vitamins and goodness is retained, but the drawback is that the texture of foods can be lost. Make sure you label and date your bags and containers, and that you organize your freezer well, with the newest produce at the back, or you will end up losing last year's frozen peas until the next ice age.

Fermenting (see pages 178–179) not only preserves fruit and vegetables for several months, but also gives complex and unique flavours to your produce. Some of our favourite ferments are sauerkraut, kimchi, and cucumber pickles, but we also love to ferment our root vegetables and garlic to make the most of bumper crops.

1. We have an army of pickles on hand to garnish meals and add a dash of colour to our dishes. **2. Curing salmon** by packing it in a dry salt rub makes it last a little longer. It's also delicious!

SEASONAL PRESERVING

SPRING

March
RHUBARB Stems of rhubarb bottle successfully, retaining much of their flavour.
THYME Store stems using the traditional drying method, and make salt rubs for barbecues.

April
ASPARAGUS If you have an abundance of spears, blanch them for 2 minutes and then freeze them.
SPINACH You could freeze it, but do so as soon as possible after harvesting.
SAGE AND ROSEMARY Drying is the quickest and most effective method for the majority of herbs.

May
BEETROOT Either pickle them or place in sand trays ready for roasting or turning into mash.
RADISH Slice and pickle them in vinegar.
GOOSEBERRIES Frozen berries retain their original, tart taste.
DILL We store dill alongside salmon when we make our own gravadlax (see page 252).
MINT Consider drying it to make a tin of tea leaves for a refreshing cuppa.

SUMMER

June
BROAD BEANS These are best blanched and then thrown into the freezer.

CARROTS You could place them in a sand box or a clamp – or freeze them.
ONIONS Dry on a wire tray and then plait to hang up and dry. Don't leave them in the dark or they will start to sprout.
PEAS No contest: freeze them if you have a glut.
REDCURRANTS Freeze redcurrants or turn them into a deliciously sweet jelly.
STRAWBERRIES Making strawberry leather is a nice way to preserve strawberries for a little longer (see page 169).

July
CUCUMBER Cucumber pickle is very tasty and worth a try.
CHILLIES We either preserve our fiery chillies in spicy chutneys or dry them on lengths of string then store in sealed jars.
POTATOES These are best stored in a clamp or in a big paper sack in a dark and dry place **(2)**.
RUNNER BEANS Top and tail them, slice diagonally into thin strips, blanch for 2 minutes, and when they have been drained and cooled, freeze them.
TOMATOES We add them to chutneys and bottled sauces, and sun-dry them.
BLUEBERRIES These freeze particularly well.
RASPBERRIES Turn them into jam or store in the freezer if you have a glut **(4)**.
CHIVES Chopping them up and freezing in ice-cube trays works really well.
LEMON VERBENA Dry some of this aromatic herb and use it to make a relaxing tea.

August
GARLIC Heads of garlic can be stored in plaits that hang just outside the kitchen for easy use. We also pickle some cloves.
GLOBE ARTICHOKES We like to cook them with herbs and other delicious flavourings and then store in oil.
BLACKBERRIES An amazing crop to forage from your hedgerows; they freeze well.
CORIANDER SEED Simple to dry, and they add huge amounts of flavour to curries, dressings, and oils.

AUTUMN

September
HORSERADISH Around this time of year we make jars of horseradish sauce.
MARROWS Using marrow for chutney is a good way to add flavour to and preserve these overgrown beasts!
APPLES Cider is one way to preserve apples (see pages 200–201) **(1)**. Or wrap them in newspaper and keep in the store room, or slice and make sun-dried apple rings.
GRAPES The only methods of preserving grapes that we are prepared to consider are adding them to chutney, or using them to make our own wine (see pages 196–197).

October
SQUASH Just before or straight after the first frost, cut the squashes and store them in nets or on shelves in a dark place; they should last until the following year.

CRANBERRIES We are traditionalists when it comes to cranberries. We try to make our own cranberry sauce and keep it untouched until Christmas day to enjoy with our home-reared turkey.
SWEET CHESTNUTS AND WALNUTS These nuts are very easy to freeze.

WINTER

November
CAULIFLOWERS Make your own piccalilli year on year **(3)**.
PARSNIPS Best stored in clamps or in sand boxes.

December
BRUSSELS SPROUTS These can be blanched and frozen, but we try to time the growing season so they are harvested fresh on Christmas morning.
CABBAGES Pickling is an effective preserving method, especially for red cabbage.

January
MUSHROOMS Dry in a solar dryer or oven and then store in jars in a dark cupboard. They will need rehydrating before using for cooking, but they last for ages.
SWEDE Set up a clamp or store in a sand box.

February
LEEKS Freeze them, but they are happy left in the ground – the simplest storage of all.

MAKING PICKLES AND CHUTNEYS

The difference between pickle and chutney is quite simple: pickles are pieces of vegetable or fruit preserved in vinegar, whereas chutneys are made from fruit or vegetables cooked in vinegar.

Vinegar is used because its high acidity inhibits the actions of harmful microorganisms that cause foods to go bad, and it adds a tangy flavour to your preserves.

PICKLES

Generally, pickling is quicker and easier than making chutney, and makes use of young, undamaged fresh fruit or veg. When pickling vegetables, it is important to first ensure that they are moist, and to then soak them in brine or pack them in salt for 24 hours to draw out some of their water. To make brine, mix 100g (3½oz) of salt with 1 litre (1¾ pints) of water. Then place the vegetables into sterilized jars (see step 4 opposite) and completely cover them with vinegar. Distilled vinegar is the strongest preservative, but also one of the most expensive; the more you pay for your vinegar, the tastier the result will be. However, you can also make your own vinegar (see page 180), flavoured with your chosen herbs and spices.

Pickled eggs are a great option if you find yourself with more eggs than your family or friends can eat. Hard boil the eggs, then shell them and pack into large jars with about 1 litre (1¾ pints) of spiced vinegar to every dozen eggs. Close the lids tightly. The eggs should be ready to eat after about one month. Add some beetroot for a fun and tasty pink version.

Pickled onions are another popular option. To achieve a sweet pickled onion, add a little sugar to the vinegar (we use white sugar to keep the pickle clear and light). Before peeling the onions, soak them in a brine solution for 12 hours, then skin them, and submerge in fresh brine for another two days before packing them into a jar. Cover them with spiced, sweet vinegar and seal the jar. Resist eating the onions for at least two months after pickling them. Use red onions and cider vinegar, with a little sugar, for a brighter version.

CHUTNEY

The chutneyfication process is one of the most satisfying and exciting aspects of growing your own food. We turn huge volumes of our produce into chutneys and they can be made from almost any fruit or vegetable. We've used marrows, apples, grapes, radishes, rhubarb, red and green tomatoes, chillies, aubergines, pears, squashes, and turnips – to name but a few. Then there are all the various spices that add those distinctive chutney flavours: cumin, coriander, allspice, cloves, ginger, peppercorns, paprika, mustard seed, and garlic are some of our first-rate selections.

BRINING OLIVES

We grow our own olives in the polytunnel, but raw olives need to be pickled to make them palatable. To remove any impurities, cut each olive to the stone, place in sterilized jars, and cover with water. To keep the olives submerged, place a small ramekin or glass weight on top. Change the water every day for a week and then pickle them in brine – 75g (2½oz) salt to 1 litre (1¾ pints) of water – for five weeks. Add a thin layer of olive oil for an airtight seal.

1. Pickled onions should be both sharp and sweet. Try our pink onion rings for a colourful version. **2. Clearly label** your chutneys, pickles, and ferments with the name, date, and even a few notes on your recipe. **3. Crunchy vegetables** like cucumbers or cabbage can be pickled in cold vinegar with a little salt and sugar.

RECIPE Spiced pickling vinegar

There is a wide variety of vinegars suitable for pickling, such as malt, wine, or cider vinegar. You can also add your own spices and herbs to clear distilled vinegar. For the contents to remain recognizable and look attractive, use whole spices, as ground spices make the vinegar go cloudy.

YOU WILL NEED

Muslin squares	Cinnamon sticks
String	Star anise
Dried chillies	Cloves
Juniper and sumac berries	Bottle of distilled vinegar

1. Lay two squares of muslin on top of each other on a work surface. Crush the chillies and berries, and then put the spices in the centre of the fabric.   **2. Gather up the corners** and tie with string to form a spice bag. Ensure the string is long enough to loop over and hang from the handle of the saucepan.   **3. Put the bag** and vinegar in the saucepan and bring it to the boil. Remove from the heat and leave the vinegar and spices to cool for a couple of hours. Take out the spice bag. The flavoured vinegar is now ready. Use it to pickle garlic, onions, cabbage, or eggs, for example.

RECIPE Garden chutney

The key to making chutney is to cook the fruit and vegetables for a long time, so that most of the moisture evaporates and it has a thick, jam-like consistency. The colours will change and the flavours intensify in the process. A good chutney also benefits from bold, contrasting flavours, so be experimental and don't be afraid to mix together fruits and vegetables.

YOU WILL NEED

Pestle and mortar
6 large cooking apples
1.5kg (3lb 3oz) tomatoes
5 onions
1 tsp chilli flakes
1 tsp allspice
1 tsp mustard seeds
1 tsp red peppercorns
1 tsp paprika
1 tbsp finely chopped fresh root ginger
2 garlic cloves, finely chopped
Handful of sultanas
1kg (2¼lb) caster sugar
4 litres (7 pints) red wine vinegar

1. Peel and roughly chop the apples, tomatoes, and onions. Crush the spices with a pestle and mortar.   **2. You can remove** the tomato seeds for a drier, faster-cooking chutney.   **3. Place all the ingredients** in a large pan, mix, and bring to the boil. Turn down the heat and simmer for 3 hours or until the chutney has a jam-like consistency and a wooden spoon drawn across the base of the pan leaves a trail.   **4. Spoon into** sterilized jars, seal, and label. To sterilize a jar, wash in hot water and put in a cool oven at 140°C (275°F/Gas 1) for 15 minutes, in a dishwasher on the hottest setting, or microwave for 2 minutes with 4 tbsp of water.

MAKING JAM AND JELLY

Jam-making is really enjoyable and captures those delicious fruity summer flavours for use all year round. Our pantry is lined with rows of gleaming jars and top-quality condiments that are expensive to buy in shops, but are made from fruits like greengage or quince that are easy to grow or forage for. Jams and syrups also make use of your surplus crops, ensuring that nothing goes to waste.

JAM-MAKING INGREDIENTS

Good jam-making relies on three key ingredients: good-quality fruit, pectin, and sugar.

Prime-quality fruit is better for jam than over-ripe fruit. If you don't have your own fruit trees or soft fruit canes, visit a local "pick your own" or forage the hedgerows for a bumper crop of fresh blackberries.

For jam and jelly to set properly you need the correct balance of pectin, acid, and sugar. Fruits that are rich in pectin (see the table, opposite) set well without any added ingredients, but those with medium to low pectin levels will need some pectin stock (see box, right). Alternatively, you can add granulated or liquid pectin, or use jam sugar, which has pectin included in it. Lemon peel contains pectin, offering another option, but be wary of adding too much as this can affect the final flavour of the jam or jelly.

Sugar preserves the fruit by inhibiting the growth of yeasts. The sugar content must be more than 60 per cent to stop the jam from fermenting. A rough guide is to add 550g (1¼lb) of caster sugar to every 450g (1lb) of any fruit that you know is rich in pectin; 450g (1lb) of sugar to 450g (1lb) of fruit that is fairly rich in pectin; and only 350g (12oz) of sugar to 450g (1lb) of fruit that is low in pectin.

Any sugar can be used to make jam. Preserving sugar is more expensive, but very easy to use as it dissolves quicker and needs less stirring. Brown sugar produces darker jams with a slightly different flavour.

MAKING JELLY

Jellies are made using a similar process to jam, but you don't need to throw away all skins, stones, and pips before heating the fruit. This is because they are removed, along with any pulp, when you strain the mixture to obtain the juice.

ACHIEVING A SET

To test for the setting point, spoon a little jam onto a cold plate and allow it to cool. Push your finger across the surface of the jam and if the jam wrinkles, remove the pan from the heat. Alternatively, remove some jam from the pan using a metal spoon. If you can trace your finger through the jam without it running back, it is ready.

You can also make pectin stock. This helps to set fruits that have a medium or low pectin content (see the table opposite). Put 1kg (2¼lb) chopped, unpeeled (the peel contains the pectin) cooking apples into a preserving pan and just cover with water. Cover the pan, bring to the boil, then simmer for 20 minutes or until soft. Remove the pulp by straining the apples over a clean pan. Simmer the juice gently once again until reduced by half. Pour into small freezer pots. Generally 150ml (5fl oz) of pectin stock is sufficient to set 2kg (4½lb) of fruit.

RECIPE Plum and cinnamon jam

Plums contain particularly high pectin levels (see box, opposite), especially when they are not quite ripe, so there is no need to add a setting agent to this recipe.

YOU WILL NEED

Preserving pan
2.5kg (5½lb) dark plums or damsons

2 cinnamon sticks
Peel of 1 orange
3kg (6½lb) caster sugar

1. Halve and pit the plums, and put them into the pan with the cinnamon sticks, orange peel, and 1 litre (1¾ pints) of water. Heat gently for about 20 minutes until the fruit is really soft and the syrup is reduced. **2. Remove from the heat,** add the sugar, and stir until dissolved. Bring to a rapid boil for about 10 minutes, stirring occasionally, until the setting point is reached. **3. To test,** trace a finger through the jam. If it doesn't run back, spoon into sterilized jars (see page 175). Seal and label.

RECIPE Apple and blackberry jelly

Try this easy jelly recipe – you don't need to add pectin, as the apples contain enough to make it set. Remember when making jams and jellies to always remove the pan from the heat before adding the sugar, and warm the sugar before mixing it with the fruit so that it dissolves more quickly. Take care when boiling your jam too. Boil it for too long and it will set rock hard; boil it too little and it will be a runny mess.

YOU WILL NEED

Preserving pan	500g (1lb 2oz) each cooking apples and blackberries
Jelly bag	
Fine sieve or muslin jam bag	450g (1lb) caster sugar for each 600ml (1 pint) of jelly liquid
Baking parchment or cling film	

REDUCING THE FRUIT

1. Peel, core, and chop the apples. Place in a large pan with the blackberries and 1.5 litres (2¾ pints) of water.   **2. Bring to the boil,** turn down the heat, and leave to simmer for about an hour.   **3. Remove any scum** and then test the jelly consistency on a plate. It should be a pulpy mass. If not, simmer for another 10 minutes and re-test.   **4. Ladle the liquid** into a jelly bag suspended over a pan. Leave to drip overnight. Resist squeezing the bag or the jelly will turn cloudy.

MAKING THE JELLY

5. In the morning, remove and compost the pulp. Measure the juice and weigh the correct amount of sugar for the volume of liquid. Gently heat the juice. Remove from the heat, and stir in the sugar until it has dissolved.   **6. Boil the jelly** for 10 minutes and remove any scum from the surface. Then, test it for a set. You should be able to trace a finger through the jelly without it running back.   **7. Strain the jelly** through a fine sieve or muslin jam bag into sterilized jars (see page 175). Fill them to the top as the jelly will shrink.   **8. While the jelly is hot,** seal the jars with a layer of baking parchment or cling film. Screw the lids on tightly.

PECTIN LEVELS

Fruits high in pectin	Fruits with medium pectin levels	Fruits low in pectin
Blackcurrants	Apricots	Blackberries
Citrus fruits (mainly in the peel)	Cranberries (ripe)	Blueberries
Cooking apples	Grapes (unripe)	Cherries (dessert)
Crab apples	Loganberries	Figs
Cranberries (unripe)	Medlars	Grapes (ripe)
Gooseberries	Morello (cooking) cherries	Melons
Plums (unripe), damsons	Plums (ripe)	Nectarines
Quinces	Raspberries	Peaches
Red- and whitecurrants		Rhubarb
		Strawberries

FERMENTING

One of the oldest forms of food preservation, fermentation is having a comeback. Fermented foods are undeniably good for us, helping to maintain our gut flora which in turn provide a variety of health benefits, from promoting a healthier digestive tract to boosting immunity. We ferment for flavour but also as a practical method to preserve the harvest using minimal tools and equipment.

EXAMPLES OF FERMENTED FOOD

While fermentation might be most closely associated with the likes of sauerkraut (see opposite), fermentation is actually a key process in many foods that we enjoy on a regular basis.

Pickled vegetables are now often preserved using vinegar and sugar, but in traditional lacto-fermentation you depend on the beneficial bacteria on the surface of vegetables, such as lactobacillus, to do the fermenting for you. This means you can lacto-ferment pretty much any vegetable, from carrots to radishes. All you need is a clean jar and 2 per cent sea salt for delicious and healthy results.

Kimchi is a Korean dish similar to sauerkraut (see opposite). It is made through the process of lacto-fermentation. Depending on how many chillies you use, you can make your kimchi extra spicy.

Cheese, butter, and yogurt may not immediately spring to mind as fermented foods but they are among the most common. There are several types of cheese and cultured butter or yogurt that you can make easily at home.

Sourdough bread has become very popular, and there's something magical about making your own from a starter (see page 216). Your bread will be truly local as it will pick up on the naturally existing "wild" yeast in the air of your kitchen.

TOP TIPS

Use clean glass jars for fermenting. Avoid using plastic, as this can more easily contaminate the flavours.

Use 1–3 per cent salt (relative to the total weight of the other ingredients) for your ferments. This will provide the appropriate environment for lactic acid bacteria to grow and suppress unwanted bacteria. It will also help protect the cell structure of your fruit and vegetables, retaining the texture.

Fermentation requires an anaerobic environment, so keep it oxygen-free by submerging your vegetables in their brine. Cover with a sheet of baking parchment, a glass weight, or baking beans to keep the food below surface level whilst fermenting.

Leave a little space in your sealed jars for expansion and release of CO_2 as a result of the fermentation process. You can also buy fermenting valves that can be screwed on as lids, allowing gas to escape but not enter.

Wait for a minimum of 10 days for the fermentation process to work. If it's all done properly there is no maximum.

If the food looks or smells wrong then trust your senses and don't risk it. Any mould growing on the top of the ferment is likely caused by oxygen and yeasts on the surface and is not generally harmful, but it's best to remove before eating.

1. Add ingredients to your ferments for more flavour. A few spices won't mask the natural taste but can offer extra colour. **2. A sourdough starter** is a living ferment that can be used for making many different breads. **3. Fermented butter** has an intense tangy taste that boosts its natural flavour and is easier to digest.

RECIPE Sauerkraut

Simply made of cabbage and salt, sauerkraut is one of the easiest fermented foods to make. Essentially, you pack all the ingredients into a clean jar and ferment for at least 10 days. Since it's a fermented food, sauerkraut will keep for several months.

YOU WILL NEED

Large bowl	Baking parchment
Lidded glass jar	1 cabbage
Weight	2 per cent sea salt

1. Remove any damaged outer layers of the cabbage and remove the hard core. Keep one clean good leaf intact and slice the rest thinly (around 5mm/⅕in). **2. Weigh your cabbage.** Calculate 2 per cent of its weight and measure out this amount of salt. Put it in a large bowl and sprinkle the salt over the cabbage a little at a time. **3. Massage the cabbage** with salt. Continue until all the salt has been added. Press the cabbage down inside the bowl with a small plate, place a heavy weight on top, and leave for a few hours or overnight. This will form your fermenting brine. **4. Then place** your shredded cabbage in a clean jar and press down until compact. **5. Pour over** your brine until the cabbage is covered. If the cabbage needs extra brine then make a 2 per cent brine (20g/¾oz of salt dissolved in 1 litre/1¾ pints of water) and top it up until completely submerged. Cover with a sheet of baking parchment and seal the lid. Leave at room temperature for at least 10 days. Once opened, store in the fridge and consume within a week.

RECIPE Kombucha

Kombucha is a cool and refreshing fermented tea. To make it, you'll need a SCOBY ("symbiotic colony of bacteria and yeast"), which feeds on the sugary tea and ferments it. Top up your kombucha with fresh sweetened tea whenever you pour yourself a glass, and the SCOBY will continue to live and do its work.

YOU WILL NEED

Wide-rimmed jar	50g (1¾oz) caster sugar, plus extra for topping up
800ml (1¼ pints) caffeinated green or black tea, plus extra for topping up	SCOBY "mother" culture

1. Brew a strong tea at 85°C (185°F) and allow it to infuse for 10–15 minutes. **2. Dissolve the sugar** in the tea and allow it to cool to room temperature. **3. Pour the tea** into a large wide-rimmed jar, and place the SCOBY on top. **4. Cover the jar** with a square of muslin and secure with an elastic band. Store at room temperature for at least 3–4 days. **5. To harvest,** pour up to two-thirds of the liquid into a separate container. Chill the harvested kombucha in the fridge before serving. Top up the SCOBY jar with more sweetened tea mixture (see step 1) to continue the fermentation process. The SCOBY should float on the surface. If, after a couple of weeks, the SCOBY starts to sink, pour away some of the liquid and top up once more with sweetened tea.

HERBS TO GROW

Herbs usually do extremely well in pots and containers and are a great starting point for growing your own produce. Grow herbs that you like eating and remember you are more likely to use them if they are close to the kitchen. Here are some of our favourites.

Basil
Ocimum basilicum

HOW TO GROW Basil can be grown as a perennial in warmer countries but in cooler climates it's usually grown as an annual. The fact that we have been growing basil from seed every summer for many years says something about how tasty we think it is. Basil grows very well in a sunny position, provided it is watered regularly. It is good practice to pinch out the growing tips to help preserve the plant's bushy shape and to remove the flowering shoots to ensure a decent supply of young leaves.

HARVEST, COOK, STORE There's nothing better in summer than a snack of our own tomatoes and basil on fresh bruschetta. The big mature leaves are the tastiest. Basil doesn't really dry and store well so the leaves are best used fresh.

Bay
Laurus nobilis

HOW TO GROW Bay is an evergreen that requires very little maintenance. It grows extremely well in pots or containers and is a great herb for an urban space.

HARVEST, COOK, STORE You can dry bay leaves to store them but as we have a bay tree, we pick and use the fresh leaves at any time of year. They are crucial for a good casserole.

Chives
Allium schoenoprasum

HOW TO GROW Chives are hardy perennials and a member of the onion family. They grow in small clumps. Sow seed outside in spring when the soil warms up in a warm, sunny position, or propagate by dividing existing clumps and replanting in good compost. Chives will grow in almost any soil but they like humidity so we tend to plant them next to our ponds. If seed is sown under cover, leaves can be harvested in just a couple of weeks.

HARVEST, COOK, STORE The hollow, grass-like leaves are great snipped over

potato salads and omelettes. We also use the pinky-purple flowers as a tasty garnish in summer salads. Yes, chives can be frozen or dried but nothing beats the taste of the freshly picked herb.

Coriander
Coriandrum sativum

HOW TO GROW We love coriander. Sow seed from late spring into a fairly rich soil in a sunny position. We succession sow coriander to guarantee a ready supply. We are also lucky enough to be able to grow coriander and other annual herbs under cover in winter, thanks to our greenhouse heat sink (see pages 130–131).

HARVEST, COOK, STORE The leaves and seeds can be used in a huge range of dishes, including lots of spicy Asian cuisine. Follow the instructions for dill, right, for harvesting the seeds.

Dill
Anethum graveolens

HOW TO GROW Dill is a hardy annual with highly aromatic dark brown seeds and lively tasting leaves. Sow dill seeds in situ

Chopped herbs add real depth of flavour to marinades and rubs. Try classic combinations like rosemary and orange, or create your own.

and be sure to keep the area very well weeded as the seedlings don't thrive with competition. To ensure a good seasonal supply, succession sow seed from late spring to early summer.

HARVEST, COOK, STORE The leaves are delicious with fish and our favourite recipe has to be the Scandinavian salmon dish gravadlax (see page 252). They are also excellent with cucumber or fresh in salads. To harvest the seeds simply cut the base of the plant when the flower heads turn brown and tie the stems into loose bunches. Cover the heads with brown paper bags and hang upside down somewhere airy for a few days. Open the bag and you'll find plenty of dry ripe seeds ready to store in a glass jar. If you want to dry the leaves for use in winter, make sure you cut them when the plants are about 30cm (12in) tall and before they start flowering.

Rosemary can be used fresh but also works well as a dried herb. Try combining with thyme, marjoram, and oregano for your own *herbes de Provence* blend.

We often use rosemary when cooking meat or roasted vegetables and almost always as a fragrant addition to the summer barbecue.

Sage
Salvia officinalis
HOW TO GROW Sage is an aromatic perennial. It grows to about 60cm (2ft) in the right sunny position with well-drained soil. For best results, grow from cuttings taken in late spring. Cut bushes back at the end of the summer to prevent them from becoming too woody.
HARVEST, COOK, STORE Sage is great for flavouring stuffing and makes a very tasty pesto mixed with garlic and pine nuts. Cut leaves and hang up to dry at the end of summer – be warned, they take a long time to dry.

Tarragon
Artemisia dracunculus
HOW TO GROW Tarragon is a very untidy but vibrant perennial herb that usually comes in two varieties: French and Russian. We recommend growing the French variety rather than the coarser Russian. Take cuttings and keep them indoors as French tarragon is susceptible to cold weather and may die over winter. Drainage is important: plant it on sloping ground in a sunny position, or near the top of a herb spiral.
HARVEST, COOK, STORE Leaves can be dried but they are better frozen. Add to chicken dishes or infuse a sprig in vinegar to use in salad dressings.

Thyme
Thymus species
HOW TO GROW Thyme is another great perennial that thrives in a well-drained, dry position. It's best grown from cuttings taken in early summer, rather than from seed. For best results cut shoots about 15cm (6in) long from the tips of the plant, avoiding the woody base stems. Prune it after flowering to stop it getting too straggly.
HARVEST, COOK, STORE Thyme can be dried successfully but as it is evergreen we usually carry on picking it even in winter. It's great in all tomato recipes.

Marjoram
Origanum majorana
HOW TO GROW There are many different varieties of marjoram and to make things even more confusing some of them are called oregano. As a general rule, marjoram varieties are hardy perennials while oregano varieties (*Origanum vulgare*) are usually tender plants. Gourmets claim the latter have the finest flavour. Both do well in pots or in a sunny spot in the garden.
HARVEST, COOK, STORE Both oregano and marjoram are vital herbs for pasta sauces and for sprinkling over pizzas. The leaves dry successfully (see pages 168–169) but you should be able to keep a pot or two going on a windowsill in the kitchen or on a shelf in the greenhouse over the winter.

Mint
Mentha species
HOW TO GROW There are lots of different varieties of mint and most of them are tasty. Mint spreads very quickly and is best planted away from other herbs, either against a wall or in a bucket, or box in the herb bed to control its roots – take the bottom of the bucket out to allow drainage. It can tolerate a degree of shade and needs lots of water.
HARVEST, COOK, STORE We use our mint to make everything from mint sauce to go with lamb to tzatziki, to help us get through our summer cucumber glut. We also enjoy it freshly picked in a cup of mint tea. Mint can usually be harvested all year round, but it's worth potting up a few roots for winter and bringing them indoors to a kitchen windowsill, just in case.

Parsley
Petroselinum crispum
HOW TO GROW Parsley is one of the most well-known herbs and has made its mark in culinary circles with its unique taste and versatility. It is also a very good source of vitamin C and iron. We sow parsley each year into rich soil that has a fine tilth; it likes to be kept well watered.
HARVEST, COOK, STORE Dry the leaves at high temperature by blasting them in a microwave for a few minutes. Add fresh parsley as a garnish: it loses its flavour in slow cooking.

Rosemary
Rosmarinus officinalis
HOW TO GROW Rosemary is a delightful evergreen shrub: some species can grow to about 1.5m (5ft), so choose carefully. You might consider using it for a hedge. It likes light and dry soil. Sow seeds in spring or alternatively take some cuttings from a neighbour's plant (ask first!). Take them in late summer from non-flowering shoots and push into a pot of compost. Keep in a frost-free environment until the following spring when they can be planted out. Prune established plants in spring after the first flowering.
HARVEST, COOK, STORE Late summer is probably the best time to pick leaves for drying but as rosemary can be harvested all year round, we don't tend to bother.

EDIBLE FLOWERS

We like to liven up our salads with a scattering of edible flowers. While most people are familiar with borage flowers floating in their summer Pimm's, they are often surprised to find nasturtiums, calendula or pot marigolds, wild garlic flowers, and heartsease in amongst the lettuce. Chive flowers add a nice touch of pink. As with all ingredients for a salad, shake flowers free of insects before adding.

Borage
Borago officinalis
With a distinctive blue star-shaped flower borage looks beautiful in the vegetable bed and it's superb for attracting bees to your plot to boost productivity. The flowers can be crystallized for a sweet garnish or frozen into ice cubes for a summer cocktail. The seeds are a rich source of gamma-linolenic acid which is believed to help lower blood pressure.

Crab apple
Malus sylvestris
The leaves of a crab apple are inedible but the flowers – along with the fruit (see page 187) – are worth trying. You can make a delicious spring blossom cordial with the petals.

Crocus
Crocus species
It's not the petals of this stunning flower that are edible, but the delicate stigma tips, which make saffron. Generally considered to be the world's most expensive spice, saffron provides an aromatic and gently spiced flavour, but beware: it can dye your food – and clothes – bright orange.

Dog rose
Rosa canina
One of our favourite edible flowers is the common briar rose or wild rose. The flowers are edible and you can also harvest the fruit – rose hips – from mid-autumn to early winter (see page 188). In cooking, the rose petals are excellent for adding flavour to desserts such as Turkish delight or dried in sweet tea.

Elder
Sambucus nigra
Native to most of Europe and North America the elderflower can be found in wasteland, hedges, and scrub. It flowers from early summer and berries in mid-autumn. The flowers give off a heady scent and they make fantastic cordials (see pages 198).

Gorse
Ulex europaeus
There is an old saying that "when gorse is out of bloom, kissing is out of fashion", as it seems it is always in bloom. Find it in coastal areas, heathlands, and the edge of forests. The flowers can be eaten raw and used for decorating cakes. They also infuse well and can be turned into a gorse flower wine or a cordial that has a sweet coconut aroma. But never collect gorse flowers wearing just a short-sleeved T-shirt! The fragrant flowers are protected by hundreds of prickly spines, so wear long sleeves and gloves to grab them.

Hawthorn
Crataegus monogyna
Growing in many hedgerows and wooded areas, hawthorn is easy to identify. The flowers are edible and can be used for their medicinal properties. Try making a syrup or tincture.

Herb flowers
Chives, wild garlic, rosemary, fennel, and thyme all have particularly delicious edible flowers too. These herbs attract beneficial pollinating insects to your plot as well as providing delicious flowers for garnishing your food. Consider replacing your traditional flower beds with more edibles for a garden that provides a palette of edible colour all year round.

Nasturtiums
Tropaeolum
The bright orange, red, and yellow hues from nasturtiums brighten up many gardens. The flowers are quite peppery to taste, but add a real splash of colour to summer salads. We also pickle the berries as garden capers to serve in a puttanesca pasta sauce or with fish dishes. The leaves and seeds are edible too.

Primrose
Primula vulgaris
You can eat the leaves and petals of this lovely early flowering plant either raw or cooked. Don't be tempted to try eating indoor flowers that look similar as other cultivated varieties taste horrible and can cause allergic reactions. We only harvest a few of our first primroses as they provide much needed food for early-rising bumblebees in search of nectar.

1. Dry petals in a cool oven or on a sunny windowsill and use in herbal teas, salads, or as a garnish.   **2. Nasturtium flowers** are not only delicious but come in vibrant orange hues.   **3. Rosemary flowers** have a strong perfume and look delicate served on roasted meats.   **4. Dandelions** are very welcome in our garden. The flowers, roots, and leaves are all tasty.   **5. Primroses flower early** in borders and hedgerows; remember to leave plenty for early season bumblebees.   **6. Calendula flowers** are delicious in salads, or use them in a hand scrub with some sea salt and oil.

FORAGING FOR WILD PLANTS

Foraging was once vital for the survival of hunter-gatherers. Now it's a popular pastime – even in cities people gather free food from parks or footpaths. Before picking and eating any wild plants, however, be certain that you've correctly identified them. We have lost the skills needed to scan a hedgerow for edible treats. Get yourself a plant identifier and, if in doubt about a plant, don't eat it!

WHAT YOU CAN FORAGE

Depending on the season and your location, you can find many appetizing wild foods – from berries to seaweed. Always check you have permission first.

Roots

There are some delicious roots that you can forage for but you can only uproot a plant in the UK if you have the landowner's permission and it is not rare or an endangered species. When you decide to dig up a plant, remember that it will then die, so only pick if there are enough growing in that spot. We like to use a digging stick (see page 186).

Fruits and berries

Plants bearing fruits and berries are some of the easiest to identify and foraging for them can be tasty and sweet. That said, avoid relying on the colour of the fruit for identification, but check the foliage of the plant, too, to make sure you have properly identified it. Also remember to leave plenty of fruit and berries for the local wildlife and only harvest what you will eat.

Leaves

When you take leaves from an edible plant, ensure that you leave enough for it to continue to photosynthesize. Take a few from each plant rather than stripping one clean. If you are harvesting leaves to eat, check they are clean and not damaged. Don't just rely on your eyes to identify leaves, as edible leaves can often look similar to toxic varieties. Check whether the leaf feels shiny or rough. Is its distinctive aroma present? If in doubt, double check with someone who's been foraging for a long time – or leave it.

Flowers

Pick flowers when it's sunny for the best flavour and fragrance. Only pick what you need and leave plenty on the plants for pollination and to feed insects.

Seaweed

Only harvest seaweeds that are connected to rocks rather than washed up on the shore – it should look in perfect condition. Use scissors rather than pulling it off rocks so the plant can regrow.

SAFE FORAGING

The single most important thing to remember is to be sure of what you've picked before you eat it. For some terrains, it's also best to plan in advance.

Mushrooms

The wealth of fungi to be enjoyed when out foraging is extremely exciting, but collecting mushrooms takes much more experience than we can cover in this section. We recommend going on a mushroom foraging course or really reading up on the subject with some expert foraging guide books. It is surprisingly easy to pick something that is toxic by mistake, so, as with all foraging, you need to be very sure of what you are picking before you eat it.

Terrain

Wherever you forage, respect the terrain. There are many rich foraging spots on the coast, for example, but the cliffs can be hazardous. Be sure to forage on steady ground, and take care not to damage the environment – or yourself.

Tides

Foraging along the seashore can offer plenty of wild treasures but remember to respect the sea and be aware of the tides. Buy yourself a tide timetable for your local area and keep an eye on the tides to avoid getting cut off. As a precaution always have your escape route planned out. When foraging at low tide, wear suitable footwear to avoid weever fish or buried sharp objects that could cause injury.

FORAGER'S CODE

Eat only what you have identified as being edible.
Leave plenty for others, and that includes wildlife – pillaging is bad foraging etiquette.
Gather fruits widely from scattered plants, not just from one. Leave some for plants to set seed.
Be selective – pick only the young tender leaves and the best-looking fruits.
Avoid roadsides because of car fumes and don't pick low-growing plants next to footpaths, where dogs might have sprayed urine.
Take a bag whenever you go on a walk – just in case.
Give as well as take. Volunteer to help lay a hedge or put up bird boxes.

1. **Make a foraging basket** from willow, but only take what you'll eat. 2. **You will find** all sorts of treats growing on walls and in hedgerows. 3. **Organized foraging walks** are a good source of both food and expertise. 4. **A small pair** of scissors is a useful bit of kit. 5. **Foraging for samphire** is very satisfying. Enjoy served with fish.

1

4
5

2
3

WILD PLANTS TO FORAGE

Part of the fun of foraging is that you never quite know what you'll find when. While it's possible to give a general idea of when you're more likely to find certain foods over others, this depends on many factors and will vary from year to year. Our advice would be to enjoy the fact that even if you don't find what you're looking for, you'll almost certainly find something else that's just as delicious.

SPRING AND SUMMER SALADS, FRUITS, AND ROOTS

Spring is the perfect time to pick the first green tops of nettles and unusual early salads and to dig up various roots. Look for them in hedgerows and on field edges. On summer walks we fill our baskets with fruits, roots, and something succulent.

Beech leaves
Fagus sylvatica
Young beech leaves are a vivid bright green, oval in shape with lots of distinctive pairs of veins. Pick the early softest leaves and try them raw in a cheese and tomato sandwich or lightly steamed as a vegetable.

Birch
Betula pendula
Found in deciduous forest and gardens, birch can be tapped in early spring to make a sweet sap wine or syrup (see page 198). Avoid if you have heart or kidney problems.

Burdock
Arctium lappa
A tall, thistle-like plant with round, purple flowers – but it's the root you want, which is absolutely delicious. Peel and cut into chunks to add to stews, or cut into thin sticks, fry in oil and serve like chips.

Dandelion
Taraxacum officinale
You've probably got some growing in your garden and you'll find them on wasteland. The roots are very tasty in stews or stir fries, and you can roast and grind them to make a coffee substitute. Mix young leaves in with salads. They're tricky to uproot and a digging stick is a great help (see box, below).

Edible bamboo
Phyllostachys edulis
You can find edible bamboo in gardens and parkland in moist areas of dappled shade. The best shoots for eating appear in the spring and although you can eat them raw, they taste better cooked. Peel off the tougher outer layers as it's too fibrous to eat, then wash and finely slice before cooking.

Hawthorn
Crataegus monogyna
A small tree with deeply lobed leaves and very abundant, strong-smelling blossoms that open in May. The tender young leaves make an interesting addition to salads. As the name suggests, watch out for thorns!

THE DIGGING STICK

One of the forager's oldest tools is a digging stick made from strong wood – hazel is ideal. Give it a sharp, bevelled, chisel end and harden it by heating in the embers of a fire. Excavate a deep hole alongside a plant. Dig all the way to the bottom of the root and on all sides and lever out the root. Fill in the hole afterwards!

We trust the old phrase "grasp the nettle". Be firm when handling nettles and their stings shouldn't bother you, otherwise wear gloves.

Wild garlic leaves and flowers are edible. Salt and pickle the flower buds and use as capers.

that it is excellent for adding a distinctive taste to panna cotta or milk-based cocktails. It can also be drunk as a tea. Sweet woodruff grows in the damp, semi-shade of woodlands, and the flowers and leaves are both edible.

Wild garlic
Allium ursinum
Our favourite wild plants! They grow in woodland margins and you can often smell them before you see them. Eat the leaves wilted in butter as a side dish, or mixed with potatoes for a tasty fried wild garlic champ.

Wild strawberry
Fragaria vesca
These little strawberries are very tasty – more delicate in flavour than the cultivated kind but well worth collecting. Find them on the edges of woodland and shady lanes.

Wood sorrel
Oxalis acetosella
This pretty plant grows on the woodland floor in spring and looks like oversized clover, with delicate pinky-white flowers. The heart-shaped leaves have a strong citrus flavour and are great sprinkled in soups or salads – just don't eat too much as they contain oxalic acid, which can upset digestion.

Horseradish
Armoracia rusticana
The leaves look like dock leaves: large, long, and coarse. Find them in fields and wasteland. Grate the very strong-tasting root and mix with cream and some lemon to make a basic horseradish sauce.

Lamb's lettuce
Valerianella olitoria
A great ingredient for bulking out salads. The leaves can be a bit bitter when they are older, so pick them young when the flowers have a delicate sweetness. Often found growing in wheat fields or dry soil.

Marsh samphire
Salicornia europaea
A green succulent plant, marsh samphire will often carpet areas around salt marshes or mud flats. Delicious pickled, then served with ham on toast. Like many wild greens, the taste is hard to describe.

Nettles
Urtica dioica
Excellent cooked or made into tea. Use the freshest tender top leaves, cook in a little water, and serve as a side dish with butter. Cooking destroys the sting, so your tongue won't tingle!

Pennywort
Umbilicus rupestris
This lovely succulent plant grows straight out of walls and rocky crevices. Enjoy as a snack, in salads, or even in sandwiches instead of cucumber. Harvest carefully with scissors to avoid pulling out its roots.

Rosebay willowherb
Epilobium angustifolium
Steam the young shoots, or boil in a very little water, as you would asparagus.

Sweet woodruff
Galium odoratum
This wild plant was traditionally used to flavour milk so it is perhaps unsurprising

For a keen forager, stumbling upon wild strawberries is like walking into a sweet shop. Don't expect many to make it into the kitchen!

SLOE PRICKER

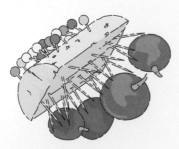

We cut a cork from a wine bottle in half lengthways and stuck lots of pins through it to make a sloe-pricking tool. Use it to pierce the skin of sloes when making sloe gin; it helps the flavour and juice to infuse. To make sloe gin you need 250g (8oz) sloes and the same weight in sugar. Fill a wide-necked bottle half way with pricked sloes and sugar and top up with gin. Shake for the first couple of days to dissolve the sugar. It'll be ready to drink after two months.

AUTUMN HARVEST AND WINTER FRUITS AND NUTS

Traditionally autumn was a time for gathering and storing nuts for winter, with people behaving very much like squirrels. It's also a good time for fungi but they are notoriously difficult to identify. We haven't covered them here; instead we recommend going on a fungi foray led by an expert.

Blackberry
Rubus fruticosus
A fruit with a rich foragers' tradition; many of us have fond childhood memories of picking baskets full. Avoid gathering next to busy roads as the pollution will spoil the fruit.

Crab apple
Malus sylvestris
This underrated fruit makes tasty jelly. It's easy to forage: just hold out a basket and shake the branches. Look for trees in mixed hedges.

Elderberry
Sambucus nigra
These hedgerow berries make very good wine. Earlier in the season we also pick some of the fragrant flowers to make our very own elderflower champagne (see page 199).

Hazel
Corylus avellana
It takes a fast-draw forager to beat the squirrels to these yummy nuts. Hazelnuts are tasty cooked in all sorts of ways. We like to store them as winter treats but make sure they are completely dry, otherwise they will rot and your crop will be wasted.

Pendunculate oak
Quercus robur
Young leaves are edible in the early summer but the acorns are what we wait for to make a dark roasted coffee alternative that is bitter and predictably nutty. You will also find oak galls, or growths, that can be used to make your own black ink.

Rosehips
Rosa canina
Rosehips are a great source of vitamin C and can be eaten raw if you scrape out the hairy seeds. They're also good for making syrup.

Sloes
Prunus spinosa
Sloes are the fruit of the blackthorn tree and look like clusters of tiny purple plums. Picking sloes is an annual ritual for us. Although they are too bitter to eat, we use them to flavour gin (see box, left).

Sweet chestnut
Castanea sativa
The nuts are contained in sheaths covered in hundreds of thin green spikes. We think that sweet chestnuts make the best stuffing for Christmas turkeys. They are also excellent roasted in front of an open fire – just make sure you pierce them with a knife first.

Chestnuts roasted over an open fire are a seasonal treat that ticks all the boxes and marks a special time of year. Just remember to leave plenty for hibernating squirrels and other woodland creatures.

1. **Rock samphire** is easy to identify and great served with seafood. **2. Sea lettuce** can be delicious fried as a crispy seaweed to accompany a stir fry.

Walnut
Juglans regia
Mature trees have huge crops. If you know where one is growing wild, keep it a secret and enjoy the benefits once a year. You'll be able to identify a walnut tree by its foliage, if you come across one out of season: they have compound leaves comprised of several pairs of leaflets. See page 165 for tips on storing and preserving.

SEASHORE SELECTION
If hedgerow pickings are thin, wrap up warm and head to the seashore instead. The footpaths near coastal areas can often be a winter treasure trove for foragers, as the warmer sea air provides good areas for winter berries, as well as better visibility through the leaf-bare trees to spot sea vegetables, such as those we've listed here.

Dulse
Rhodymenia palmata
A tasty seaweed from the inter-tidal part of the shore. Identify it by its rich red colour. You can treat dulse much like cabbage, or dry it and flake over salads for a flavoursome kick.

Edible kelp
Alaria esculenta
Kelp has long brown fronds and is found in rock pools or deeper water. Cut them into short lengths and soak in warm water for 5–10 minutes before adding to stir-fries.

Seabuckthorn
Hippophae rhamnoides
Seabuckthorn is a small shrub with stiff, spiky branches. The distinctive leaves are pale green with a silvery glimmer and the berries are bright orange and juicy. Seabuckthorn is often found growing in sand dunes and its leaves are edible when picked young and fresh. Try infused in a herbal tea. The autumn fruit makes a delicious juice or sorbet and is reported to contain, weight for weight, over 10 times as much vitamin C as oranges.

Sea kale
Crambe maritima
This plant is part of the Brassica family and can often be found on shingle beaches. You can eat the leaves, stems, and flower buds. Similar to other cabbages, it is best steamed, boiled, or sautéed with a little butter.

Sea lettuce
Ulva lactuca
Its stunning emerald-green colour makes sea lettuce easy to identify. We use it in stir fries. Easy to find in rock pools and available all year round.

Wild cabbage
Brassica oleracea
The leaves of this classic wild cabbage are very good raw or cooked and have a slight heat to them. Often found in coastal areas, it can be harvested all year round.

PROJECT Grow your own mushrooms

One of the biggest worries about foraging for mushrooms in the wild is identifying the edible ones. Growing your own eliminates the risk – you can confidently sample lots of different types without needing to be an expert mycologist. They'll be cheaper than shop-bought ones and whichever of these two methods you try, you'll be reusing a waste product.

A fun and easy way to grow mushrooms uses an old paperback book. We tried pearl oyster mushrooms. In the wild they grow on dead or dying hardwood trees but we grew them on our windowsill at home.

The other method uses logs and dates back to China a thousand years ago. Back then, they simply placed freshly felled logs next to ones that had mushrooms growing on them and waited for them to spread. These days we can buy spawn plugs – small lengths of grooved dowel that contain the mushroom's root system, or mycelium – to insert in a log. Look out for shiitake, pearl oyster, or chicken of the woods spawn plugs.

YOU WILL NEED – BOOK METHOD
Paperback book with 200–400 pages
Mushroom spawn (order online)
Reused plastic bag and elastic bands
Mister spray bottle

YOU WILL NEED – LOG METHOD
Axe
Gorilla bar
Narrow spade
Post-hole digger
Hand drill and 8mm (⅛in) drill bit
Hammer
Logs that have been felled within the last six months, about 1.2m (4ft) long and 15–18cm (6–7in) in diameter
About 50 mushroom spawn plugs per log (order online)
Metal pot

BOOK METHOD

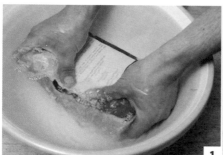  

1. Soak the book in warm water for 20 minutes. Squeeze the bubbles out of the book while it is underwater to get rid of most of the air. Take the book out of the water and squeeze it again, so that it is wet but not saturated.   **2. Carefully open the book**. Break up the mushroom spawn and gently spread some on the pages.   **3. Press** the pages together. Repeat this process every 50 pages or so throughout the book, until you have used all the spawn.

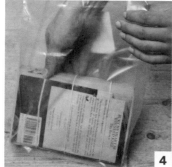  

4. Put a few elastic bands around the book and place it spine down in a see-through bag. Seal the bag and label it with the date. Keep in a warm place away from strong, direct sunlight and draughts – about 20°C (68°F) is ideal. In winter, keep it near a radiator or heat source. After 24 hours the spawn will begin to activate. In about two weeks the paper will be covered in white fuzz – the mycelium, or mushrooms' roots. When the edges of the book look white, place the bag in the fridge for two days.   **5. Remove from the fridge** and open the bag, folding it down to just above the top of the book. Put it in a light position (indirect sunlight) and water regularly with a plant mister.   **6. Harvest** oyster mushrooms when they are about 10cm (4in) across.

GROWING TIPS

Maintaining high moisture levels is very important as mushrooms are mostly water. Growing them in a book means they need a bit of attention, but no more so than any indoor houseplant, and you can expect a harvest much faster than with the log method.

You can get as many as three crops from your book. There's no need to put it back in the fridge – the mushrooms should just continue to grow as you harvest them.

Dry mushrooms in a solar dryer (see pages 168–169) then store them in airtight jars.

LOG METHOD

1. Choose a disease-free hardwood log and trim off any side branches and twigs. Remove any lichen or moss. Try not to damage the bark: it will help to retain moisture when the mushrooms start growing. Moisture is top priority for mushroom cultivation. Choose a damp, shady site, sheltered from strong winds, and dig a hole. Use a gorilla bar to get started.   **2. Finish the hole** with a post-hole digger: the hole should be about half as deep as the log's total length.   **3. Stand the log** in the hole and secure it firmly by back-filling with rubble and soil. The log will draw up water from the earth.

4. Drill holes at right angles into the log. Make them half as deep again as the length of the spawn plugs. Stagger the holes at 10–15cm (4–6in) intervals in a diamond pattern, going all around the log. **5. Push the spawn plugs** into the holes immediately after drilling – make sure your hands are clean. Then tap them in place with a hammer. Water the log thoroughly. Place a metal pot over the top of the log. This will help to limit moisture loss by evaporation. Continue to water in hot weather. If the log gets too dry, the bark will flake off; if it sprouts green mould you have overdone the watering. **6. Repeat with more logs** if you have space. In 12 to 18 months your crop will be ready. Pick mushrooms when they form clumps of about 15cm (6in) across. Use a sharp, short knife, and cut close to the bark.

Any hardwood log can be used for growing mushrooms but we've had great success with beech, oak, and birch.

TRY THIS

Tree stumps are ideal for growing mushrooms, as their roots still draw up plenty of moisture from the soil, but you must inoculate the stumps within several months of the tree being cut down. Drill only the outer ring of sapwood on the face of the stump, then insert spawn plugs.

The north side of a building can be an ideal location for your logs. The wall will keep the logs a bit warmer, but don't position them too close to the building or they won't get any rain.

Tray culture gets great results with button mushrooms. The trays should be about 20cm (8in) deep and filled with well-rotted horse manure. Sow dry mushroom spawn on the surface and cover with damp newspaper. Store the trays in a garage or shed. Remove the newspaper after about two weeks when the mycelium has grown. Cover the surface with 5cm (2in) of normal topsoil and water well. You should be picking mushrooms in two weeks.

HERBAL TINCTURES

Tinctures are an efficient and simple way to preserve herbs. Alcohol, or a mixture of alcohol and water, is used to dissolve and extract the active substances from the herbs at the same time as preserving them. On pressing, fresh herbs yield more tincture than dried herbs, so if you want to make your own at home, consider growing your own medicinal herbs.

THE INGREDIENTS

Tinctures can be made from all sorts of herbs and flowers. Some plants that make effective tinctures include echinacea for immunity boosting; nettle leaves for arthritis and rheumatism; dandelion to help with indigestion or loss of appetite; and hawthorn to help with high blood pressure. However, we would recommend that you always seek expert advice from a qualified herbalist before making and using your own tinctures, especially for more serious medical complaints.

The simplest way to make a tincture is by maceration (soaking the herbs in a solvent). Clear alcohol is the most commonly used preserving solvent in tincture making. The minimum proof required to inhibit the growth of unwanted organisms is 25 per cent. Qualified herbalists have access to ethanol, which is 96 per cent alcohol, but for home-made tinctures, 40 per cent vodka is more than adequate.

The weight of herbs to volume of alcohol used indicates the potency of the tincture: the greater the proportion of herbs, the higher the potency.

TAKING A TINCTURE

Consuming your tincture is as easy as making it. All you need to do is add a few drops to a glass of water once a day or as prescribed by a herbalist. Tinctures are stronger than infusions (see pages 194-195), so the required dosage can be much smaller.

RECIPE ## Digestive tincture

This tincture is ideal for those who suffer from burping, bloating, or indigestion. If you have ever wolfed down a meal too quickly and regretted it, the time spent making this remedy will be more than worthwhile. Take it before or after eating to stimulate your digestive enzymes. Do not use it during pregnancy.

YOU WILL NEED

Large jar	100g (3½oz) rosemary
Muslin	85g (3oz) bay leaves
Press	10g (¼oz) wormwood
Dark-coloured bottle	400ml (14fl oz) vodka

 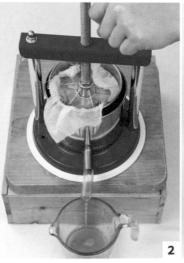 

1. Pick your herbs and remove any insects. Chop the herbs into 2cm (¾in) pieces and place in a large jar. Pour the vodka over them, close the lid tightly, and shake the jar well. Continue to shake the jar daily for at least 2 weeks. **2. Strain the contents** into a jug using the muslin and squeeze as much of the liquid out of the herbs as possible with your hands or with a small press, if you have one.
3. Transfer the tincture into a dark glass bottle, label it with the contents and date, and store in a cool, dark place. Take 1 tsp in a small cup of water up to three times a day.

Plant your own medicinal garden

It's easy to create a medicinal bed or border in the smallest garden. At our old farm we had a medicinal garden, full of herbs which we used in tinctures and infusions to help treat common minor ailments. Medicinal plants not only look attractive, they also tend to attract bees and smell gorgeous. Although many overlap in their healing properties, together they make a dazzling display.

KEY

Bites and bruises
1 Calendula
2 Comfrey
3 Lavender

Sleep
4 Chamomile
5 Passionflower
6 Valerian
7 Hops

Hangover and fatigue
8 Echinacea
9 Dandelion
10 Mint
11 Chillies

Headache
12 Rosemary
13 Feverfew
14 Mint

Coughs and colds
15 Lemon balm
16 Garlic
17 Sage
18 Thyme

Indigestion
19 Senna
20 Mint
21 Fig

QUICK STING RELIEF

If you are plagued by bites and nettle stings, find the nearest bunch of greater plantain and pick a handful of leaves. The juice from the leaves is more soothing than the traditional remedy of dock leaves.

Rub the plantain leaves firmly between your hands until they start to produce some juice. Keep going and then squeeze the cooling green liquid on to the sore area.

MAKING BOTANICAL DRINKS

Whether it's tea or beer, we certainly enjoy a brew or two, and it's all the more satisfying to know we've made it ourselves using plants from our own garden. Over the years, we've tried many combinations of herbs for infusions. Some can be an acquired taste, but provide natural health benefits, while others are delicious in their own right. We've also consumed large quantities of our own beer. Brewing your own botanical beer can not only save you money, it's a fun way to try beers that you can't get in the pub!

MAKING TEAS

Teas, or infusions, are relatively easy to make from leaves and flowers. Place a heaped teaspoon of each herb in a mug of boiling water and allow the leaves and flowers to infuse. Or to be more scientific about it, add 30g (1oz) dried herbs or 60g (2oz) fresh herbs to 600ml (1 pint) boiling water. Cover and leave the herbs to infuse for about 10 minutes. We like to use a tea pot with a central strainer. Alternatively, you can use an old-fashioned tea strainer that infuses in the mug and is then removed when the drink is to your taste.

Using teas

You can drink these herbal infusions either hot or cold and also use them topically as a wash or in a bath, as a douche or enema, or use them as a mouthwash or gargle – mint is especially effective here.

Drink 200ml (7fl oz) three times a day if you are a bit under the weather or 150ml (5fl oz) every 2 hours to relieve particular symptoms.

Infusions keep in the fridge for 24 hours and can be taken cold, but be aware that cool infusions can stimulate urination.

GREAT HEALERS

As well as our favourite calming herbs listed to the right, try drinking ginger infusion for travel sickness; feverfew for reducing the intensity of a migraine; elderflowers for helping ease cold symptoms; and nettle leaves – wear your gloves when picking them – to reduce stress.

Chamomile makes a great lawn and has the reputation of soothing the central nervous system, leaving you super calm. It also combines well with mint. As well as drinking the flowers as an infusion, place dried chamomile in a muslin bag in the bath for extra relaxation. Mix it with any of these calming herbs: comfrey, hyssop, lemon balm, passionflower, or valerian.

Lavender flowers are highly attractive to bees and as a bedtime drink will encourage a good night's sleep. It is also especially soothing for a headache: use 2 tsp of fresh flowers or 1 tsp dried and infuse.

As well as being calming, chamomile and lavender are, of course, also very attractive additions to the garden – and bees love them, too.

After an afternoon's hard work in the veg patch, you can't beat a refreshing glass of mint tea, with leaves freshly plucked from the medicinal garden.

RECIPE Nettle beer

Nettles make surprisingly delicious beer. Generally, we are not too keen on nettle soup or nettle tea, despite the health benefits, but nettle beer is on a whole other level. It is easy to make and tastes great. Use the fresh green tops of the nettles, as they make the best beer, and ensure all your equipment is scrupulously clean (see page 256).

YOU WILL NEED
Large pan
Small funnel
Bottles
1kg (2¼lb) young nettles
2 lemons, rind and juice
500g (1lb 2oz) demerara sugar
25g (scant 1oz) cream of tartar
1 tsp brewer's yeast

1. Wearing gloves, pick the nettle leaves and compost the stems and any roots. Rinse the leaves briefly to remove any bugs.
2. Boil the nettles in about 4.5 litres (1 gallon) of water.   **3. Strain the liquid** into a large container and add the lemon rind and juice, sugar, and cream of tartar. Make up to a volume of 4.5 litres (1 gallon) with water and then stir vigorously.   **4. Transfer** to a demijohn. Then, when the brew is cool, add the yeast. Allow 3–4 days for fermentation to take place and then strain the beer into clean, sterilized bottles (see page 256).

RECIPE Ginger beer

Ginger beer has always been a great favourite of ours, and ever since we discovered that we can grow ginger at Newhouse Farm, the beverage has become a seasonal treat. To grow ginger, plant a sprouting section of root into a pot of good compost in spring and wait for the rhizome to develop. Ginger must be brought inside over winter to avoid the cold, and we harvest ours when the tall stalk withers in winter, saving a bit to replant ready for next year.

YOU WILL NEED
25g (scant 1oz) root ginger
15g (½oz) cream of tartar
1 lemon, grated rind and juice
500g (1lb 2oz) white sugar
1 tsp brewer's yeast

1. Pick your ginger and scrub it clean before peeling it.   **2. Crush the ginger** in a pestle and mortar and then mix it with the cream of tartar and some grated lemon rind in a bowl. Add to a bowl of boiling water and then mix in the sugar. Stir until the sugar has fully dissolved and allow to cool. Add the yeast and the lemon juice and make the mixture up to 4.5 litres (1 gallon) with water.
3. Cover with a tea towel and leave in a warm position until fermentation starts. Remove the surface scum.   **4. Transfer** to a demijohn. Bottle and cork after 2 days and drink within a few days after that.

MAKING WINE

Pursuing the self-sufficient dream can be a lot of hard work. Indeed, sometimes getting out of bed early in the morning to feed your animals or braving the cold weather to weed the vegetable beds can feel monotonous or deflating. However, on a cold day after labouring outside, what could be better than relaxing in front of a roaring wood fire with a glass of aromatic, home-made wine?

SUCCESSFUL WINE MAKING

When you make a good bottle of wine there is not much in the world that can beat the feeling of satisfaction, pride, and, after a glass or two, relaxation. For a small investment it is easy to make lots of drinkable wine, and you also get the chance to try some exotic recipes that are not available at any wine merchants we know of. To make a fine wine, however, requires good ingredients, overall cleanliness, and loads of patience.

Wine making can be approached as a scientific experiment with accurate measurements and meticulous preparations, or treated as a vibrant and fun activity, as we do. We are not too bound by a strict regime and prefer to let the process evolve naturally. Having said that, there are a few basic rules.

ESSENTIAL TIPS

All wines consist of water, flavourings, sugar, acid, tannin, and yeast. The quantity of each ingredient varies greatly and the wine improves or degrades accordingly. Time is another key component and the need to exercise patience during the process cannot be overemphasized.

Keep your bottles and all other apparatus scrupulously clean. Warm bottles in the oven, and then fill with boiling water. Leave for a few minutes, pour out the water, and hang them upside down to dry. Or use a sterilizing solution. Campden tablets clean all types of equipment. Read the instructions and always leave for the full designated time so they can do their job properly.

1. Use secateurs to cut bunches of grapes from vines. **2. The traditional way** of making wine is excellent fun. Get your socks off and give your feet a very good clean. Then squish the grapes under your feet and between your toes until they are reduced to pulp and swills of grape juice.

Our quantities of sugar for making wine come from old-fashioned advice. We use 1kg (2¼lb) of sugar "in the gallon" to make dry table wine ("in the gallon" means you put the sugar in the container and add wine up to the volume of a gallon/4.5 litres). Use 1.25kg (2¾lb) of sugar for a medium wine; anything over 1.35kg (3lb) makes a very sweet wine.

Use good-quality wine yeast, granulated or liquid, available from wine-making shops or online. You can use baker's yeast, but it often makes a poorer quality, frothy wine. For a good strong wine you can also add yeast nutrients available from most wine-making shops.

More acid may be needed for wine that you make from flowers or grain. Buy citric acid from a wine shop, or use lemon juice, which works well.

Two fermenting stages are required. First, the frothy, active stage where the yeasts are multiplying and need some air around them. Leave the jar or demijohn three-quarters full during this stage. For the second stage, top up the jar with water and place a fermentation lock on the top. This allows the gases that have been produced by the wine to escape, while simultaneously keeping the air out. The lock also prevents the bacteria that turn wine into vinegar from forming, and stops any fruit flies that carry vinegar bacillus from entering the jar and contaminating your precious wine.

Temperature is all-important, and you must try to keep your wine at temperatures favourable to the vital yeasts. During the first fermentation, aim for 24°C (75°F). Do not allow the temperature to rise over 27°C (81°F) or the yeast will start to die; allow it to fall below 21°C (70°F) and it will be too cold for the yeast to work.

RECIPE Wine from grapes

You can make wine out of just about anything! Some of our more unconventional favourites are pea pod, blackberry, nettle, parsnip, and rosehip. We have tried and enjoyed all of them, but in our opinion it's hard to beat the taste of a traditional grape wine. We have struggled to make a heavy-bodied red wine, but have been very successful using our own grapes to make rosé, as shown here. Grapevines can thrive even in our colder climate (see page 161). Try growing them against south-facing walls, under cover in a greenhouse, or outside on a warm, south-facing slope.

YOU WILL NEED
Wine press
Mesh cloth
Demijohn, funnel, and fermentation lock
Campden tablets
Bottling and racking equipment
Red and white grapes
Sugar (for quantities, see opposite)
Wine yeast
Citric acid

1. Place a mesh cloth on the press within a stainless steel frame and cover it with a layer of grapes; they can stay in bunches. Add the sugar and yeast to a demijohn (see opposite for quantities). Then fit a funnel on top and place beneath the press ready to catch the juice.
2. Wrap the mesh cloth around the grapes, ensuring none can fall out. **3. Remove** the stainless steel frame **4. Repeat** for up to three layers, separating each layer with a sheet of wood. **5. Squeeze the grapes.**

6. Don't fill the demijohn right up to the top or it may explode during the first fermentation stage; the yeasts also need air to multiply. Leave it about three-quarters full. **7. When the wine is no longer frothy,** indicating that it is in the second stage of fermentation, top up the jar with water and place a fermentation lock on top. When the wine is six months old, it will be time to rack it and bottle it (see right). **8. The by-product** from the pressing stage of wine-making is a flat, fairly dry cake of grape skins and pips. We feed this to our pigs as a little treat.

RACKING AND BOTTLING

To "rack" means to siphon your wine off from the yeast (or "lees") that has settled at the bottom, so it doesn't spoil the flavour. We use a short plastic tube to siphon the wine into a clean container. Repeat a month later and, if you have the patience, do it again three weeks after that. Some people leave their wine in a cold place to hasten the settling of any sediment. After six months the wine should be ready to seal in sterilized bottles without risk of exploding. Leave about 2.5cm (1in) at the top for the cork. A corking gun is a good investment, or use a wooden mallet. Cork and label the bottles and leave them for a year, stored on their sides in a dark place, at about 13°C (55°F).

Flower, vegetable, and hedgerow wines

Country wines are fermented drinks made out of all sorts of different ingredients other than grapes. Vegetables also make interesting country wines, but need extra table sugar or honey added to them. Not everyone lives in a location suited to growing grapes, but everyone can make some type of country wine. The only disadvantage is that these wines don't store for long, and are best consumed within a year of bottling.

FLOWER WINES

Flower wines are extremely easy to make, tasty, and beautifully aromatic. Simply pour 4.5 litres (1 gallon) of boiling water into a large fermenting vessel, preferably made of metal, and add the flowers (see below for ideas). Bring the water and the flowers to the boil and simmer on a low heat for 15 minutes, stirring occasionally. Once the mixture has cooled, add 1.8kg (4lb) of caster sugar and the juice of 3 lemons. When it has cooled to about 24°C (75°F), add 1 teaspoon of yeast. Strain and pour the wine into a demijohn, and put a fermentation lock on top. Leave it to ferment, and bottle when ready (as described on page 197). Leave for a few months, then drink.

Some of our traditional flower wine favourites are listed below.

Gorse flowers make amazing, fragrant wine, which smells a bit like coconuts and is very sweet. You will need at least 1.7 litres (3 pints) of gorse flowers for every 4.5 litres (1 gallon) of water.

Dandelion flowers can be turned into a lovely light wine. Collect the dandelion flowers from mid to late morning and then pick the petals off the flower heads. You will need 1.2 litres (2 pints) of petals. Wash them in running water to remove any bugs or unwanted residue. Then leave in 4.5 litres (1 gallon) of water to infuse overnight. The next day, make the wine as shown on the opposite page.

OTHER TRADITIONAL FAVOURITES

Pea pod wine is particularly remarkable because it is made from the bit you normally throw into the compost bin. You can use any sort of pea pods, or even bean pods, as long as they are still greenish. Add 1.8kg (4lb) of empty pea pods, plus the rind of an orange and a lemon, to 4.5 litres (1 gallon) of water. Bring to the boil and simmer for at least 30 minutes, then allow to cool. Strain the liquid over 900g (2lb) of sugar, 1 teaspoon of yeast, and the juice of the lemon and orange, and then stir until everything is dissolved and mixed. Ferment and bottle (see page 197), and leave for 6 months before indulging in one of the great mysteries of the vegetable plot.

Parsnips produce a very pleasant, light white wine. Pick and scrub 1.8kg (4lb) of parsnips. Then slice them, without peeling them, and place in a large pan with 4.5 litres (1 gallon) of water. Cook until tender and strain the liquid into another vessel. Remove the parsnips and make a warming winter soup with them. For the wine, reheat the infused

Dandelion wine is a country classic and well worth brewing at home.

BIRCH SAP WINE

If you like German dry white wine, find some mature birch trees to make a country alternative. Tap the sap in midspring from a few trees with trunks at least 30cm (12in) in diameter, so that no tree gets over-tapped. Use a cordless drill to make a 2.5cm (1in) diameter hole in the trunk. Insert a short offcut of pipe into the hole, hang a bucket beneath it, and wait for the rising sap to flow out. Make wine as for pea pods, boiling 4.5 litres (1 gallon) of sap (adding water if you don't have quite enough) with 900g (2lb) of sugar. Drink after a month.

You can collect about 4.5 litres (1 gallon) of sap in a day from a mature tree.

liquid with 1.35kg (3lb) of sugar, plus the juice of a lemon, its rind, and 1 teaspoon of yeast. Ferment and bottle, and drink within 6 months.

Elderberry wine can be made in late summer or early autumn. With a fork, strip 1.8kg (4lb) berries from their stalks. Crush in a large bowl with the end of a rolling pin. Add 4.5 litres (1 gallon) of boiling water, stir and, once it has cooled to 20°C (68°F), add 1 teaspoon of yeast and the juice of 2 lemons. Cover the mixture and leave in a warm place for a few days, stirring daily. Strain the liquid through a sieve into a pan with 1.35kg (3lb) of sugar and stir until it has dissolved. Pour from the pan into a large demijohn. After the first ferment has finished, put on a fermentation lock, and leave to continue fermenting. Transfer to dark bottles and wait at least 6 months before opening.

RECIPE Elderflower champagne

Elderflower champagne is definitely at the glamorous end of self-sufficiency. It is made simply by fermenting the natural yeasts in the flowers, which then turn into a refreshing bubbly drink that is only mildly alcoholic. Elderflower champagne is cheap, delicious, incredibly fragrant, and quick to make. Be sure to store it in plastic bottles, though, as the fermenting yeasts can cause gas to build up in the bottles and they may explode if you leave the champagne to ferment for too long, spraying sticky, floral liquid all over the place. Luckily, this can be easily avoided by slowly releasing some of the gases when the bottles start to bulge.

YOU WILL NEED

Large metal container that holds 4.5 litres (1 gallon) water

Zester

Jug

Sieve

Funnel

Plastic bottles

Campden tablets

About 600ml (1 pint) compressed elderflowers

675g (1½lb) white sugar

2 lemons

2 tbsp white wine vinegar

Pick elderflowers on a hot day while they are in full bloom and strongly scented. Remember not to strip a tree bare, otherwise there will be no elderberries to use later in the season!

BREWING

1. Once you have collected enough flowers, mix them together with 4.5 litres (1 gallon) of water and stir thoroughly.   **2. Add the sugar** and continue stirring until most of the sugar has dissolved.   **3. Remove the zest** from the lemons with a zester.   **4. Cut the lemons** in half and squeeze out the juice.   **5. Add the zest** to the elderflowers and water, then add the lemon juice and white wine vinegar.

BOTTLING

6. Cover the container and leave in a warm place for 24 hours.   **7. The next day,** strain the liquid into a clean jug.   **8. Pour** into sterilized bottles. We like to use 2 litre (3½ pint) plastic fizzy drink bottles as they allow you to release the explosive gas by turning the screw top a little. Sterilize using Campden tablets rather than boiling water, which could melt the plastic (see page 196). Leave the bottles capped for a fortnight.   **9. Drink** before the bubbly is a month old.

MAKING CIDER

Cider making is a really social activity. Every year we collect apples from our orchard, but we also do a bit of scrumping. This involves making the most of any under-appreciated fruit trees in the area and going around the village with a trailer attached to a mountain bike – of course, we ask permission first. We then make the cider from the fermented juice of our apple collection.

KNOW YOUR APPLES

Any variety of apple can be used to make cider, and even dodgy-looking windfalls can be turned into good "scrumpy", as cider is known in the West Country. Some varieties of apple store well, like Cox's for example, but on the whole, apples don't last too long unless they are picked when ripe and are not bruised or damaged in any way. So, the advantage of cider is that you can successfully preserve huge quantities of fruit and save lots of money on alcohol by doing so.

Unless you are growing specific cider-making apples, such as Langworthy, Foxwhelp, or Crimson King, which can be brewed on their own, the trick to making a truly tasty cider is to use a mixture of different varieties. Combine about one-third each of bitter-sweet, sweet, and sharp apples for a good brew. Try the following:

Bitter-sweet apples are low in acid but high in tannin, such as Dabinett, Somerset Red, and Yarlington Mill.

Sweet dessert apples have a medium acidity and are low in tannin, such as Cox's Orange Pippin, Golden Delicious, and Worcester Pearmain.

Sharper apples include Royal Russets and Herefordshire Costards.

We have planted all sorts of traditional apple trees for eating, cooking, and for making cider, but until the trees are mature and fruiting well, we tend to use whatever apples we can lay our hands on, and the resulting cider always does the job.

FERMENTING

Once you have your fresh apple juice stored in demijohns or a larger fermenting bin, it's time for the natural yeasts to work their magic. We allow the yeasts from the apple skins and "wild yeasts" in the orchard to do the fermenting, but if you want to guarantee success, add a few teaspoons of yeast to the mix.

Many people dislike roughly made scrumpy and add sugar or syrup to their cider to make it taste sweeter. We don't mind the dryness of pure cider and find that it is also great to use in cooking, particularly with rabbit loin, mustard, and some root vegetables. Yummy...

Once fermentation has ceased (anytime from ten days to a month), you can rack off the cider straight into bottles for storage, using a plastic tube to siphon (see page 197). You'll often find a secondary fermentation takes place in the bottle. We once bottled some apple juice after trying to pasteurize it and were surprised to find it had turned into a bubbly cider similar to the fizzy drinks of Normandy. Of course, we didn't reveal the mistake to our guests, who loved it.

There is a stage in the fermentation process of cider when it is described as being "hungry". We have heard countless stories from locals about cider makers throwing in bits of beef and even the odd rat to add strength to the brew. Undoubtedly, most of this will just be hearsay, but it pays to be wary of unknown scrumpy.

MAKE CIDER VINEGAR

When cider turns into vinegar by mistake it's a disaster, but when made specially, it is great for cooking and in chutneys and pickles. To make vinegar, soak some wood shavings in a vinegar you like the flavour of, then place them in a large barrel with a tap at the bottom. Put a wooden or plastic plate, drilled with pin-sized holes, on top of the shavings, and pour over your cider so it drains slowly though the holes. Cover the top of the barrel with muslin. As the cider drips down, it is exposed to air and the active bacteria *Acetobacter*. Drain off the liquid and leave in an open container; it will turn into vinegar in a week.

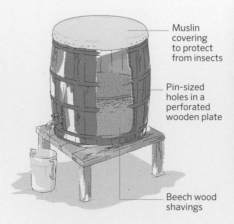

Muslin covering to protect from insects

Pin-sized holes in a perforated wooden plate

Beech wood shavings

Collect windfalls as soon as possible or they will start to ferment naturally on the ground. Don't worry if they are a little bruised.

Apple cider

Before you crush your ripe apples, leave them in a heap for 2–3 days to soften, or pick up your windfalls, which will already be very juicy. Try to avoid using apples that have serious pest problems – feed them to the pigs if they are maggoty. Generally, most apples will be usable; even if they don't look perfect they can still be turned into cider. The quantity of juice apples produce in the press varies greatly, but on average we'd expect about 4.5kg (10lb) of apples to make just under 4.5 litres (1 gallon) of cider. Make sure all of your equipment is scrupulously clean and sterilized (see page 196). As with many seasonal jobs, it can be almost a whole year since you last used your tools, so a good cleaning is essential. Use this same method for making perry – a sweeter drink than cider and made from pears.

YOU WILL NEED
Apple crusher
Apple press
Mesh fabric or muslin
Sterilized demijohn or fermenting bin
Apples (or pears if making perry)

1. Pour your apples into the crusher, keeping your hands away from the rotating blades, having first positioned a container to catch the apples. (You can also chop the apples in a wooden box using a spade, as described below.) The crushed apples will appear in the container at the foot of the machine. **2. Place the mesh fabric** over the apple press base, leaving plenty of spare fabric around the edges. Spread a layer of crushed apples on top. **3. Cover the apples** with the fabric and place the demijohn under the press.

4. Wind down the pressing mechanism until all the juice has run into the demijohn. Unwrap the leftover pulp and feed it to the pigs or put it in the compost bin. Then refill your press with more crushed apples and repeat the steps until you have pressed them all. **5. The juice looks cloudy** initially, but it clears throughout the fermenting process. Leave to ferment in the demijohn for 10–28 days, then bottle.

CRUSHING BY HAND

We have invested in an electric apple crusher that's quick and easy to use, and allows us to make large volumes of cider. But you can crush the apples with just a strong wooden box and a spade. Fill the box almost to the top with apples. Make sure that they are closely packed together so that they can't move easily, but don't overfill the box or the fruit will jump out and spill onto the floor. Use a sharp clean spade to chop the apples into small pieces before laying them on the fruit press.

WORKING LARGE AREAS

If you are lucky enough to have a large plot to work on you'll also have lots of big decisions to make. This chapter is intended to provide an overview of the exciting options open to you and guide you through the important process of deciding between them with a degree of caution. Our advice is to survey your land, consider what you want to do, establish what will work, and take a gradual approach – don't bite off more than you can chew. It is a real privilege to be the custodian of land: remember that grandparents plant trees so their grandchildren get the shade.

OBSERVING, PLANNING, AND DESIGNING

It's hard to resist jumping in with both feet and getting stuck in, but time spent planning your plot can save you from expensive mistakes further down the line. Gathering as much local information as you can and making a detailed site map will help you to make your dreams come true without too many false starts – and you may uncover some fascinating history at the same time.

START BY OBSERVING

Ideally, you should take notes throughout the seasons before making permanent decisions. In our first year, for example, we waited to see where frost pockets developed before planting our fruit trees, and monitored wind speeds in different sites before erecting a turbine.

For a comprehensive picture of your land, speak to previous owners and neighbours or research old photographs and maps. Part of our motivation to plant an orchard was discovering a 400-year-old map that showed the whole of the valley covered in fruit trees. Both local knowledge and the nature of the land itself should influence any decisions you make. The result will be a well-designed landscape that grounds your dreams and ambitions in reality.

PLANNING YOUR PLOT

Now you've got to the exciting stage. First decide which existing elements of your plot you are going to keep, and which have to go. Then think about what you are going to produce on your plot, and why. What kinds of fruit and vegetables do you like eating? Do you want to keep poultry?

Our advice is: don't over-stretch yourself. Your energy and enthusiasm will fade if you take on too many projects and don't have time to step back, gain some perspective, and think to yourself: "This is the life".

DESIGNING YOUR PLOT

The design that you eventually come up with will never be a finished thing. Nature constantly moves the goal posts – anything from the weather to the local slug population has the power to alter your plans. But this unpredictability is what makes life fun. We often stray from the initial idea, but this is another advantage of the organic lifestyle – allowing yourself the luxury of diverging.

DESIGN TIPS

- Site your **vegetable garden** in a sunny, sheltered place near the house.
- Keep **toolshed and compost bins** close to the vegetable patch.
- Align rows of fruit and vegetables from **north to south,** to catch the sun.
- On a slope, **plant across the contours** to prevent erosion.
- Plant **fruit bushes** densely to make harvesting and netting easy.
- Use **fruit trees** as a way of hiding any ugly features.
- Earmark **south-facing walls** for delicate fruits like peaches and vines.
- Site **wind turbines** away from sources of turbulence, or plan to build them high enough to avoid problems.
- Keep **poultry** near to the house so it's easy to let them out and shut them away at night.
- **Geese** are noisy: site their pen near the entrance of your property and they'll act as watchdogs!
- A south-facing roof is ideal for **solar PV panels** or **thermal collectors.**

1. Assembling a geodesic dome is a flat-pack challenge. **2. Digging a channel** for the pipe that brought water from the spring to the buildings at Newhouse Farm. **3. The water wheel** as it began to take shape next to the barn.

Conducting a site survey

Making a map is without a doubt the best way to get to know your land. A well-observed map fits everything together spatially and helps you to visualize your plans. But resist the temptation to add your ideas onto the map at this stage – it makes the process much more confusing. Simply record what is already there, and always mark north and south.

Newhouse Farm in the early days was in need of attention, from clearing the land of brambles and gorse to fixing the roof and sorting out the plumbing.

Filling in the details

Record your site's dimensions as accurately as you can – to scale if you have the patience. Try using a printout of your plot from Google Earth on the internet and lay tracing paper on top. Then you can add extra information such as the sunny and shady areas, and different soil types. Consider drawing a separate map of your buildings at a larger scale to show details such as services, mains, and utilities.

Look beyond your boundaries. Are there tall neighbouring trees or large buildings that cast shadows? Areas such as these may be harder to utilize, so highlight them on the map.

Mapping out the options

Annotate your map with extra information, from soil types to prevailing winds and sun traps.

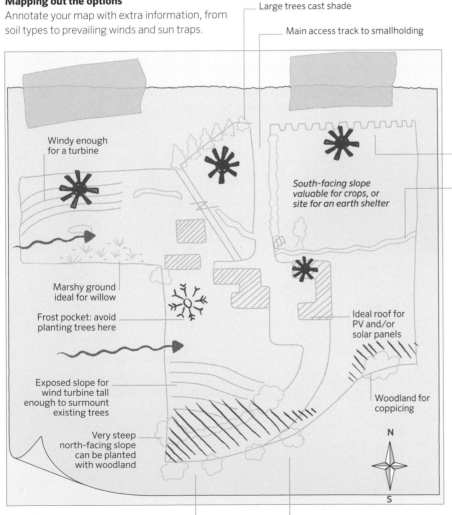

Large trees cast shade

Main access track to smallholding

Windy enough for a turbine

South-facing slope valuable for crops, or site for an earth shelter

Marshy ground ideal for willow

Frost pocket: avoid planting trees here

Exposed slope for wind turbine tall enough to surmount existing trees

Very steep north-facing slope can be planted with woodland

Ideal roof for PV and/or solar panels

Woodland for coppicing

Shady lane for mushroom logs

Secluded spot for beehives

Observe a stream throughout the year and measure the flow in different seasons (see pages 80–81).

Make a note of wildlife habitats such as ponds, wildflower meadows, and bird nesting sites.

South-facing wall ideal for fruit trees

Stream to power waterwheel

FINDING WATER

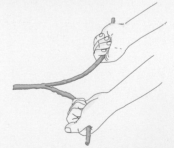

Find water on your plot by using a willow divining rod. Dowsing, or doodlebugging as it is known in the US, may sound wacky, but don't knock it until you've tried it. Finding springs using this ancient method is an amazing experience and it could even help you cut your water bills.

CREATING BOUNDARIES

Losing a crop of carefully tended vegetables to farm animals on the rampage is one of the most frustrating things that can happen on a smallholding, so on our farm we always made sure our livestock was securely contained. A mixture of permanent hedging and fencing, together with easily assembled temporary enclosures, helps to keep animals where you want them.

PERMANENT ENCLOSURES

Hedges, walls, and fences are traditional permanent boundaries used the world over to contain animals – or to keep them out. All need regular maintenance to keep them fit for purpose.

How to plant a hedge

Hedges are bushes that have been trained to form an impenetrable mass that livestock cannot push through. Once planted, they are cheap to maintain. A traditional quickthorn hedge is planted with thorn bushes (normally whitethorn, *Crataegus laevigata*). Use seedlings about 15cm (6in) tall and set them in two staggered rows, with about 45cm (18in) between each plant and 23cm (9in) between rows. Protect from livestock using conventional fencing for at least the first four years. This extra fencing is the most expensive aspect. Once the bushes are mature, you can lay the hedge.

Hedge-laying (see opposite) encourages the plants to produce dense growth that contains livestock efficiently and creates an important habitat for wildlife. It is ideal for nesting birds and for dormice.

Traditionally, a professional hedgelayer can lay a "chain" or 20m (22yd) a day. Seven years after laying a hedge, the regrowth can be as high as 4.5m (15ft).

Dry-stone walling

Dry-stone walls are built by laying together large, closely fitting stones without using mortar. The old walls on our boundary were animal proof but needed regular repairs. We also built a wall from scratch. The trick is to dig a level foundation trench and use flat angular stones that fit together snugly. You can buy stones from a salvage yard or building suppliers, but it is not cheap. It makes sense to build dry-stone walls only if you've already got plenty of stones on your land.

Wire fencing

Wire is expensive but relatively durable and makes an effective boundary. A good wire fence needs to be tensioned, so buy a wire strainer to achieve this. This simple tool can also strain slack or broken wire, including barbed wire, exerting up to 2 metric tonnes (2 tons) of pull, so the pressure on fence posts

THE STONE HEDGE

This form of hedging is basically a cross between a stone wall and a living hedge. It is widely used in the far south-west of England.

Building a stone hedge

Use large rounded stones to construct a pronounced batter – that is, two walls that lean towards each other. Make the walls around 1.2m (4ft) tall. Fill the gap between them with rammed earth and plant a quickthorn hedge on the top. Plants will also grow in gaps between the rocks. Stone hedges are valuable habitats for small mammals and reptiles, as well as for plants.

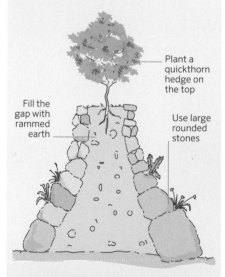

Plant a quickthorn hedge on the top

Fill the gap with rammed earth

Use large rounded stones

A typical stone hedge should be as wide as it is tall, and the width at the top of the wall should be half the width of the base.

can be considerable and they will need anchoring (see box below).

Match the fencing height to your livestock. Goats need fencing that is at least 1m (3ft 3in) high so that they can't jump over. You can increase the height of standard stock fencing (square netting) by adding two extra lines of wire. Put up fencing about 1.15m (3ft 8in) high for cattle, and 80cm (2ft 6in) for pigs and sheep. To keep deer out, you need a fence that is at least 1.9m (6ft 2in) high.

Post and rail fencing

If you have oak or chestnut trees to coppice, you could make your own post and rail fencing from stakes. Split them in half to make the uprights and into quarters to make the cross rails. Post and rail fences are good enough for most livestock and for marking out footpaths, but obviously won't keep rabbits out of a vegetable plot.

TEMPORARY ENCLOSURES

When grazing sheep, cattle, or pigs on fodder crops, you need a way of confining them to one area at a time that is quick and easy to set up and dismantle. Such temporary enclosures are also important when you need to corral sheep for shearing or lambing, or to inspect their feet.

Electric fencing

With electric fencing it's easy to create a secure temporary boundary in no time at all. Electric fencing is ideal for pigs and poultry. It has the added benefit of protecting poultry from potential

WIRE FENCING

Wire fencing is only as strong as the posts that hold it. Use a post rammer to set them firmly in the ground. If necessary, add box anchors to stop posts being pulled straight out again under tension when you use a wire strainer to take up slack in the wire.

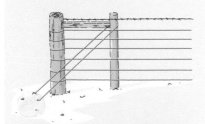

Bury a boulder or large stone as a box anchor. Corner posts need two, to take the strain in each direction.

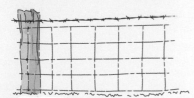

Square-meshed stock fencing with a single line of barbed wire on the top will stop pigs and cows straying.

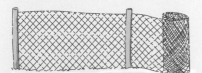

Temporary wire fencing such as diamond netting will contain sheep. Roll it up when not in use.

TWO TYPES OF HURDLE

Lightweight gate-style hurdles can be used for instant pens; woven wattle hurdles make short-term fences. Make gate-style hurdles out of any wood that splits easily, such as ash or chestnut; thin stakes make lighter hurdles. Sharpen the end stakes, then use mortices to join horizontals. Nail cross braces to the horizontals for strength.

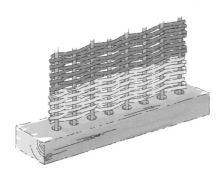

Use a frame to make a wattle hurdle. Space holes 23cm (9in) apart and weave between stakes with willow or split hazel.

To assemble an enclosure place solid posts in the ground at intervals and attach hurdles between them with twine.

Wattle hurdles make good temporary fences, but they don't last more than a few years. They also look good in the garden, but take quite a long time to make. See pages 232–233 to find out more about basic weaving techniques.

predators if the voltage is high enough and the fencing is in good condition.

Electric fencing is also effective with cattle, but it doesn't always work with sheep as their thick fleeces don't conduct electricity easily. You can up the voltage to high levels of 4,000–5,000 volts with three strands of wire, or consider using electric netting, when your sheep are lambing to protect them from foxes. But under normal circumstances, the best way to secure sheep temporarily is with diamond wire netting or hurdles.

The simplest electric fence uses one strand of hot wire. Position it at nose level for pigs and cattle. For pigs that's about 30cm (12in) off the ground.

If you set up a small solar panel to trickle charge your battery, you won't have to take it back to the workshop every time it needs recharging. The whole set up can be rolled up and put away in minutes.

Hurdles

Lightweight and easy to move around, hurdles are ideal if you have a ready supply of coppiced wood (see pages 220–221), and are ready to build your own (see opposite). Alternatively, you could make more permanent, woven wattle hurdles, which make good, cheap, short-term fences (see opposite).

Temporary wire fencing

Wire netting that can be rolled up is particularly useful as a temporary fence. Diamond netting is easiest to work with, as square-meshed netting may get bent out of shape.

PROJECT Set up electric fencing

Electric fencing is cheap and effective, and can be set up or taken down in moments. It can be adapted for a range of livestock by varying the height of the hot wire – the ideal position for any animal is nose height. Pigs, for example, are remarkably intelligent animals, and it only takes one belt on the nose for them to learn it's best not to cross the line.

> **YOU WILL NEED**
> Plastic or wooden stakes
> Electric wire
> Control unit and battery
> Watering can
> Waterproof crate

1. Clear the grass around the perimeter to reduce "bridging" – losing electricity into the ground where grass touches the wire. Set out fencing stakes. Space them close enough so the wire doesn't sag. **2. Feed the wire** around the stakes, setting it at an appropriate height. **3. Check again for grass** touching the wire. In this set-up all the strands are "live".

4. Firmly push the spike of the control unit into the ground. This acts as an earth, so wet the ground for good conductivity.
5. Connect the control unit to the fence and the positive and negative terminals to a fully charged battery. You can then cover the battery with a waterproof crate. **6. Test the fence** by touching a piece of grass to the wire. If it's working, you'll know (get someone else to do this if you have a weak heart or a pacemaker). Remember to tell visitors that the fence is live.

Lay a hedge

A hedge needs to be at least 2.5m (8ft) tall for laying. The technique is usually carried out when most of the leaves have fallen and the hedge is starting to look gappy. The stems to be laid are called "pleachers" and should be the straightest, best-looking stems in the hedge. Remove dead or bent stems as you work your way along. This is the Blackdown style of hedge-laying, specific to a region of the West Country.

YOU WILL NEED
Stafford bill hook (1)
Tenterden bill hook (2)
Short-handled slasher (3)
Kent pattern hand axe (4)
Hedging maul (5)
Hedging rake (6)
Leather gloves (7)

1. Clear brambles and nettles from the base of a mature hedge using a bill hook and slasher.   **2. Select a pleacher** and, using a sharp bill hook, make a vertical cut through the stem so that it is pliable enough to be laid down. Although the technique looks drastic, the pleacher continues to grow.

3. Trim the stob – the cut-away stump – and any whiskers of wood, to reduce rotting.
4. Lay the pleacher down horizontally.   **5. Roughly weave** the pleacher into those that have already been cut and laid, to hold it in place.

6. Cut a crook from the hedgerow – look for a post with a natural hook at the end, to hold the pleachers in place. Sharpen the base with the bill hook.   **7. Hook the crook** over the pleachers facing inwards and use the maul to hammer the sharpened base into the ground.   **8. The finished hedge.** The pleachers throw up vertical growth along their length and the cut stump will also send up six to eight stems.

FODDER CROPS TO GROW

Rearing livestock on a small scale is a great way of becoming more self-sufficient. However, animals often need more food than they can obtain from grazing and this extra feed can be expensive. In the past this was why most livestock was slaughtered before winter when fresh food could not be spared. Agriculture then developed so that planting crops to feed livestock through the harder months became the norm. It's now perfectly viable for a smallholder to subsidize winter feeding with crops specifically grown as fodder. An added bonus is that grain can be milled for flour and oats made into dairy-free milk (see pages 218–219). For some you will need to use a seed drill for best results; others can be sown broadcast by scattering handfuls of seed.

FODDER CEREALS

Cereals are not just a crop for farms with hundreds of acres; growing them on a smallholding can easily be achieved with a bit of hard work. Most cereals grow quickly and should smother any weeds.

MAKING HAY

Leave grass to grow in a field or orchard. Cut with a scythe in early summer when the grass is tall, either just before or just after flowering, but before it goes to seed. To make a haycock, erect a 1.8m (6ft) tripod of wooden poles tied together at the top and around the structure. Place a bent piece of tin on the ground facing a windward side; this will keep the air flow circulating through the haycock. Cover the structure with hay and leave to dry. When dry, store in a hay shed and feed to livestock such as cattle, goats, and sheep in autumn and winter.

Barley
Hordeum distichum

HOW TO GROW Barley will grow in surprisingly poor conditions. It does like a finer seed bed than other cereals and should be sown in spring, from March onwards, when the soil is warming up and drying out. Sow at 247kg (544lb) per hectare (100kg/220lb per acre). Then harrow the ground, which involves pulling – either by manpower, horse, or tractor – a spiky "bush" or attachment called a harrow to cover the seeds and mix them into the ground. When some of the barley reaches 15cm (6in) tall, harrow again to kill weeds and open up the surface.

HARVEST Cut barley late when the ears are bent over, the grains have gone pale, hard, and yellow, and the straw is dry. Store loose in a stook until it's completely dry. To make a stook, secure together two sheaves – grab-sized bundles tied around the middle – standing them upright so the grainy heads stick to each other. Add more pairs so they won't blow over. Leave for about three weeks to dry thoroughly, bring under cover, then thresh to separate the grain. This can be done by bashing handfuls against the back of a chair or by beating it with a home-made flail. Feed the grain to pigs or cattle, but soak it for at least 24 hours first. Use the straw as food and bedding for livestock.

Maize
Zea mays

HOW TO GROW Maize is another excellent fodder crop. Sow in spring into light soil, about two weeks after the risk of frost has passed. Sow at about 37.5kg (83lb) per hectare (15kg/33lb per acre) at a depth of 7.5cm (3in) in rows 38–75cm (15–30in) apart. Put up a scarecrow to stop rooks eating the freshly sown seed.

HARVEST Ripe maize cobs are great for cattle or you can feed them the whole plant in summer like grass. Maize can even make good silage: cut it when green, crush, and store it in plastic sacks so it ferments. Maize is one of our favourite fodder crops – especially as we love eating it too! To harvest the cobs simply walk along the rows, rip out the ears, and then drop them into a sack.

Oats
Avena sativa

HOW TO GROW Oats grow well in damp, cold places. In wet areas they are best sown in spring. In drier and warmer conditions, they are better suited to winter sowing – but make sure you have a scarecrow ready at this time of year as birds like to snack on them. The seeds should be sown into a coarse seed bed with small clods about the size of a child's fist. Ground that has been turned over by pigs using their snouts is ideal. Sowing is most easily done in a biblical fashion by holding a bag full of seeds and broadcasting them over the ground. Sow oats at 100–125kg (220–276lb) per hectare (40–50kg/88–110lb per acre) and harrow in the same way as barley.

HARVEST Cut the oats when there are still green bits in the straw. Tie into sheaves and make a stook (see left). Sheaves can be fed directly to cattle alongside their grass diet, at a rate of one sheaf per beast per day. Oat straw is generally accepted as the best cereal straw for feeding.

Rye
Secale cereale

HOW TO GROW Rye grows well in cold, dry conditions and light sandy soil. Plant in a similar way to oats. Sow in autumn for best results as it grows fast. Rye is known as a catch crop due to its quick-growing habit. The other excellent thing about rye as a fodder crop is that birds don't like it much so the success rate is high.

HARVEST If you plant rye straight after a crop of potatoes, it will be ready to be grazed off by sheep or cows in spring when other greens are in short supply. This space can then be planted with another spring crop. It also makes good straw bedding or animal feed.

FODDER ROOT CROPS

Roots are useful as winter fodder for animals, as well as being delicious in the kitchen. Root vegetables store up energy during the summer and retain all this nourishment as they lie dormant over winter. They are packed with goodness and can supplement your animals' diet.

Fodder beets and mangolds
Beta vulgaris

HOW TO GROW Sow seeds at 10–12kg (22–26lb) per hectare (4–5kg/9–11lb per acre) into a fine seed bed at the beginning of April. Make rows 55cm (22in) apart and thin them to around 23cm (9in) between plants.

Maize has a high energy yield compared to many fodder crops so makes good feed.

HARVEST Mangolds can yield huge crops but are not as rich in protein as fodder beet. Pull both in autumn before the first frost and place in piles covered with their leaves until you have time to clamp them (see page 167). Wait until after New Year's Day to feed them to pigs and cattle: this reduces the risk of any toxins being present. Neither crop is nice enough to eat – like bland-tasting tough old beetroot – but some people do grow mangolds specifically to make wine.

Turnips and swedes
Brassica rapa and *B. napus*
HOW TO GROW Turnips and swedes do extremely well in wet and cold conditions. Sow into a fine seed bed in May in drills at a rate of about 1kg (2lb) per hectare (0.5kg/1lb per acre) and then thin out to one every 23cm (9in). Try to hoe between plants a couple of times as they grow to reduce the competition from weeds.
HARVEST Lift both turnips and swedes some time after Christmas and store in clamps (see page 167). Alternatively, try grazing sheep directly on the growing crops in fenced strips, which you extend each day. After the sheep have finished, let your pigs out into the field to root up all the leftovers. Don't forget to lift and store a good supply of roots for use in the kitchen, too.

FODDER BRASSICAS
These tend to be nutritious, hardy crops used as autumn and winter feed for lambs or cows; fodder brassicas are also good for suppressing weeds and so they clear the ground ready for future plantings.

Cabbage
Brassica oleracea Capitata Group
HOW TO GROW Sow cabbage seeds in early April in drills 50cm (20in) apart, in well-manured soil.
HARVEST You get double-value feeding when you grow fodder cabbage. You can cut and cart the cabbages to cows when they're ready to harvest, usually around October. Then you can let your pigs out on to the ground where they will turn up all the edible roots with their snouts.

Kale
Brassica oleracea Acephala Group
HOW TO GROW Sow seed as for cabbage above. Kale tends to have a higher yield and is rich in protein.
HARVEST As with cabbage, you get double-value from kale as you can feed livestock on both leaves and roots. Start grazing animals on kale from early autumn onwards, then let pigs into the field.

Mustard and rape
Brassica juncea and *B. napus*
HOW TO GROW Sow mustard from March onwards by drilling at 10–12kg (22–26lb) per hectare (4–5kg/9–11lb per acre). Rape is sown in summer by drilling at 6kg (13lb) per hectare (2.4kg/5lb per acre); it can be ready for your animals to graze in 12 weeks and will carry on producing until Christmas. If you keep bees, be aware that they will harvest rape nectar voraciously but that the honey they produce is not at all palatable.
HARVEST Like turnips and swedes, both mustard and rape are ideal for strip grazing. Rape seed can also be pressed to make excellent oil.

OTHER FODDER CROPS
Sunflowers and Jerusalem artichokes look stunning – the artichokes are also a member of the sunflower family. More importantly, they will fatten up your pigs in winter.

Jerusalem artichokes
Helianthus tuberosus
HOW TO GROW Plant Jerusalem artichokes any time after Christmas. Set the tubers every 30cm (12in), in rows 90cm (3ft) apart. They are prolific and you will find that the following year, despite thorough harvesting, tubers will sprout up again in the same place.
HARVEST Instead of wasting time digging up your Jerusalem artichokes, simply put your pigs in an enclosure with them, giving them access to a few plants at a time, like strip grazing. They will root them up and happily eat them for weeks, while you save plenty of money on your normal feed costs.

Sunflowers
Helianthus annuus
HOW TO GROW Sunflowers are a great dual-purpose crop. The large heads of seeds can feed the chickens – and us – while the pigs forage on the stems and leaves. Sow 2.5–4cm (1–1½in) deep at roughly 4kg (9lb) per hectare (1.5–2kg/3–4lb per acre).
HARVEST The seeds are ready to harvest when the back of the flower head turns brown. We scrape out the seeds and dry them. Leave some heads for wild birds.

UNUSUAL CROPS TO GROW

Climate change is a reality. We saw a noticeable change in conditions during our time in Cornwall and evidence shows the same is happening all over the world. Wherever possible, we embrace changes to our advantage and try to focus on the glass half-full perspective. If the weather conditions are warmer in the summer, we see it as an opportunity to grow unusual crops and reduce the amount of food we import. It also happens to be a lot of fun!

Bamboo
Phyllostachys species

HOW TO GROW Bamboo is an incredible crop. It grows quickly, is very strong, and is excellent in damp areas. Some species are even edible. Bamboo can be grown all over temperate regions. It needs little care and attention, thriving naturally and spreading profusely through its root system. Choose species carefully; some bamboos are rampant and will spread rather too efficiently. We planted *Phyllostachys nigra,* which has black canes, and *P. glauca,* which forms good clumps and spreads well. We grew them at the bottom of our plot where the land was fairly wet but drained well. There it acted as a windbreak for the vines further up the hill and thrived in the moist conditions.

HARVEST Cut canes regularly to remove older or decaying culms (stems) and let in enough light for the rest of the plant to flourish. Harvest canes annually near the end of summer. We love the idea of being self-sufficient in as many materials as possible, so never again needing to buy a single bamboo cane for the garden was reason enough to plant bamboo. As plants mature, larger growth can also be woven together to build some form of garden shelter or seating area.

Flax or linseed
Linum usitatissimum

HOW TO GROW Flax is one of the UK's oldest crops – used to make linen and also linseed oil, a rich source of omega-3. Flax seed grows best in well-drained soil, rich in organic matter. Sow in drills in late spring and cover lightly with soil.

HARVEST The seed capsules are ready to harvest in late summer when most have turned brown. Mow the crop, rake it up into sections, and let it dry. The dried seeds can be pressed to make oil or milled for flour. The husks make a rich oil cake for cattle and the rough straw a base layer for animal bedding.

Hemp
Cannabis sativa

HOW TO GROW Hemp is a versatile crop and can be made into all sorts of natural products, such as insulation, fabric, paper, and animal bedding. It grows well on most soil, but avoid poorly drained or sandy ground. Prepare a weed-free seed bed and sow from late April to early May when the risk of frost has passed. Make drills at a depth of 2.5cm (1in) and leave 13–25cm (5–10in) between rows. Hemp plants tend to be very competitive and once mature they will smother other plants with their tall, dense canopy. The environmental benefits of hemp compared to crops such as cotton are unquestionable. It grows fast, needs little water, and no pesticides or chemicals. It is also important to understand that hemp is not the same as wacky baccy! The agricultural variety has almost negligible levels of cannabinoids.

HARVEST Harvest the fibres when the plants start shedding pollen, and the seeds when the crop is about 16 weeks old. The plants can clog machinery, so for larger crops consider hiring professional help.

Hops
Humulus lupulus

HOW TO GROW After a hard day working the land, there are few things more enjoyable than going to the local pub for a well-deserved pint of beer. By chance we prefer our local brewery's ale to imported lager, so the ensuing "beer-miles" are not too bad. However,

Bamboo provides good protection in windswept parts of your plot, but it can spread so be careful where you introduce it.

Like most plants in the *Camellia* family, tea plants produce attractive flowers. You can use the young leaves to make black, green, or oolong tea.

if you are serious about becoming more self-sufficient then growing your own hops is worth trying. Hops are relatively easy to grow in well-drained soil with plenty of manure. First, remove any perennial weeds and grass growing in the area and then lay bits of hop roots, from 30–60cm (1–2ft) in length, under a good layer of manure or compost. Space these bits of root 1m (3ft) apart. Male and female flowers are produced on separate plants. It's the female "cones" you need for brewing, and for best flavour they should be unpollinated, so dig up any male plants that grow.

HARVEST Cut the flowers when they are blooming and the inside is full of a yellow powder that will be the key ingredient in your very own beer (see page 272). Dry them thoroughly by hanging from wires over a stove or wood burner. Store the

TRAINING HOPS

To make picking easier and keep your plot neat, train hops on a traditional-style frame with 3m (10ft) tall strings; the hops race up them when they burst into life in spring. If you have space, try a "maypole" with strings running from the top of a central pole – or train hops on a pergola instead of a grapevine to create a shady place to sit.

hops in a hessian sack until you are ready to use them.

Rice
Oryza sativa

HOW TO GROW Although better known as an Asian crop, rice is in fact grown in large quantities in America and Southern Europe. What makes rice tricky to grow further north is that it requires summer temperatures of over 20°C (68°F) for most of its growing season, which can be up to four or even five months. Rice also needs plenty of water. We recommend growing it under cover (with a minimum temperature of 10°C/50°F in spring) and using a ram pump to supply the irrigation system (see pages 84–85). Prepare a seed bed that has warmed up in spring and rake the seed in well after sowing. Flood the bed with water once. As the rice plants grow, keep the bed well watered, making sure that the water is below the top of the shoots. When the plants are about 20cm (8in) tall, thin them to about 10cm (4in) between plants and continue to keep your small-scale paddy field wet. Transplant the thinned seedlings to a second bed. Seeds can be difficult to source: we recommend growing mountain rice.

HARVEST Cut the plants at the end of summer, when the rice spikelets are hard and yellow-coloured or turn from green to brown. Drain the growing patch and allow the rice to dry for 2–4 weeks in the greenhouse.

Tea
Camellia sinensis

HOW TO GROW Tea is native to China and India but has more recently been grown in other parts of the world. It is an evergreen plant from the *Camellia* family. Plants are best propagated from cuttings and grown under cover for the first couple of years. Outside on a south-facing slope, a standard tea bush will grow to about 1m (3ft) tall, and needs selective pruning to make it easy to pick the young leaves. Position tea plants about 1.5m (5ft) apart in rows, with 1m (3ft) between each row. We grew our own tea thanks to the climate in the south-west of England. It so happens it is very similar to the climate in Darjeeling. Best go get the kettle on...

HARVEST Bushes take six years to mature before they can be picked. Pick the tender young tips every 7–10 days in the growing season.

STORING CROPS ON A LARGER SCALE

When you're growing crops for animals as well as humans you need more storage space. A root cellar is designed to keep vegetables fresh for months on end. We think there is something very satisfying about eating a stored carrot in spring or having turnips to feed to our pigs on cold winter days. And if you've made hay in summer, you'll need somewhere weatherproof to keep it.

THE HAY SHED

Once hay is harvested and dried (see page 210), you'll need a covered hay shed to keep it dry. Wet, mouldy hay breaks down quickly and is not only unhealthy for animals to eat, it's a fire hazard – hay heats up as it rots. If it gets too wet, mix it into your compost bins or use as a mulch.

Site the shed close to livestock housing so you don't have far to lug the hay. Make the sides from vertical wooden slats about 7.5cm (3in) apart, to allow the hay to breathe but stop it getting wet. Design the door so there is enough headroom to walk in while holding a pitchfork of hay – this will avoid backache later! The shed should be solid and wind-resistant but don't waste large sums of money on what is essentially a rain-cover for dried grass.

GRAIN STORE

Grain that has been threshed to separate it from straw needs to be kept safe from vermin. Metal bins with lids are a good option or try a traditional grain store bucket raised off the ground. Keep straw under cover in the same way as hay.

Some cereals, such as oats, can be fed to livestock without threshing. Store these cereals in sheaves (see page 210) under cover.

STORING ROOT VEGETABLES

Root cellars are designed to store large quantities of produce all through the year without refrigeration. Their design is based on the fact that most root crops keep best in cold, humid, dark conditions, with plenty of room for air to circulate. Traditional root cellars were built wholly or partly below ground. The ideal site is carved into a north-facing slope. They don't need windows – roots must be kept in the dark, just as they were when they were growing. Site your root cellar close to the house. The last thing you want in winter is to get cold and wet fetching produce. Keep a wind-up torch hanging inside the exterior door to the cellar so you can see what you're doing.

Temperature

Put a thermometer in the cellar and check it regularly. The ideal temperature range is between 0–4.5°C (32–40°F). Install an intake vent near the bottom of the cellar wall to draw in cool night air and, at the high point of the cellar, another vent to let out warm air. Use the vents to control the temperature: open at night if the cellar gets too warm.

Humidity levels

High humidity is best: about 90–95 per cent will stop root vegetables from shrivelling. Use a hygrometer to measure it accurately. The easiest way to increase humidity is by pouring water on the floor. Earth floors maintain higher humidity levels than concrete.

Ventilation

Keeping air circulating in the cellar also helps remove ethylene gas produced by vegetables as they ripen. Left to build up, it can cause other veg to ripen quickly, and make them smell bad or start to sprout. Use the cold and warm air vents to control circulation and temperature.

Undercover storage space is invaluable when storing animal bedding and feed in bulk.

Building a root cellar

If you haven't got a natural slope to house a root cellar, you can achieve a similar effect by building it partly below ground and heaping soil over the top to create an earth berm (see pages 48–49). You will need advice from a structural engineer to create a framework strong enough to hold the weight of earth. Make your cellar at least 1.5m x 2.5m (5ft x 8ft).

Fitting out the cellar

Add plenty of shelves for storage. Make them from a durable hardwood such as oak that won't need treating with wood preservatives, which can taint stored produce. Spread a thick layer of gravel on the floor to stop your feet getting wet when you pour water on the ground. Or use large shallow pans of water to adjust the humidity instead.

Traditional root cellars in Canada built into the slope of a north-facing hill, with turf roofs and solid doors for insulation.

Making use of natural factors

The design of a root cellar relies on the earth's insulating properties combined with the tendency of warm air to rise.

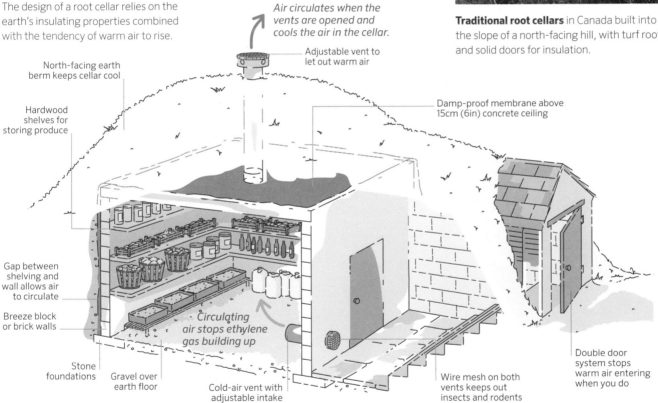

Air circulates when the vents are opened and cools the air in the cellar.

North-facing earth berm keeps cellar cool

Adjustable vent to let out warm air

Hardwood shelves for storing produce

Damp-proof membrane above 15cm (6in) concrete ceiling

Gap between shelving and wall allows air to circulate

Circulating air stops ethylene gas building up

Breeze block or brick walls

Stone foundations

Gravel over earth floor

Cold-air vent with adjustable intake

Wire mesh on both vents keeps out insects and rodents

Double door system stops warm air entering when you do

MINI ROOT CELLAR

If you have an old wooden barrel, don't even think about cutting it in half and turning it into two planters. Sink it into the ground against a wall close to the kitchen and turn it into a mini household root cellar.

Converting the barrel to a store

Dig a slight dip in the ground next to the wall and start layering up rocks so that the barrel can be positioned at a 45° angle with the lid facing outwards. Fill the barrel with harvested root crops, packing straw around them for insulation. Back fill with earth behind the barrel, add a layer of straw and another layer of earth, except above the barrel opening. Add a wooden board for a door.

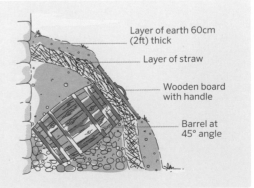

Layer of earth 60cm (2ft) thick

Layer of straw

Wooden board with handle

Barrel at 45° angle

BAKING BREAD

Even if you're a long way from the self-sufficient life, the smell of freshly made bread is reason enough for baking it regularly at home instead of buying it. There is an amazing number of breads you can make in the comfort of your kitchen, and whatever sort you decide on, and whichever type of grain you use, the process is simple. So experiment, be inventive, and have fun.

SOURDOUGH

Sourdough bread is made using a natural fermenting process so you don't need to add yeast. However, you do need to make a "starter" 2–3 weeks ahead of when you plan to start baking sourdough for the first time.

Making the starter
Mix equal quantities of strong white flour and water, say 500g (1lb 2oz) of flour and 500ml (16fl oz) of water, in an open container. Cover with a cloth and leave somewhere out of harm's way at room temperature.
After 2–3 weeks the starter will start to ferment and increase in volume. You will also see air pockets in it (see below).
When you use some of the starter, you have to replace it – called "feeding" it. For example, if you use 400g (14oz) of starter in your recipe, replace it with 200g (7oz) of flour plus 200ml (7fl oz) of water, whisked in to the original starter.
If you are not baking every day, keep your starter in the fridge and take it out 24 hours before you need to use it. This will bring it back to room temperature and kick-start it into action.
You can also freeze it. To bring it back to life again, let the liquid defrost and then leave it at room temperature for a day before topping it up to feed it.

Air bubbles and a sour aroma indicate that the wild yeast culture has formed and it's ready to use.

YEASTS

Bread can be broadly divided into two types, plus sourdough, based on whether or not they use yeast. They are:
Leavened breads, such as white or wholemeal bread, rye bread, and bagels, each of which use yeast as a rising agent.
Unleavened breads, such as pitta breads, chappatis, and rotis.
Sourdough (see box, left) is a rising bread that is made using a "starter".

The most commonly found breads are those leavened by yeasts; they rise when the yeast converts the sugars in the flour into alcohol and gases. After the alcohol has evaporated, carbon dioxide makes bubbles in the dough giving it an unmistakable fluffy texture. Yeast is a temperamental ingredient that needs a warm environment to rise, but make sure it doesn't get higher than 35°C (95°F), or you'll kill it.

Fresh yeast looks like soft putty and has a strong yeasty smell. Keep it in the fridge or freeze it in 2.5cm (1in) cubes. We prefer it to dried yeast, but if you are using dried, use half the weight stated for fresh in the recipe.

KNEADING

Pummelling your dough is not only a very important part of the bread-making process, it is also a superb stress-buster. While you're dissolving any pent-up anger or tension, you are distributing the yeasts evenly through the dough and developing the gluten strands, without which the bread simply wouldn't have the elasticity and strength to be able to rise properly. It can be treated harshly and will love you for it. Push and pull the dough until it becomes silky and elastic. After a while, do the window pane test: stretch a portion of dough between your hands until it forms a thin sheet that you can see light through. (If it tears, it's not ready.)

Now put the dough aside until it has doubled in size. You'll know the bread is fully risen if it springs back to its original size when you stick your finger in it.

EXPERIMENT WITH FLOURS

Try traditional breads from your local bakery and a variety of flours in your baking, instead of buying cotton wool loaves from a supermarket. Here are some of our favourite flours:
Wheat flour is a rich source of fibre as well as protein and vitamins B and E. It is also rich in gluten, rises very well, and produces a workable, stretchy dough.
Rye flour makes a traditionally dark, heavy bread with a slightly sour taste and an excellent chewy texture. It's a favourite in much of eastern Europe. It is low in gluten and a popular choice for people with wheat intolerance.
Barley flour is extra delicious if you toast the barley flour first. It makes a lovely, sweet-tasting bread. We mix one-third barley flour with two-thirds wheat flour for a good loaf. Barley is also low in gluten.
Spelt flour is made from the earliest known grain and is especially high in protein, vitamins, and minerals as well as being naturally low in gluten. The loaf can be fairly dry so add some wheat flour if you want it to rise well.

RECIPE Bread rolls

Dividing your dough to make bread rolls is a useful skill to learn and makes eating "on the go" easy. We also highly recommend a dough scraper for getting dough out of the bowl, cutting and dividing it, and cleaning hands and work surfaces.

YOU WILL NEED

500g (1lb 2oz) strong white flour

350ml (12fl oz) water

10g (¼oz) dried or fresh yeast

10g (¼oz) salt

Milk or egg for brushing

Seeds (optional)

1. Mix together the flour, water, yeast, and salt and knead until smooth and springy. Place in an oiled bowl, cover with a damp tea towel, and prove in a warm place for 1–2 hours. **2. Divide the dough** into equal portions, then shape the rolls by tucking the edges into the middle. Repeat until you have a ball shape. **3. Gently roll** each ball to make a neater shape. **4. Place the rolls** on a lined baking tray, then leave to prove again, covered, for another 45 minutes. Brush with milk or egg, then sprinkle with seeds. **5. Bake at 200°C (400°F)** for 20 minutes until golden brown.

RECIPE Rosemary and Cornish sea salt foccacia

Many breads need to be buttered or used to make a sandwich to be a satisfying snack, but focaccia makes a delicious meal all on its own. That said, we do enjoy dipping this bread in a good olive oil or some balsamic vinegar. Once you've mastered the basics, have fun adding your own flavours.

YOU WILL NEED

500g (1lb 2oz) strong white flour

15g (½oz) fresh yeast

20g (¾oz) semolina

330ml (11fl oz) water

50ml (1¾fl oz) olive oil

10g (¼oz) fine salt

3–4 sprigs of rosemary, leaves stripped from the stems

1 tsp sea salt flakes

4 tbsp olive oil

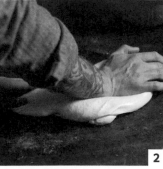
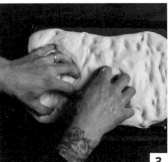

1. Mix together the flour, yeast, semolina, water, olive oil, and salt. **2. Knead for 10 minutes** until smooth and elastic, then form into a ball and prove, covered, for 1–2 hours. Turn out the dough into an oiled baking tray with deep edges and knock it back gently so that it fills the tray. Leave covered for a further 45 minutes to prove. **3. Use your finger tips** to knock back the dough, making shallow indentations. Leave to prove for a final 30 minutes. **4. Push the sprigs of rosemary** into the indentations and sprinkle over the sea salt flakes. **5. Bake at 200°C (400°F)** for 25 minutes, then leave to cool on a wire rack and brush with more olive oil. Enjoy!

MAKING FLOUR AND ALTERNATIVE MILKS

An added advantage of growing your own nuts or fodder crops like barley and oats (see pages 210–211) is that you can use them to make flour and non-dairy milks. What's more, it's widely considered that a lower-dairy diet can drastically reduce your carbon footprint. For us, it all ties in with our ethos of buying locally produced food and learning to grow our own. The key is to try it and see if it works for you.

NON-DAIRY MILKS

Dairy-free milks can be yummy and healthy alternatives to cow's milk that are still rich in protein. They are also surprisingly versatile: they can be used for baking, in hot drinks, or mixed with some coconut milk to make delicious dairy-free ice creams and desserts. They are simple to make at home from oats or nuts, and because they will keep for 7–10 days, can be made in big batches to reduce the amount of work needed.

RECIPE Oat milk

We've been using oat milk for several years. It's fun to make at home and a great alternative for those with nut or dairy allergies. If you like, add a pinch of sugar or vanilla to sweeten the taste, or mix in cinnamon or nutmeg for a warming drink.

YOU WILL NEED
50g (1¾oz) oats
500ml (16fl oz) water, plus extra for soaking

1. Soak the oats in water for between 30 minutes and 12 hours, then pour them into a sieve and rinse thoroughly; otherwise, the end product will be too claggy. **2. Blend the oats** with 500ml (16fl oz) water for 2–3 minutes until smooth. **3. Strain the mixture** through a nut bag or muslin into a jug or bowl. Discard the solids, or save them for pancakes, smoothies, or porridge. Store in a clean jar in the fridge for up to 1 week. Shake well before use.

RECIPE Nut milk

Nut milk can be made from many different kinds of nut, so if you grow your own, it's definitely worth giving nut milk a try. Cashew milk, shown here, is creamy in flavour and free from saturated fats and cholesterol.

YOU WILL NEED
50g (1¾oz) cashews
500ml (16fl oz) water, plus extra for soaking
Pinch of salt
½ tsp of vanilla extract (optional)

1. Soak the nuts in water overnight to soften them.
2. Drain and rinse the nuts, then place in a blender along with the water, salt, and vanilla (if using). Blend for 1–2 minutes until smooth. **3. Strain the milk** through a nut bag or very fine muslin to remove any small chunks. Keep in a sealed jar in the fridge and use within 10 days.

MILLING GRAINS FOR FLOUR

Milling your own grains produces really nutritious flour, and if you've grown your own cereal crops (see pages 210–211), then it's great fun to go the whole way through this ancient process and end up with an entirely home-made loaf of bread. That said, you can also try milling a wide range of ancient grains, such as amaranth, quinoa, buckwheat, millet, teff, or kamut. These grains tend to be easier to digest than modern cereals and have many health benefits.

Why grind your own flour?

Freshly ground flour is tastier (think of the difference between pre-ground and freshly ground coffee beans), noticeably more moist, and nuttier.

Variety is the spice of life, and we find it can often be easier and more affordable to get hold of ancient grains like spelt, kamut, or teff as whole grains rather than as specialist flours.

Whole grains are cheaper to buy in bulk and store ready to grind, especially the speciality types.

GRINDERS

There are several options for grinders that mill flour, ranging from a standard coffee grinder to miniature mills with different grades for milling.

Coffee grinders that are cheap and easy-to-clean work well for very small batches. However, the blades aren't as tough as grain grinders, so may break.

Hand mills are a fairly cheap, easy-to-store option. Hand mills can attach to a kitchen counter, are pretty quick to use, and can grind just about anything, but they can't produce very fine flour.

High-powered blenders can grind harder grains and the strong blades will last well, but you may have to alternate between grinding and sifting the flour.

Food processors can grind nuts, slightly larger batches of grains, and oats into flour. But watch out: they can quickly turn nut flour into butter!

If you're new to grinding flour, try a coffee grinder. They're best for small batches that you'll use straight away.

Grinding machines are more expensive, but they are great for large batches, and you can usually control the coarseness of the flour. If you plan on grinding lots of grains – whether you are a keen baker or looking for a source of gluten-free flour – then it could be worth investing in this specific piece of kit for the job.

It's greener to grow your own grains. Even buying grains locally with less packaging and milling at home can significantly reduce your environmental impact. It's also more self-sufficient.

It can be healthier. Many commercially milled flours may have chemical caking agents added as a preservative and have the nutritious germ removed during the milling process. Milling your own flour means you know exactly what you're getting and you can keep the germ, which contains lots of vitamins B and E, as well as nutritious oils.

PROJECT   # Making spelt flour

We love using flour ground from this ancient grain for baking. If you don't grow your own spelt, you can find it in shops or online. Finer milled flour tastes fresh and is worth the extra time spent grinding. You can also use the leftover bran as a topping for your sourdough loaves or added to oatmeal for porridge.

YOU WILL NEED

1kg (2¼lb) spelt grains
Hand mill or grinder (see box, above)

1. **Place a large bowl** or tray under your grinder and, if possible, select the grade of grinding to determine how fine or coarse you want your flour to be. **2. Turn your grinder** carefully while keeping the hopper filled with spelt grains. Grind a second time if you want to achieve a finer flour. **3. Sieve the flour** before storing in an airtight jar or bag. Use while fresh for best results.

MANAGING A WOOD

Productive woodlands have been carefully managed in the UK for centuries, and we believe that some form of control is really beneficial, as it helps to keep trees strong and healthy, and can extend their life. It also makes good economic sense, allowing you to become self-sufficient in fuel, as well as providing a shelterbelt for crops and playing an important part in nature conservation.

WOODLANDS FOR FUEL
Small woodlands are most often managed to provide a source of firewood for the home. You can season the wood by storing it under cover with good air circulation for at least a year – the longer you store it, the better it burns. If you have firewood to spare, you can sell it or make charcoal (see pages 228–229).

CONSTRUCTION AND CRAFTS
Roundwood is wood that's in the same state as it was when felled. It isn't worth much money, but is useful for building simple structures, like sheds and barns. Coppiced wood is ideal for fencing, basketry (see pages 232–233), and green woodworking techniques (see pages 222–223).

You may need permission from the Forestry Commission to fell trees. Some are protected by preservation orders or may be growing in a conservation area – check with your local authority. Trees with trunks less than 8cm (3in) in diameter when measured 1.3m (4ft 3in) above the ground are fine to fell yourself. For all others, call in a professional.

LIVESTOCK AND WOODLAND
Pigs kept in woodland enjoy eating acorns and beech mast.

Cattle and sheep allowed the occasional foray into the woods can help to regenerate the vegetation.
Chickens are woodland birds.
Pheasant shoots, held occasionally, bring in extra income.

MANAGING A COPPICE
Coppice comes from the French word "couper", meaning to cut, and it's a sustainable way to harvest large quantities of timber from your trees. Plant the trees close together and, when the trunks are about 23cm (9in) in diameter, cut them in winter to just above the ground. Tall, straight poles will grow from the stumps, and you can then repeat the procedure every few years and harvest the wood. Ash, hazel, and willow are ideal.

If you have an existing coppice that has been left for several years, divide it into "coupes", or areas to be coppiced. The diameter of a coupe should be about 1.5 times the height of the trees. For fencing and firewood, cut the trees every 10–15 years. Harvest one coupe at a time within this cycle; don't cut all the coupes in one go, as this will disrupt the natural ecology. A short cycle of 3–4 years is known as a "short rotation coppice". Use willow for this, and turn the slimmer poles into woodchip biomass for heating (see pages 92–93).

1. Coppiced hazel trees planted close together grow tall and straight towards the light. **2. Chickens** will enjoy foraging for food in a willow and dogwood coppice. **3. Using a log splitter** reduces large logs from mature trees to a manageable size for a stove.

Creating a new wood

Planting a mixture of conifers and broad-leaved trees helps to ensure biodiversity and provides the best environment for wildlife. Your plot will benefit too, as woodland creates an effective windbreak for crops.

Getting started

Siting a new wood is easy. Choose land that is not good for much else, such as north-facing slopes, poorly drained land, and infertile soil.

Spacing is important. If trees are too far apart, they will grow with short trunks and too many branches, and if planted too close together, they will be tall and spindly. The best technique is to plant closely and then thin out the trees as they grow (see page 147). Buy bare-root trees from a nursery and plant in winter, or grow your own saplings from seed (see below).

(see page 147)

In a mixed woodland of conifers and deciduous trees, the conifers create shelter for wildlife in winter, and early nesting sites for birds in spring. They also retain warmth in autumn so that deciduous trees hold on to their leaves for longer, which improves their growth rates.

TOP TREES

Our suggested trees for planting a small-scale woodland for timber and firewood:
Sweet chestnut is awesome for timber, fast growing, strong.
Oak is slow growing and slow burning.
Ash is tough and resilient with deep roots. Fast growing. Excellent firewood as it splits easily and burns when green.
Larch is an unusual deciduous conifer. Very fast growing. Makes good fencing.
Cherry is a good hardwood for sweet-smelling firewood.
Elm makes great chopping boards or butcher's blocks. Almost extinct in the UK.
Silver birch is one of our favourites for burning but it must be well seasoned.

PESTS AND PROBLEMS

Newly planted saplings are most at risk from animal pests, though some can damage mature trees too.
Deer damage is easy to spot, as they eat all the leaves and nibble the twigs up to a definite, visible line. Put up fencing 1.9m (6ft 3in) high or, if you must, shoot them.
Grey squirrels munch on the bark of many broad-leaved trees and will often eat the leading shoots of young saplings. Reduce their population using traps and releasing them a safe distance from your property. Or, as a last resort, shoot them.
Rabbits and hares strip the bark of saplings, killing the young trees. Protect them with plastic guards or fencing.
Rhododendron ponticum is an invasive evergreen shrub that casts such heavy shade that nothing can grow beneath it. Keep it under control if you value biodiversity; ideally dig it all out.

Growing your own saplings from locally collected nuts and seeds ensures your new trees are truly local species.

Putting up boxes for dormice and birds is just one way of making your woodland more wildlife friendly.

Plastic tree guards protect saplings from rabbits and hares. Remove them as the trees mature.

GREEN WOODWORKING

Using traditional skills that are not a million miles away from modern carpentry, green woodworking produces products with a more rustic feel. As managed woodlands are increasing in number (see pages 220–221), we are seeing a resurgence of the skills needed to turn the freshly felled green timber. Everyone can try their hand at it, and we recommend novices start by carving a spoon.

WORKING WITH GREEN WOOD

Fresh and filled with moisture, green wood is much easier to cut and carve than seasoned timber, which has an average moisture content of just 20 per cent. Our favourite green wood is wild cherry because it is both strong and not too hard to cut, and comes in a surprising range of beautiful colours. Other woods suitable for green woodworking include alder, English pine, elm, and willow. The best introduction to this type of woodworking for a beginner is to make a spoon (see opposite), which allows you to try your hand at the basic techniques before investing in specialist tools.

USING A POLE LATHE

The real magic of green woodwork lies in the pole lathe, a simple, efficient tool that has a timeless charm. It may be an old technology but at one point in history it was responsible for the mass production of a whole range of timber goods, such as tools and kitchenware.

A pole lathe is used for "turning" wood. The wood to be worked on, known as the stock, is held firmly between two points. You then use a foot pedal to rotate the wood while you shape it using a blade resting on a wooden tool rest. Using a pole lathe is like hopping on one leg uphill. It takes skill and practice but once you've got the knack you can turn out anything cylindrical.

Preparing the wood

Choose your wood carefully, and check that the shape lends itself to the object you are planning to make. The next stage is to remove the bark using a knife and shave horse, which is essentially a large clamp that holds the wood firmly while you work. You can make one from scrap wood; look up instructions on the internet. Pull a draw knife along the wood to shave off the bark. It will glide through the wood like a knife through cheese. The shavings are great for kindling, and because the wood is so moist, you won't get sawdust in your eyes.

Split green wood with a froe (see opposite), and wiggle it as you ease it down the wood. It will cut with the grain, thereby retaining the inherent strength in the fibres.

Finishing touches

Sand items with some shavings. Dry for a couple of weeks, then coat in a mixture of beeswax and oil warmed together in a pan. Once the mixture has soaked into your items, wash them in warm soapy water and dry.

1. A traditional pole lathe doesn't need an electrical power supply and can be set up anywhere, even in the woods. **2. A pine bowl** and green wood goblet turned on a pole lathe, with a hand-carved cherrywood spoon. **3. Green woodworking tools,** including a froe (top left) for splitting wood and various sharp knives.

Make a spoon

Michelangelo believed that every stone contained a sculpture. Green woodworkers know that within every length of wood lies a perfect spoon. Making a wooden spoon is a fun way to try out some of the easier green woodworking skills.

YOU WILL NEED
Short-handled axe
Shave horse
Draw knife
Froe or saw
Sharp knife and hook knife
Piece of green wood, e.g. cherry

PREPARING THE WOOD

1. Cut off any off-shoots from the wood using a short-handled axe.   **2. Remove the bark** by holding the wood in a shave horse and pulling a draw knife along it to shave off the bark.   **3. Saw the wood** roughly to length.   **4. Split the wood** to make a flat block to work with. Use a froe if you have one, and hit it with a lump of wood or a mallet. Otherwise use a saw.

CUTTING AND SHAPING THE SPOON

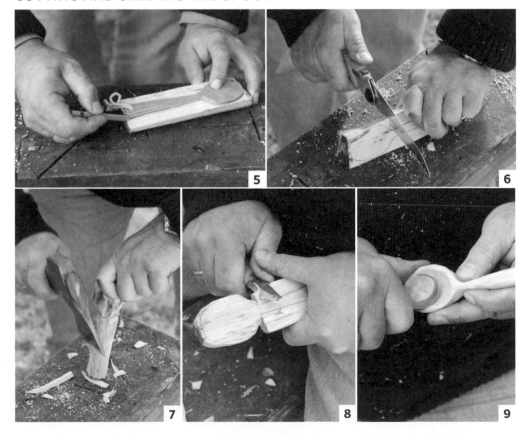

5. Draw the shape of your spoon onto the wood. We use a template so that our spoons look similar each time we make one – we didn't say identical, only similar.
6. Make some cuts from one side of the wood towards the outline of the spoon. Repeat on the other side. This makes it easier to cut out the shape.
7. Use a small axe or a sharp knife to carve the handle.
8. Pare shavings away from the spoon using a sharp straight knife. Be careful not to cut the spoon too thin near the neck. Try to cut over the edges of the grain, not up into them. Leave a bit of a rib on the back of the neck to give it some strength.   **9. Use the hook knife** to shape the bowl. Leave to dry for a couple of days, then use it.

WOOD-FIRED COOKING AND SMOKING

The popularity of smoked meats and fish and barbecue reflects changing attitudes towards rustic cooking techniques, including wood-fired kitchens and slow food. If you really love smoked food and want to take your home cooking to the next level, then consider building your own smoker or buying a commercial one. Whatever technique you use, remember that the wood itself is a key ingredient.

COOKING WITH WOOD

When it comes to cooking with wood, there are two main techniques: barbecue (or campfire) cooking, which involves cooking directly over hot embers, and smoking, which uses – you guessed it – the smoke to cook your food.

BARBECUE OR CAMPFIRE COOKING

You've probably already had some experience of barbecuing, most likely using pre-prepared charcoal (you can make your own – see pages 228–229). Cooking over wood is a very different – and infinitely more satisfying – experience. All you need is the right wood (see box, below), a grill, and something to cook. Before you light your fire, make sure you choose a well-ventilated spot that is clear from flammable materials and won't upset any neighbours.

You can light your fire in one of two ways. For the log cabin method, start by laying two logs parallel to each other 30cm (12in) apart as the foundation. Then bridge these two with two more parallel logs. This will help protect the fire from the wind whilst allowing air to flow. Lay four pieces of kindling across the logs and some tinder in the hollow. Then repeat, maintaining an open structure. Lighting the tinder will set a fire from the inside out and set the "cabin" alight. Gently blow to encourage the flames.

Another way to start a wood fire is in a chimney starter – a metal cylinder designed to light charcoal easily. Fill the chimney with hardwood chunks and light as you would charcoal. You can also use it as an "under fire" to bring wood to flame.

However you light your fire, allow plenty of time – up to 45 minutes – for it to mature and burn down to embers.

Then, with a shovel or long-handled grill hoe, rake the glowing orange embers underneath the grill grate. Like charcoal (see pages 228–229), the deeper the pile, the higher the heat.

SMOKING

This technique started out as a way of preserving food in the days before refrigeration. Like brining, it works by drawing moisture out of food so bacteria can't survive. Smoking coats food with antibacterial and antioxidant compounds found in the wood smoke, while at the same time adding a delicious, smoky flavour. Smoking only penetrates the surface of the meat or fish – the centre won't benefit from the preserving effect – so it is essential to start off by salting or brining it (see pages 252–253). This does most of the preserving; smoking helps and adds more flavour.

TOP TIPS FOR WOOD-FIRED COOKING:

Always use seasoned hardwoods like oak, alder, ash, beech, hickory, maple, pecan, birch, walnut, fruitwoods, or mesquite.
More unusual options include: olive, wine barrels, and grapevine clippings.
Avoid softwoods like pine and fir, which produce a resinous smoke that generally spoils the flavour of food.
Wood burns faster than either lump charcoal or charcoal briquettes. Be prepared to replenish the embers every 20 to 30 minutes.
Keep a fire extinguisher, water bucket, or a pile of sand and a spade nearby to keep the fire from spreading out of control. Extinguish the fire completely once you are finished with it.

1. Smoke jerky or biltong to make a delicious meaty snack. **2. Let the flames** die down and cook over hot charcoal for best results. **3. With a little welding,** an old oil drum can become a respectable smoker or barbecue.

TYPES OF SMOKING

Smoking is generally either hot or cold. The technique you choose will depend on the materials you have available and your desired results.

Hot smoking

This method is really more of a cooking technique that gives food a smoky flavour. You can do it on a hob or in a DIY smoker (see pages 268–269). Hot smoking uses very hot smoke, and the heat cooks the food while simultaneously smoking it. Hot smoked food doesn't store for long – just a few days in the fridge – but, unlike cold smoking, it's ready to eat straight away.

Cold smoking

This is our preferred method of smoking food. It's an outdoor activity, and to do it successfully, you need a slow-burning fuel and an enclosed heatproof container of some sort, with enough space for the smoke to circulate around the food (see page 227). The fire smoulders rather than burning fiercely and produces clouds of aromatic wood smoke. The

Attach a fire box to the main smoker to create an indirect heat source. This allows the smoke to flow, but keeps the temperature low.

Types of wood

Choosing which fuel to use for smoking your food is as important as selecting the right ingredients for a dish, and should be treated with the same care. We decide which fuel to use based on the time we have available and the flavours that marry well with the dish. We like to use a blend of local lump charcoal (see pages 228–229) for reliable fire management and a hardwood to give more flavour to the smoke and a more hands-on feel to the cooking. For us, charcoal briquettes may have a place in smoking food, but we're drawn to the unpredictable character and magic of using wood in its natural form.

For the best flavour, we always season wood for at least 12 months before using for smoking or cooking.

Hardwoods

Most hardwoods are deciduous, and in the UK, we have a good supply of oak, beech, cherry, birch, alder and apple. They tend to be slower growing than softwoods and, as a result, they are denser with a more intense heat, which sustains a good plateau for cooking. Wood varies hugely from species to species; oak will burn very differently from something like chestnut, for example. It is not an exact science, so experiment and try to source seasoned wood for the best results.

You can buy specialist wood for smoking and as a flavoursome fuel. Hickory wood chunks are available from many barbecue suppliers and whisky oak barrels are becoming more popular in the UK. But in the spirit of self-sufficiency, we'd suggest offering to pick up branches from a local orchard for some good fruit wood, or visiting traditional boat builders, carpenters, and furniture makers to ask for offcuts. Green woodworking will also give you some useful shavings for your smoking (see pages 222–223).

Softwoods

In general, softwoods such as pines, spruce, or firs are evergreen conifers. Their lower density and highly flammable resins mean they burn more easily and quickly. Softwood can be very useful for lighting a fire, but the resins exude an acrid smoke and produce a bitter flavour that can destroy your smoked food. Whilst we wouldn't usually recommend using them for cooking, there are exceptions. When soaked in advance to release their essential oils, cedar and juniper are lovely for light smoking on planks.

You can also use dried pine needles to hot smoke shellfish. The needles burn intensely for a short period of time, and if you have enough, they will create a sweet smoke and cook the fish perfectly.

Seasoned wood

Like heating your home, using seasoned wood, rather than fresh or green wood, makes a big difference to the cooking process. Green wood is often still supple and contains much more moisture (see pages 222). Seasoned wood is dry, lighter, paler in colour, and splits with a completely different sound. It is referred to as "seasoned" because it has been left for a few seasons to dry out. Wood like oak, for example, needs at least two years' seasoning for best results. Choose open-air seasoning over kiln-dried if possible, as seasoning outside uses less energy. Ask around to find local suppliers.

heat from the fuel doesn't actually reach the food, so the food doesn't cook; instead, it absorbs all of the flavour from the smoke.

Food that has been cold smoked needs to be left for at least 24 hours before eating for the flavour to develop fully. After that, it will normally keep for at least a week in the refrigerator but this all depends on the type of food and the extent to which it has been cured before smoking – if in any doubt, don't eat it! Cold-smoked salmon can be eaten raw, but other smoked foods such as bacon, kippers, cod, and haddock will require cooking afterwards.

CHOOSING A SMOKER
Once you've decided on a method of smoking, the next thing to consider is the smoker's eventual location. Whilst you can hot smoke reasonably well on a hob (see page 269), most traditional smokers are intended for outdoor use and are designed with vents or chimneys for ventilation purposes.

Be considerate of regulations, as well as your neighbours and any air vents. Conduct a full site survey before getting too carried away and assess what is feasible.

WHAT TYPE OF SMOKER?
Smokers vary drastically in size and the best option for you will depend on your needs. Building your own can be good fun if you have the time and space (see opposite), but there are many ready-to-buy options available too. Some are intended to be static, while others are designed and equipped to be pulled on a trailer. If you plan to smoke on a commerical scale, you will need something bespoke.

Ceramic smokers
Made famous by brands such as the Big Green Egg and Kamado Joe, this iconic barbecue and smoker is heavy duty yet small enough to fit in a garden or well-ventilated kitchen. It diffuses internal heat thanks to a ceramic plate

that holds heat consistently and can keep a good, even temperature for cooking over long periods of time. You can also use it as a tandoori oven, grill, or barbecue. The downsides of ceramic smokers are that they tend to be quite expensive and are not very portable.

Tennessee smoker
This style of smoker epitomises the southern USA's pit barbecue style. It has an offset barbecue, with a smaller chamber for burning your charcoal and wood. The air passes through a vent and cools in the main chamber. You can use it as a low-and-slow smoker, which is perfect for briskets, ribs, and pork shoulders. We like being able to use its side firebox to keep charcoal hot and then top up the main chamber for hotter smoking and grilling.

Electric smokers
These smokers are capable of both hot and cold smoking. They come with an accompanying smoke generator in

1. Curing and then cold smoking your own bacon is well worth the effort. And if the meat is home-reared, then so much the better. **2. A thermometer** that shows the temperature in your hot smoker is an essential piece of kit.

which flavoured bisquettes are burned for 20 minutes, producing a steady temperature. This eliminates high-temperature gases, acids, and resins which can distort the flavour of smoked food. The result is clean-tasting food, without any aftertaste, but for us, the process can lack creativity and fun.

Commercial smokers

Smoking food isn't just an enjoyable (and delicious) pastime – for some, it can be a viable business. Cabinet smokers are an excellent option for commercial smoking. Their precise temperature control gives you the consistency required in a catering environment while maintaining the authenticity of smoking with a real fire and wood smoke. However, if you want to be able to move your smoker easily – to attend events, for example – a barbecue trailer is a practical choice.

PROJECT DIY cold smokehouse

In a permanent cold smokehouse, the smoke is produced by a fire in a pit. It passes through a flue and into the smoke chamber, where the food is arranged on shelves or hanging in the space.

The smoke chamber

Almost anything large enough, made of fireproof material, and with no paint or enamel that could burn off and contaminate food, will do – an old refrigerator is ideal. Cold smoking won't damage the plastic lining, which is easy to clean. Do not use galvanized metal, copper, or brass oxides for shelves as they taint the food and are a serious health hazard.

Getting the temperature right

In cold smoking the temperature should be around 32°C (89.6°F) and should never go higher than 50°C (122°F). Use an oven thermometer to check the temperature in the smoke chamber.

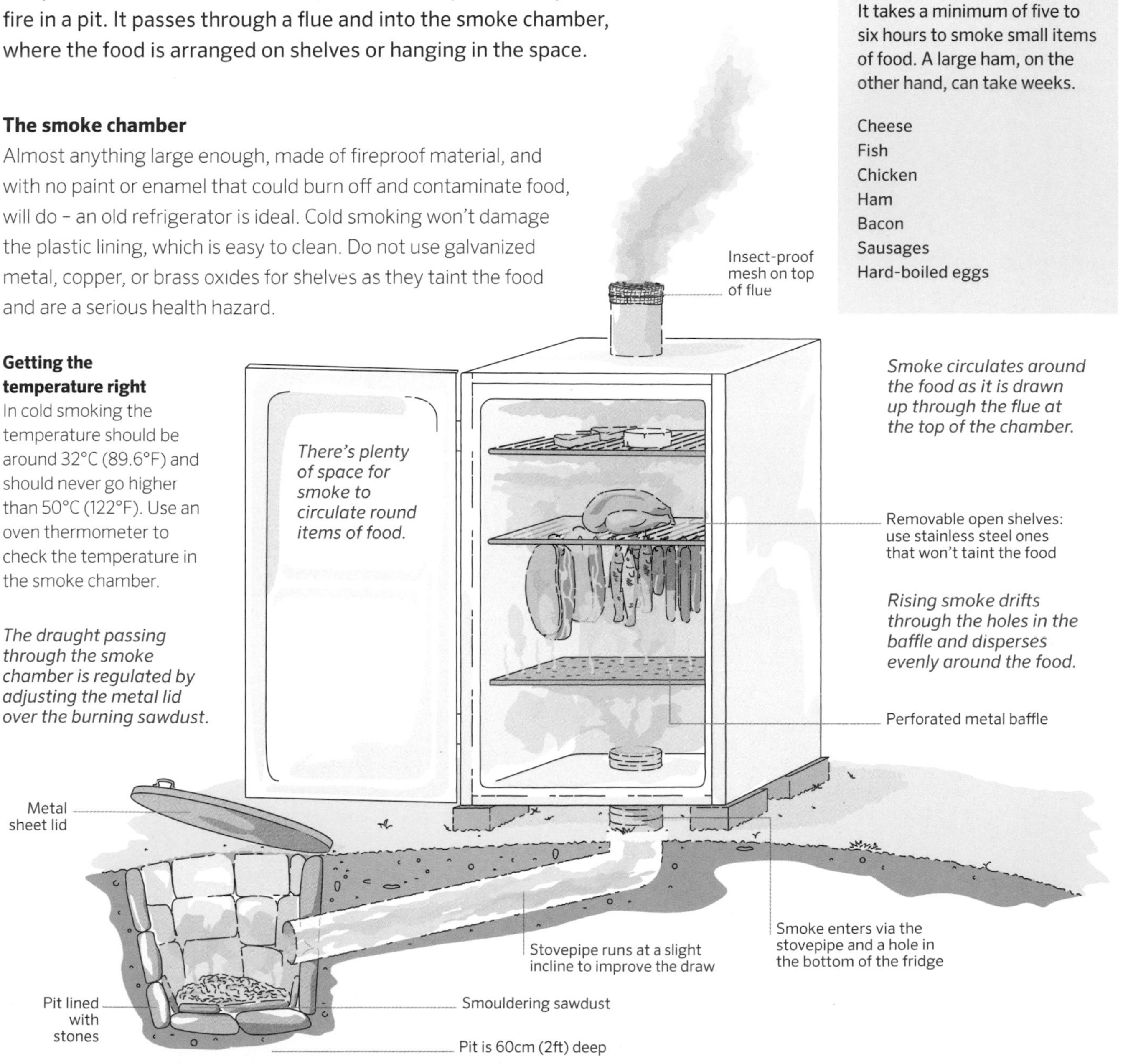

The draught passing through the smoke chamber is regulated by adjusting the metal lid over the burning sawdust.

WHAT TO SMOKE

It takes a minimum of five to six hours to smoke small items of food. A large ham, on the other hand, can take weeks.

Cheese
Fish
Chicken
Ham
Bacon
Sausages
Hard-boiled eggs

Insect-proof mesh on top of flue

Smoke circulates around the food as it is drawn up through the flue at the top of the chamber.

There's plenty of space for smoke to circulate round items of food.

Removable open shelves: use stainless steel ones that won't taint the food

Rising smoke drifts through the holes in the baffle and disperses evenly around the food.

Perforated metal baffle

Metal sheet lid

Smoke enters via the stovepipe and a hole in the bottom of the fridge

Pit lined with stones

Smouldering sawdust

Stovepipe runs at a slight incline to improve the draw

Pit is 60cm (2ft) deep

MAKING CHARCOAL

Of the thousands of tonnes of charcoal that are burned on barbecues in the UK each year, most comes from tropical rainforests and endangered mangrove swamps on the other side of the world.

Yet charcoal has been made in this country for centuries, and it's easy to build a kiln and make some on a small scale, so why not give it a go and enjoy truly satisfying summer barbecues?

WHAT IS CHARCOAL?

Charcoal is made by heating wood with insufficient oxygen to burn, but to a temperature high enough to drive out water and volatile gases. Charcoal burns slowly to a very high temperature. In Europe it was used for thousands of years for smelting metals until coal took its place, and traditional blacksmiths still prefer to use charcoal in their forges. It is also popular with artists for sketching, but today its main use is for barbecues.

Its slow-burning properties make charcoal perfect for outdoor cooking, but never use it on an open fire indoors, as it produces toxic carbon monoxide when it burns.

BUY LOCAL

Often, a bag of lump charcoal will contain a mix of hardwoods from South America or Asia. We opt for sustainably managed local charcoal, so we know the harvesting of the wood has not been responsible for habitat degradation and deforestation.

Wood from British trees is also less dense than timber used for imported charcoal, making it far easier to light and quicker to heat up for cooking.

Local forestry volunteer groups often make charcoal as an annual fundraiser, using the money for woodland management. If everyone in the UK decided to buy locally produced charcoal, we'd all get better quality as a result.

When we can't buy locally, we buy bags with the Forest Stewardship Council (FSC) logo. This non-profit organization promotes conservation of the world's forests and guarantees that products come from well-managed sources.

TYPES OF CHARCOAL

Charcoal has different properties depending on how it's made and the type of wood used.

Lump charcoal

Hardwood lump charcoal, also known as char, is one of our favourite fuels to use for smoking. It provides a good plateau of heat and, although it doesn't burn

A TRADITIONAL CHARCOAL CLAMP

The clamp was built with logs arranged in a criss-cross pattern around a central hole. More wood was stacked around the logs to form a mound about 3.7m (12ft) across and topped with layers of leaves and turf, with a few vents at the bottom. Once lit the collier tended the fire constantly for two weeks, making sure it smouldered but didn't break into flames, until blue smoke signalled the charcoal was ready.

Make a modern trench version
A charcoal pit is a good idea if you have the space and want to make a large batch. Dig a decent-sized trench and fill it with wood. Set fire to the wood, and when it is fully ablaze, cover the trench with a couple of sheets of corrugated iron. Then quickly shovel earth on top, burying the sheets of metal and cutting off the oxygen supply. Leave the heat to work its magic. After several days, open up the trench and remove your newly made charcoal.

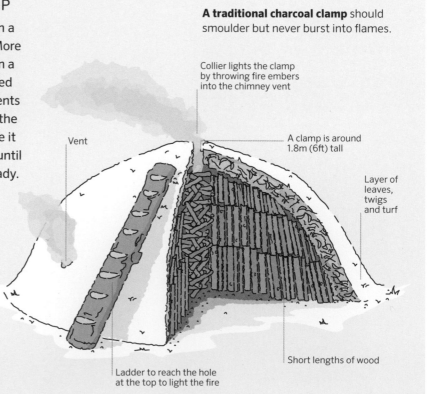

A traditional charcoal clamp should smoulder but never burst into flames.

Collier lights the clamp by throwing fire embers into the chimney vent

A clamp is around 1.8m (6ft) tall

Vent

Layer of leaves, twigs and turf

Short lengths of wood

Ladder to reach the hole at the top to light the fire

1. **When choosing** charcoal opt for sustainably produced lump charcoal.
2. **Lump charcoal** lights and heats up quickly, but briquettes have a more constant temperature for cooking with.

uniformly like shop-bought briquettes, it also doesn't contain anything extra like sawdust or petroleum that can taint your food. Today, the process of making lump charcoal typically begins by stacking wood logs in underground trenches and covering them with sheet metal and dirt (see box, opposite).

Traditionally used for Japanese yakitori grilling, Binchotan is a type of lump charcoal primarily made from holm oak. The tree clippings are stacked vertically in a large oil drum and charred at a low temperature for two weeks. They are then heated to a high temperature until the smoke is clear and all the impurities have burnt away. The result is nearly 100 per cent pure carbon which burns steadily as a fuel without smoke. Where it surpasses other types of charcoal is in the burning time: Binchotan can burn for up to five hours, and can be extinguished and started again up to three times.

Charcoal briquettes

These are basically charcoal dust with a starch binder mixed up with additives.

It is claimed that they release a stronger heat for longer. Frankly, there's little in the heat strength argument, but they certainly last longer and the important point is: they give off a consistent level of heat. This means that, with practice, you can forecast when food will be ready and minimize fuel waste by using a certain quantity of briquettes with a fairly standard level of heat. Another positive is that there's less ash and cleaning up is a little easier.

Self-lighting charcoal

Avoid the stuff that comes wrapped up in a flammable brown paper bag for lighting. Generally it works, but it doesn't burn well, and because it is impregnated with fuel, it will taint the flavour of your food. Another negative is that it creates a lot of light ash which can rise up and stick to your food.

COOKING WITH CHARCOAL

Grilling and smoking food over a lump charcoal fire (see box, right) requires care and attention. Lump charcoal lights and heats so quickly that you can get a

burst of heat within 5–10 minutes by using a chimney starter or adding a few unlit coals. For a longer heat plateau, we often put some oak logs onto a charcoal fire to boost the middle section of cooking time and maintain a stronger, steady heat source.

USING CHARCOAL ON A BARBECUE

To speed up the cooking process try this method of lighting. Take an off-cut of a metal pipe and stand it on top of the barbecue tray to make a "chimney". Stuff the bottom with paper, add some kindling, and light it. Once the kindling catches, add charcoal via the top of the pipe. When the charcoal starts to burn, put on some oven gloves and carefully lift the pipe upwards so that the glowing charcoals spread across the tray.
Don't start cooking until the bed of hot embers has been glowing for 10–15 minutes.
When the charcoal is ash grey it's reached the right temperature.

WORKING WITH WILLOW

Willow is an amazing plant that can grow at a phenomenal rate – more than 2.5cm (1in) a day in summer. We harvest its long straight shoots, or "withies", annually and use them around the farm to make fencing and wigwams for climbing plants, as well as weaving them into baskets and hurdles (see page 208). Nothing is wasted – even the scraps are turned into broom heads.

GROWING WILLOW

The roots of willows (*Salix* species) are renowned for absorbing moisture, which is why they are often planted beside streams or rivers to help stabilize banks and keep them intact, or on waterlogged ground where little else will grow. But willow's tendency to draw moisture from the soil also means that you should not grow it, or construct a living willow structure, too close to vegetable plots, drains, or buildings. We planted our willow in a marshy valley and around the edge of the duck pond.

Different species have different coloured stems, from yellow to black. We mixed in some dogwood (*Cornus* species) too, adding reds and greens to the colour spectrum, and used the cut stems in the same way as willow.

You can increase your stocks every year by replanting some of the withies when pollarding (see box opposite). Planting them close together encourages new growth to compete and stretch upwards to the light to produce more withies to work with.

When planting willow withies, bury at least a third of the stem in the ground. We made our own Derby dipper (an old fork with a 30cm (12in) spike welded to the bottom) to make perfectly sized planting holes. Once in the ground, your job is done – the willow will sprout roots, even if you plant it upside down!

WILLOW CRAFTS

Living willow rods planted into the ground can be shaped and woven into domes, tunnels, and sculptures. We have great fun creating our own structures and weaving baskets (see pages 232–233).

Living boundaries

To make a willow hoop fence, take straight withies and rub off any side shoots. Then push one end in the ground at an angle, bend the willow over into a hoop, and stick the other end into the ground. Overlap them and tie the hoops together at the top with soft twine if you want the stems to graft together. Each winter take the new vertical shoots and bend them down and back into the ground to continue the hoop pattern. This fence needs very little maintenance, but in the early stages remove any grass and weeds around the base, as they will compete for water and nutrients. Use the

1. A living willow fence grows thicker and bushier each year. If you bind stems together, eventually they will fuse.
2. Wigwams are perfect for beans and other climbing plants. **3. Weave willow** into baskets and decorative items, and turn shorter off-cuts into brooms.

Pollard and plant willow

Pollarding is an agricultural method that has been practised for centuries. It is a form of pruning that increases a tree's productivity. New shoots grow into straight lengths, known as "withies", which can be used for weaving baskets and hurdles, or to make living structures. Another key reason to pollard willow is to keep it under control and prevent it from growing into a massive tree. Pollard in the winter, between November and March, when the tree is dormant.

YOU WILL NEED
Pruning saw
Loppers
Buckets
Derby dipper for planting

1. Allow the tree to grow and then cut the trunk off at a height of 1m (3ft) or 2.2.m (7ft) with a pruning saw. This image shows the new stems that have sprouted from the cut in one year. **2. Remove any damaged,** diseased, crossing, or dead stems first and put aside for burning. Use loppers to remove any thin withies too. **3. Use a sharp pruning saw** to cut thicker branches. Remove all the current year's growth from the pollarding point. Rub off any thin shoots that are sprouting lower down. The finished tree should have a flat top – it could be dangerous to fall on to sharp spikes. **4. Plant some long straight withies** to make more trees. **5. If you're not planting** all the withies immediately, keep them soaked in 15cm (6in) of water in a bucket and store in a sheltered spot outdoors.

fence to make a decorative edging, as it takes several years before it is stock-proof.

We have found that planting willow alongside conventional wire fencing is a great overlap system. Eventually, the willow will outgrow, and can replace, the wire fencing. (See page 208 for making woven willow hurdles.)

Wigwams for climbing plants
Push several lengths of tall dry withies into the ground in a circle and tie the top tightly; you can use willow for this too but we tend to cheat and use a bit of wire. Weave around the base with flexible withies to bind the uprights together. Once you've woven to a depth of about 15cm (6in) around the bottom,

start to spiral upwards to strengthen the wigwam (see wigwams opposite). Lift the structure out of the ground and dry it out before you use it or it will start to sprout roots and turn into a living willow wigwam – not ideal for crops.

USING UP OFF-CUTS
The by-products of working with willow are twiggy little branches and off-cuts. But, you guessed it, all of these have uses too. You can make a useful brush. All you need is a broom handle sharpened into a point at one end and a large jubilee clip. Bundle a selection of twiggy off-cuts together and use the jubilee clip to fasten them tightly around the broom handle. We find that dogwood off-cuts make extra colourful brushes.

Alternatively, use your dry willow twigs as pea sticks to support your crops (see page 152) or to make charcoal for drawing.

Make a rooting compound
Willow bark contains auxins, a type of plant growth hormone. You can make your own rooting compound from willow off-cuts and use it to encourage cuttings of other plants to send out roots. Chop thin stems into 8cm (3in) lengths and place in a pan. Pour 4.5 litres (1 gallon) of warm water over the willow, cover, and leave overnight. In the morning, strain and bottle the rooting extract; seal it and store in a cool place. When you take cuttings (see page 147), dip them in a cup of the extract before potting up.

PROJECT Make a willow basket

We think everyone should have a go at making a basket. Book yourself on a course, or try this basket made from buff willow, which has been stripped of its outer bark. Soak stems in water for 20 minutes, then wrap in a damp cloth overnight. Keep them pliable by spraying with water as you weave.

YOU WILL NEED

Willow ring (1)
Weight (2)
Block of wood (3)
Rapping iron (4); bodkin (5)
Sharp curved knife (6)

Straight knife (7)
Secateurs (8)
1.2m (4ft) willow bundle sorted into 6 thick base stakes, 42 thin weavers, 24 thicker rods

MAKING THE BASE

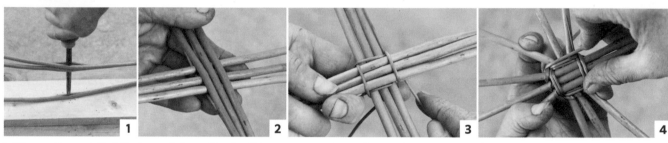

1. Skewer through 3 of the base stakes with a bodkin, protecting the surface with some wood.   **2. Insert the other 3** in the gap.
3. Weave around the centre of the stakes with 2 thin weavers. Insert the thin tips into the split stakes, then use a pairing weave (see step 5). Weave twice around the centre.   **4. Pull the stakes apart** on the third round and weave them.

5. Use the pairing weave, take one weaver behind one stake and over the next. Then bring the other weaver in front of the first stake and behind the next, and so on.   **6. Join in 2 more weavers** when the first pair run out. This time start with the thick butt ends. Continue pairing.   **7. Trim the butt ends** when the weavers run out. Tuck in the tip ends.   **8. Cut 8 thin weavers** at the butt end to the same length and insert 2 pairs of 2 weavers, tips first, opposite each other on the base. Work with one pair until it reaches the opposite pair; drop them and start weaving with the other pair – the chasing technique.

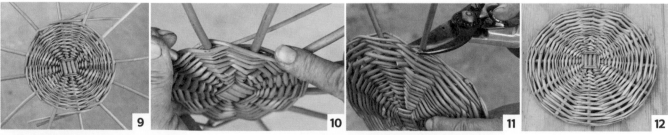

9. The pairs finish opposite each other.   **10. Add the remaining 2 pairs** of weavers – this time starting with the butt ends. Continue pairing. Tuck in the tip ends.   **11. Cut off the base stakes** using secateurs.   **12. The completed base.**

WORKING LARGE AREAS

232

MAKING THE SIDES

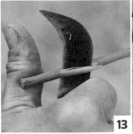

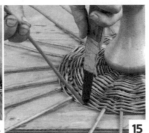

13. Select 24 thick rods. Make a sloping cut on the butt ends. **14. Insert the rods,** one on each side of the base stakes. Add a band of waling (see box). **15. Use a heavy weight** to steady the base. Use the back of a knife to kink each rod up at 45°. To do this, bring the rod upwards while pressing on the base of the stem with the knife. Let the rod drop back.

16. Gather the kinked rods and secure with a willow ring. Trim 12 weavers to length. **17. Insert the tip ends** of 3 weavers and work a wale around half of the basket. Insert the tips of another 3 weavers and use the chasing technique (see step 8). **18. Join in** the remaining 2 sets of 3 weavers when the first set run out. Continue waling. Repeat with another set of 12 weavers.

WALING

Waling is a strong weave, often used at the bottom and top of the basket to make it sturdy.
- Take 3 weavers and insert tips into 3 consecutive spaces.
- Take the left-hand weaver in front of 2 uprights and behind the third, working anti-clockwise.
- Take the weaver that is now on the left and work in the same way – in front of 2 uprights, behind the next one.
- Carry on in this way, always using the weaver on the left.
- Using the chasing technique in waling (see step 8), ensures the basket is an even height.

Waling is used to strengthen a basket.

19. Tap down the sides evenly using a rapping iron. **20. Make a simple border.** Soak the basket in water first. Use a straight knife to kink each upright down to the right. Take it over the next two uprights, weave it behind the third upright, and trim it between the third and fourth upright, so that it lies against the third one. Thread the last few uprights into the first ones. **21. The finished basket.**

ANIMAL HUSBANDRY

By keeping animals you take a huge step towards leading a self-sufficient life. Even just a couple of hens can give you a real sense of independence. If you choose to produce your own meat, however, there are stark facts to be faced. Meat does not come pre-sliced and wrapped in cellophane, and we know lots of people who have problems killing their livestock. To give your animals the best possible life, and then to dispatch them with the least amount of stress, seems to us the sensible way to produce meat.

KEEPING CHICKENS

People have been keeping hens for thousands of years. The Ancient Egyptians, Greeks, Chinese, and Romans all enjoyed the benefits of rearing chickens for eggs and meat. Nowadays they are still ideal domestic animals, whether you live in an urban or rural area. However you like your eggs in the morning, if they come from your own free-range chickens, nothing compares to their great taste.

CHOOSING THE BREED

We generally try to mix our small flocks to include pure breeds and some hybrids. Hybrids are used commercially and you can expect lots of eggs. Traditional breeds look more interesting and are useful to have as broody hens if you want to rear your own chicks. Light Sussex and Rhode Island Red are ideal for both eggs and eating but are ignored by commercial farmers because of their small eggs. We tend to opt for bantams, which are smallish breeds: they are relatively low cost to feed and you don't need much space. For a small plot, especially an urban one, we would recommend getting only a few hens and not bothering with a cockerel (see box opposite). As long as you keep fewer than 50 birds in total, including other poultry, there's no need to register your flock with DEFRA (see www.defra.gov.uk).

WHAT AGE TO BUY?

Day-old chicks are very cheap to buy but are unsexed, so you won't find out which are cockerels until later. This is good if you want some birds for meat but bad news if you simply want plenty of eggs. There's also the expense of heating their enclosure when they're young and giving them extra feed until they're ready to lay.

Pullets

These are birds that can be anything from about eight to 20 weeks old. You can easily buy them in batches of all females – no worries about cockerels. They take a little more feeding before they start producing eggs but are a good age to buy.

OUR FREE-RANGE PHILOSOPHY

The key for us has always been to provide our chickens with as much free roaming space as possible. On our suburban plot we let them out into the garden in the morning and lock them away at night, and on our smallholding they had a substantial area to roam. Chickens are traditionally woodland birds and appreciate being in willow coppice. They're also at home in open fields, where they supplement their diet with insect pests and weed seeds. However, there are disadvantages to the free-range lifestyle. Hens will start laying eggs in hidden places instead of their nest boxes. If it becomes a problem, keep them in their run till midday, until they start laying inside again.

Check whether an egg is good to eat by putting it in a bowl of water. A fresh egg lies on its side at the bottom (1); rotten eggs float (2).

Point-of-lay pullets

At 20 weeks old, hens start laying eggs. Their first season tends to be their most productive. You have to spend a bit more to get point-of-lay pullets, but we have occasionally found it worthwhile, especially when our existing hens are getting a bit old and we want a few more daily eggs.

Year-old ex-battery hens

Commercial battery farms kill their hens after their first year of laying. After they've reached their productive peak they no longer have value. There are huge animal-welfare issues associated with battery hens. Often they spend their whole life in an area no bigger than an A4 piece of paper; they lose their feathers, never learn how to scratch, and generally look a sorry state. We buy battery hens and give them a second chance. They are normally hybrids and are excellent layers. It's a humbling experience to see a maltreated animal rediscovering its instincts and enjoying its first dust bath or struggling to eat its first worm. The colour of their combs changes from a sickly pale pink to a healthy red in just a few weeks. Not only is it a feel-good thing to do, but battery hens are also productive and cheap.

COCKEREL PROS AND CONS

A cockerel is a great way of keeping a free-range flock together; he will shepherd them and protect them from predators. One of our cockerels, William, lost his tail feathers defending his ladies from a fox and walked around with an unsightly stump for a few months afterwards. Another reason to keep a cockerel is that it gives you the opportunity to rear more chicks, at little expense, if you let a hen go broody. The disadvantage in urban areas is their tendency to act as an alarm clock, so apologies to all our neighbours who have been woken up over the years.

1. Marans are one of our favourite breeds. They tend to be very hardy and lay large, deep-brown eggs. **2. Light Sussex** are a lovely old English breed and are great layers as well as good table birds. **3. Buff Orpingtons** are placid birds that lay smallish eggs. They are excellent hens for fostering chicks. If you don't plan to breed from your flock, don't choose these hens as they have a tendency to go broody. **4. Rescued hybrid battery hens** adapt quickly, soon thrive in their new free-range environment, and are reliably good layers.

Housing and basic care

Whether you keep chickens in an urban garden or on a smallholding, they have the same needs: a cosy nest box to lay in, a patch of ground on which to scratch and forage, and – most important – somewhere to sleep at night, safe from predators. We built our big chicken coop inside a shed with an outside run attached. This made cleaning them out easy, even on a rainy day. We also had a movable ark (see pages 192–193) down in the field. A movable system is ideal for gardens: it gives hens access to grass and you can change its position when the land is looking a bit hen-pecked. All chickens need somewhere to perch during the day and to roost at night, constant fresh water, and, of course, tasty food.

Chickens like to perch as they are woodland birds – ideally on a round branch to avoid damaging their feet.

KEY TO CHICKEN RUN

1	Outside run
2	Dust bath under shelter
3	Mesh roof on outside run
4	Indoor run within shed
5	Pulley system to operate door
6	Door to outside run
7	Staggered perches
8	Nest box with access hatch
9	Hen access to nest box
10	Vermin-proof feed store
11	Access for humans
12	Sawdust on floor
13	Rat-proof feeder
14	Bowls of grit and oystershell
15	Automated water supply

Dust baths are a vital way to get rid of parasites like lice; make one by filling a shallow wooden box with fine silver sand and keep it somewhere covered from the rain

Give grit and oystershell in small pots to aid digestion and avoid calcium deficiency

Chickens must have clean water and an automatic watering system saves a huge amount of time – if you fill up their water manually, do it regularly

Chicken arks that are easy to clean and that have their own run attached are excellent for smaller suburban gardens.

Remember to change your hens' bedding regularly so it's nice and clean. Finding fresh eggs in the morning is the ultimate reward.

A simple movable run with secure, sheltered accommodation for night-time. Move to fresh grass every few days.

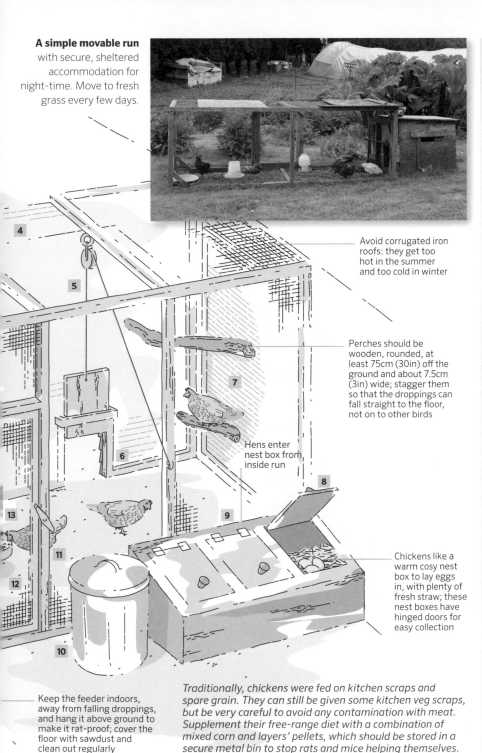

Avoid corrugated iron roofs: they get too hot in the summer and too cold in winter

Perches should be wooden, rounded, at least 75cm (30in) off the ground and about 7.5cm (3in) wide; stagger them so that the droppings can fall straight to the floor, not on to other birds

Hens enter nest box from inside run

Chickens like a warm cosy nest box to lay eggs in, with plenty of fresh straw; these nest boxes have hinged doors for easy collection

Keep the feeder indoors, away from falling droppings, and hang it above ground to make it rat-proof; cover the floor with sawdust and clean out regularly

Traditionally, chickens were fed on kitchen scraps and spare grain. They can still be given some kitchen veg scraps, but be very careful to avoid any contamination with meat. Supplement their free-range diet with a combination of mixed corn and layers' pellets, which should be stored in a secure metal bin to stop rats and mice helping themselves.

Let your chickens out to roam during the day, even in frost. The only time you'll need to keep them shut in is when there is deep snow on the ground. Remember, egg production will drop in winter – it's controlled by the hours of daylight.

POTENTIAL PROBLEMS

Keeping chickens is generally straightforward, but you may need to use some of the following techniques and solutions.

Trimming a cockerel's spurs

This will stop your hens from getting hurt during mating. Wrap the cockerel in a towel and hold him firmly. Use wire cutters to trim the spurs a little at a time. File them smooth when you're done.

Avoid the vein when making the cut: stop at the first sight of blood below the surface.

Red mites

Parasites that emerge from the woodwork at night. To deal with them, keep the coop clean and get a remedy from a vet.

Scaly feet

Caused by burrowing mite. Isolate affected birds and consult a vet.

Broody hens

If one of your hens has gone broody and you don't want any more chicks, place her in a small slatted coop for a couple of days. The cold air circulating underneath will deter her from sitting.

The slatted floor in this coop stops the broody hen sitting comfortably. A wire mesh flooring works just as well.

PROJECT Build a chicken ark

A movable chicken ark is ideal for small gardens, as it keeps your chickens where you want them while giving them access to grass. When a patch wears out, you can simply lift the ark and move it on to pastures new. Raising the nest box above the run makes maximum use of the space available for the chickens to move about. This ark houses two or three hens comfortably.

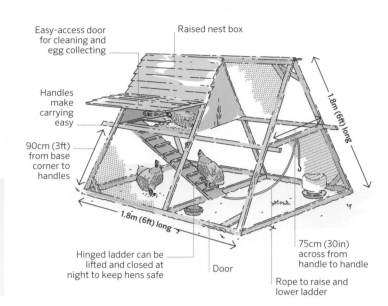

Easy-access door for cleaning and egg collecting

Raised nest box

Handles make carrying easy

90cm (3ft) from base corner to handles

1.8m (6ft) long

1.8m (6ft) long

75cm (30in) across from handle to handle

Hinged ladder can be lifted and closed at night to keep hens safe

Door

Rope to raise and lower ladder

YOU WILL NEED

Handsaw
Circular saw, jigsaw
Power drill/driver
Tape measure, pencil
Wire cutters
Hammer

Pressure-treated planks, battens, and struts
Plywood, feather boards
Chicken wire
Screws, pins, staples
Hinges, latch, clasp, rope

MAKING THE FRAME

1. Cut planks to size and screw them together. We made two 1.8m (6ft) square frames that were a convenient fit for chicken wire of 90cm (3ft) width. Screw angled brace supports into the corners of each frame, except for the corner where the door will be located.
2. Fit the final corner brace on the inside edge of the frame, so that the door to the ark will fit flush in the corner. Rest one frame on the other while working for stability.   **3. Fix a vertical strut** into the centre of each frame. Attach longer horizontal struts to the inside of the frame, protruding at least a hand's span each side of the ark. These will be used to carry it, so fix them at waist height and sand the ends to make them comfortable to handle.

4. Work out the ideal width to maximize space for the chickens while making the ark easy to carry and prop the frames against each other. Take measurements. Ours was 1.5m (5ft) wide at the base and 75cm (30in) wide at handle height. Cut lengths for horizontal central and base struts.   **5. Mark angles** and cut ends to fit.   **6. Fix base struts** to inside edge of the long verticals, and central struts to the inside edge of the horizontals. Fix an additional horizontal strut higher up, to attach the nest-box door frame (see step 10).

MAKING THE NEST BOX

FEATHER BOARDING

To fit the boards, secure screws without over-tightening, lift the edge and slide the next layer under. Once in position, tighten top screws and repeat process.

7. Measure the floor for the nest box, and cut out from plywood using a circular saw or handsaw. Cut out a hen-sized access hatch using a jigsaw. Screw the floor to the supporting struts. Measure, cut, and attach plywood end pieces for the nest box. **8. Make a ladder** from a 0.3 x 1.2m (1 x 4ft) plyboard plank. Pin on rungs made from battens, using a block of wood as a spacer. Hinge the ladder to a section of plank and screw the plank to the underside of the floor, level with the access hatch. **9. Clad one side** of the nest box with feather boards (see box); fit them to about halfway down on the other side.

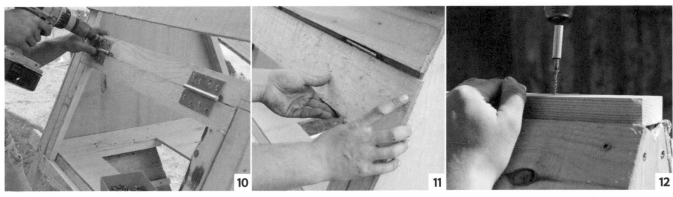

10. Construct a door big enough for you to collect the eggs; aim to make the height such that the bottom of a feather board will fit over the top part of the hinges. Fit the door flush with the main frame and let it overhang the nest box floor to avoid draughts. Add hinges. **11. Complete the feather boarding,** attaching those covering the door to its frame. You can also add a simple clasp for a padlock. **12. Attach a capping timber** to the top of the ark to stop rain dripping through the roof joint.

ATTACHING CHICKEN WIRE

13. Make another door to fit one of the bottom rectangles. Strengthen the door with two corner braces at opposite ends. Attach hinges to one of the vertical sides. Cut chicken wire to size and staple it in place. Screw the door hinges in place and attach a latch to keep out predators. **14. Continue** attaching chicken wire over the whole frame, taking care not to leave any loose edges. Snip off any sharp ends. **15. Install a drinker** and feeder, and introduce your chickens!

How to increase your flock

Which comes first: the chicken or the egg? Well, we have always started with chickens and increased our flocks from their eggs – that's why we keep a cockerel (see page 237). But for a beginner we recommend buying fertilized eggs (do this online) and incubating them yourself. It's harder work than leaving it to nature, but it's the least risky way to build a flock.

HATCHING YOUR OWN

An electronically controlled incubator takes care of the temperature and humidity, but you'll still need to turn the eggs three times a day so that the embryo does not get stuck on the shell (see box). Incubation is about 21 days. After seven days, "candle" the eggs to check they are fertile (see box). Stop turning them on day 19.

After hatching, leave the chicks in the incubator for the first 24 hours and don't worry about feeding. Then move them to a heated brooder, where they will live for the next six weeks. The brooder must be rat-proof, with wood shavings on the floor and an infra-red lamp above. Ideally, the temperature should be 35°C (95°F); then reduce it by 3°C (5.4°F) a week.

THE NATURAL WAY

It is easy to tell when one of your hens is broody. She won't leave the nest box and fluffs up to twice her size when you peer in. She will make loud noises, and if you move her, she'll go straight back to her eggs.

If you want her to hatch chicks, you must move her to her own house. Firstly, so that the chicks are safe and secondly, so that she doesn't discourage all the other hens from coming in to the nest box to lay.

Her house should be warm and cosy, with food and water nearby. We find it easiest to move broody hens at night. After that, there is very little to do – other than provide chick crumb and water for your new chicks.

DIY HATCHING

An electronic incubator allows minimal handling, which can cool the eggs or introduce bacteria.

Warm it up before the eggs arrive. In a still-air machine the ideal temperature is 39.4°C (102.9°F), measured 5cm (2in) above the eggs. Set the humidity to 75–80 per cent.

Label eggs with the date they were added to the incubator.

Turn each egg through 180 degrees using your pencilled date label as a helpful marker.

Candle the eggs to check how they're developing. Cut two holes in opposite sides of a cardboard box. Hold an egg in front of one hole, shine a torch through the other, and examine the yolk. Discard infertile eggs and those with the blood ring of a failed embryo.

Once the chicks hatch, give them some time to dry off in the warm incubator before you handle them.

Look for these signs (from left): failed embryo, infertile egg with clear yolk, fertile embryo.

Show your chicks where the water is by dipping their beaks in it, and drop food in front of them like the mother hen would.

Use an infra-red lamp and adjust the heat as needed: if it's too hot, chicks move away from the lamp; too cold and they huddle below.

Broody Buff Orpingtons are great foster hens. Give them a clutch of fertilized eggs to sit on and they will hatch the chicks and raise them as their own.

PROJECT Prepare chickens to eat

Killing one of your chickens and preparing it for a meal is relatively simple. The first time you do it, it's a good idea to get an experienced person to help so that the chicken is rendered unconscious immediately and doesn't suffer.

KILLING AND PLUCKING

DO
Make sure that the bird has not been fed for 24 hours so that its digestive system is empty.
Get someone to help.
Stay relaxed.
Pluck the chicken straight away when it's still warm.
Wash your hands between plucking and drawing.

DON'T
Panic.
Use too much force.
Kill a bird in front of other chickens.
Tear at the feathers or you'll rip the skin.

1. Take both legs firmly in your left hand and hold the head cupped in the palm of your right hand, with your index finger and middle finger either side of its neck. **2. Stretch** the neck down so that the head bends backwards. Stretch until the neck is broken: don't go too far as the head will come off. **3. Tie the feet** and hang the chicken. Start plucking immediately. It is far easier to do while the bird is warm. Pluck the wing feathers, followed by the leg feathers, then the body.

DRAWING

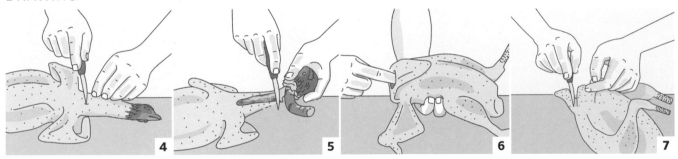

4. Cut from the bottom of the neck up to the head. **5. Sever** the neckbone and remove it, along with the head. **6. Put your finger** inside the gap and loosen the innards by moving your fingers inside the cavity. **7. Turn** the chicken and slit between the vent and tail, being careful not to pierce the rectum or you'll contaminate the flesh. Cut all round the vent.

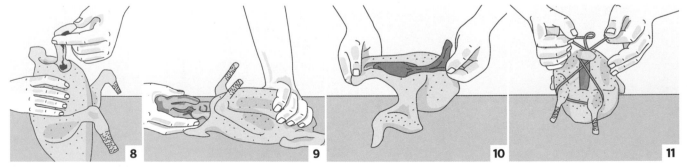

8. Pull the vent carefully away from the body; it should come away quite easily now that you've cut it free. **9. Draw out** the intestines, gizzard, lungs, liver, and heart with the vent as you keep pulling it steadily. **10. Remove the crop** from the neck end of the chicken, being careful not to pierce it. Tuck the flaps of skin back inside at either end of the bird. Gently singe the skin with a flame to burn off any remaining fluffy feathers. **11. Truss the bird** up ready for roasting. Cut off the feet and tie the wing tips to the legs, then wrap the string round the hocks and the end of the tail.

KEEPING TURKEYS

The first Christmas at Newhouse Farm we ordered an organic turkey from our local butcher. It was a lovely bird, but it had a huge price tag. Next year we did the sums and reckoned we could rear them for about a third of the price. So we fattened a dozen, giving them lots of room and organic feed, and it was a resounding success. Watching them grow is fascinating and they make great presents.

CHOOSING WHAT TO BUY

Buy birds that are 5–6 weeks old rather than day-olds, because it cuts down on the hassle of providing lamps and looking after them when they are at their most fragile. We would often buy 12 Norfolk Black turkeys, which were too young to sex at a glance, and would expect to be able to tell which were stags and which were hens by the time they were about 12–13 weeks old. As long as you keep fewer than 50 birds in total, including other poultry, there is no need to register your flock with DEFRA (see www.defra.gov.uk).

The best breeds to choose from are below.

Bourbon Reds are an American breed that originated in Kentucky and Pennsylvania and can grow to a monstrous 15kg (33lb). Their feathers are a reddish-chestnut colour and their tail feathers are white.

Bronze are named after the unusual colour of their feathers, which are a green copper-brown in the sun. Bronze turkeys often require artificial insemination to reproduce due to the large size of their breast meat. We have found them a calm bird to keep that can grow very large.

Buff turkeys have reddish buff-coloured feathers and the hens can be good egg producers. The added advantage of a buff turkey is that the pin feathers are nearly white, so the bird looks great when plucked.

Narragansett are extremely tasty birds that are hardier than most other breeds due to their mixed genetic selection. Narragansett turkeys are traditionally known for having a calm disposition and making good mothers.

Norfolk Black are our favourites and possess striking black plumage. They are slower to mature than many

CLIPPING WINGS

If you want to restrict the movement of your turkeys, clip one of their wings. Clip both wings and they will put lots of effort into flying and eventually take off, whereas one wing means they are lopsided and spin back to earth. When clipping, simply cut off their primary feathers with a sharp pair of scissors at the point where they line up with the next layer.

Get someone else to hold the turkey as you clip the wing feathers.

1. Turkeys don't have great eyesight, so they may need coaxing into their shed at night. **2. Give them organic feed** and clean them out regularly. **3. Turkeys will thrive** with plenty of space to enjoy.

commercial breeds, but this is more than compensated for by their really tasty flavour. Having said Norfolk Blacks are slow to mature, the stags can still reach a good 11kg (24lb)!

HOUSING AND BASIC CARE

Turkeys must be kept separate from chickens to guard against the fatal blackhead disease. We kept our turkeys outside in a small paddock surrounded by electric fencing to keep out any hungry foxes (see page 209 for how to set one up). They had a decent-sized shed to shelter in and a big perch in the middle of their enclosure, which they sat upon and practised their best vulture impressions.

We fed our birds with turkey grower's pellets, which have less protein in them than commercial turkey feed. We were able to use this type of feed because we usually bought birds as early as possible in the season, so that they had plenty of time to increase in size naturally. They also had the opportunity to obtain a great deal of their protein by snacking on the insects and bugs they found in their enclosure.

When looking after turkeys, it is also vital to provide a good source of grit for them to eat. This acts as an aid to digestion, enabling them to mash up their food in their crop and digest the goodness, ensuring they grow nice and big during the year.

TOP TURKEY TIPS

Keep your turkeys clean! Meaning the bedding and food containers – and give them plenty of fresh water. If you keep their house clean, there is far less chance of one of your birds getting ill.

Shut them up at night. Sounds like common sense, but free range doesn't mean providing the local fox with an early Christmas present.

Leave the ground to rest for 12 months before a new batch of turkeys return to the same spot. This helps avoid blackhead, which is a parasite that lives in earthworms and gut worms. Signs of a problem are sulphur-yellow droppings and, more often than not, a dead turkey.

Consider homeopathic worming alternatives if you want to rear your birds organically.

If you plan to keep breeding stock so you can enjoy turkey all through the year, buy some special leather saddles as protection for the hens. These sit on their backs and stop the stags from damaging them too much while mating. The saddles need to stay on in spring and come off in early autumn.

KILLING LARGE BIRDS

Slaughtering turkeys needs to be fast and efficient. Calmly take the birds one at a time to a barn, then place them upside down in a traffic cone inserted into an old chair. One person holds the legs while the other two place a broom handle on either side of the turkey's neck. They firmly squeeze the handles together and twist and push them down while the person holding the legs pulls up, which separates the vertebrae in the neck. Then immediately slice the jugular with a sharp knife. Once dead, hang the birds up by their feet and pluck them straight away (see page 243).

Doing the deed alone

To kill a turkey solo, place a broom handle over its neck, twist and press down with your feet, whilst pulling it up by the legs.

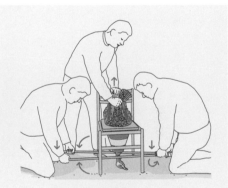

The "suspended traffic cone" method requires three people to make it as efficient and kind as possible.

Working alone requires greater coordination and strength to ensure the job is done quickly and humanely.

KEEPING DUCKS AND GEESE

Geese are absolutely amazing to rear yourself. We found that they are hardy, tough, and surprisingly self-reliant – almost the perfect self-sufficient animal – but they can be noisy and a handful if you're not used to big birds. Ducks are also extremely useful animals to keep on your plot, but you'll find they can be messy! Buy them both either as young birds or as eggs to raise your own flock.

WHY KEEP DUCKS?

Provided you have a source of fresh water, ducks are great to keep for four reasons: their delicious eggs, which make the best cakes in the world; they are tasty table birds; they are selective eaters that are less likely to scratch garden beds than chickens; and they can be useful at keeping down the population of pests like slugs.

Choosing what ducks to buy

Among the best breeds of duck to choose from are:

Aylesbury table ducks are famous as large birds for eating. They can weigh up to 4.5kg (10lb).

DUCKS TO WATER

Ducks need to be able to periodically submerge their heads to clean their eyes and nostrils. We believe that it's only really fair to keep ducks – and geese, for that matter – if you have running water on your plot or at least a pond regularly topped up with fresh water. Flowing water and a pond are ideal, but if you don't have either of these, you must ensure that you can provide them with a deep container of water that you can regularly change.

A Muscovy duck enjoys a paddle in the stream with her young.

Indian Runners don't have much meat on them but they are good egg layers – on average, you can hope for about 180 eggs a year. They have a distinctive upright stance and stroll around like penguins.

Khaki Campbells can lay over 300 eggs a year!

Muscovy ducks are heavy and placid, and make excellent mothers. They are very good fliers, so you will probably want to clip their wings (see page 244). Be careful when you pick them up as their feet have sharp claws. Muscovies are thought to descend from geese, but we class them as our favourite ducks.

Welsh Harlequins are good all-round birds. They are decent layers and are big enough to be an impressive table bird.

Housing

Ducks will live happily in a conventional hen house, easily adapted and placed near a pond or stream – water is essential to a duck's wellbeing (see box, left). You may want to provide a ramp into the house as ducks are clumsy and can easily damage themselves, but they don't need a perch. Additionally, the housing should be rat- and fox-proof, draught-free but well ventilated, and always have fresh, dry bedding. Ducks enjoy a wander, but we kept ours in a fenced-off area with a pond most of the time, to avoid stepping in duck poo each time we left the house.

Feeding

Ducks aren't grazing birds like geese (see opposite), but they will supplement their diet with grass if you give them the access. We fed our ducks corn daily to

These Indian Runner ducklings are particular favourites – we enjoy their perky natures.

increase the number of eggs they laid and to fatten them up nicely. Ducks are also partly carnivorous and will happily eat slugs, snails, worms, frogs, and insects. Therefore, in spring, before we start the next round of crop planting, we allowed our ducks access to the vegetable beds to find and eat any hiding pests. Don't get carried away and give them a free run of your beds all year round though, as they will damage young brassicas and eat peas and lettuce without a care in the world.

Slaughtering ducks

For larger ducks, we recommend that you follow the methods described on page 245. It's best to find someone to help you. Smaller ducks can be killed single-handed (see page 243). You can kill your ducks at about 10 weeks old, but we left ours for much longer.

WHY KEEP GEESE?

Geese are first-class grazers and are great guard birds, with their noisy honking serving as an intruder alert. They usually start laying enormous eggs in February and March. Soft-boiled for breakfast, you'll find you need a whole regiment of soldiers to dip in them. Put in cold water, bring to the boil for 9–10 minutes – perfect! Bear in mind, though, that geese are large birds and need to be treated as such. Respect them when you pick them up or you will receive a strong wallop from their wings.

Choosing what geese to buy

There are many breeds to consider:
Brecon Buffs not only look lovely, they also taste great!
Chinese geese are excellent egg layers and can be ready to slaughter as early as 8 weeks old.
Common English is generally the standard goose, often an Embden crossed with a Toulouse breed.
Embden is an excellent, very large table breed.
Roman is particularly good if you want a goose that can be killed when it is young, but still have plenty of breast meat to enjoy.

Housing

Provide geese with larger houses than ducks. The traditional design is an open, three-sided construction, often made with straw bales. We always protected them from foxes at night by locking them away in a small shed. Geese are vulnerable to rats and foxes when they are young or when they're sitting on eggs. We used an electric fence with poultry wire to stop them roaming (see page 209).
Geese require a bit more space for grazing than ducks or chickens. We found that our small orchard was perfect for them. Although they are officially classed as water fowl, they don't absolutely need a pond. All they require is access to water to keep their nostrils and eyes clean by immersing their heads. A large container will do until you make something a bit more permanent such as a small pond, but it must be deep and the water changed regularly.

Feeding

Geese eat grass like living lawnmowers and can be used for weeding between strawberry plants or vines, as they tend to avoid eating broad-leaved plants.
When you get your goslings, feed them chick crumbs for the first couple of months, reducing quantities after four weeks as they start to graze more on grass shoots. Then, when the geese are fully grown, occasionally substitute their grass diet with corn feed if you want to fatten them up.

Slaughtering geese

We would recommend that you follow the methods described for large birds on page 245. Geese are large birds and should be killed by a team effort at 5–6 months old.

GREEN GEESE

Traditionally, rearing geese was inextricably linked with the growth and decline of green grass. Goslings begin grazing when the grass is fresh in spring and then, when the grass slows down around September, they are slaughtered. It is a harsh natural cycle, but efficient. In the past, they were called Michaelmas geese in Britain as their butchery tied in with the 29 November, and in America geese are often bred to coincide with Thanksgiving feasts.

Keeping "green geese" makes good financial sense and mows the lawns.

Imprint yourself on goslings by being around when they hatch and for the first few days. They'll flock to you much more readily when they're bigger.

KEEPING PIGS

Pigs are often the heart of a smallholding. They plough and manure the land, as well as providing tremendous meat. These highly intelligent animals are capable of coming when they're called and of showing individual character. We are omnivores, however, and enjoy eating a bacon butty, so for that reason, our pigs at Newhouse Farm didn't have names and we never became too attached to them.

CHOOSING WHAT TO BUY

Rare-breed pigs such as Berkshire, Cornish Black, and Tamworth tend to be much hardier than specialized commercial hybrids and are better suited to outdoor smallholding conditions. Ideally, you want your pigs to put on weight fast and not have any health problems.

You can often look at the history of certain pedigree animals in the records of a pig breeder to get an indication of the size they'll grow to and whether they will be better suited for making bacon or for processing the meat as sausages.

Some old-fashioned breeds, for instance, can carry excessive amounts of fat, whilst cross breeds such as Lop and Large White are strong and healthy, with a fat layer that is perfect for rustic bacon and sausage making.

For us, there is no competition with mass-produced breeds. Mass-produced breeds are often fast-growing, produce big litters, and have less fat. This makes them profitable, but lower animal welfare is a serious issue. For example, they are often kept inside (see box, opposite).

We would not recommend buying your first batch of pigs from a local livestock market unless you are very experienced or have a knowledgeable friend to help you. Often farmers will off-load some of their inferior stock at market and you won't know what you're looking for. Instead, develop a relationship with a reputable farmer, who should then sell you good stock because, if the pigs don't do well, you won't go back to them again and they end up losing good business.

HOW TO BUY LIVESTOCK

You will need to register your property as an agricultural holding before buying any livestock. There will also be paperwork to fill out when you buy. You'll need to keep records, tag each animal, and abide by regulations when you need to transport your animals. Visit www.defra.gov.uk for more information and check with your local authority.

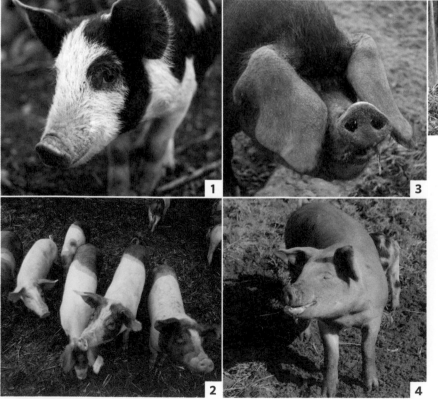

1. Berkshires are one of Britain's oldest breeds. They are early maturing pigs and have a high proportion of lean meat. **2. British Saddleback** is a hardy breed with an excellent foundation in many cross-breeding programmes. **3. Cornish Blacks** have a thick fat layer, which keeps them happily warm over winter. **4. Tamworths** are a lovely, medium-sized, red-coloured pig that is extremely hardy, even in northerly areas such as Scotland and Canada. **5. Gloucester Old Spots** were bred originally to live in apple orchards. They are hardy and make great bacon. **6. Large Whites** provide top-quality back bacon and are extremely popular commercially.

Gilt weaners

The easiest and most economical way to buy pigs is to get some 8-week-old weaners that have already been weaned for a couple of weeks and run around actively. The rear ends should be clean and watch for any signs of lameness.

We always bought gilts. These are young female pigs that have not mated or had any piglets. We opted for gilts rather than boars as the pork does not end up tasting "pissy" – a colloquialism for meat tainted by male hormones as the young boar matures.

It is best not to buy breeding stock straightaway because, as a beginner, it is nearly impossible to select genuinely good stock. Breeding pigs also involves additional equipment, attention, and some expertise. We liked to leave that side of things to local experts and concentrate our efforts on the fattening-up side of things.

Transporting your pigs

Getting your newly bought pigs home involves a bit of planning. Many countries have strict rules governing the movement of animals and penalties if you don't adhere to them, so check them out via DEFRA or your local authority.

A key piece of equipment when moving pigs is a length of plyboard with two holes cut in the top to hold it at each end. This enables you to shepherd your pigs in and out of trailers or to new pastures.

HAPPY PIGS

Animal welfare is one of our main motivations for rearing pigs. Commercial factory farming often denies pigs space to move around and their feed can be pumped with a cocktail of chemicals resulting in cheap, tasteless pork. If you can't rear your own pigs, you can still choose to eat free-range organic pork, but it can be difficult to understand how it is labelled and what different welfare conditions mean.

Organic production in the UK allows pigs to spend their entire lives outdoors, with access to straw-bedded huts or tents and large paddocks.

Free-range has no legal definition for pork production, but we believe this should mean pigs are born and reared outdoors throughout their lives, with permanent access to pasture.

Outdoor-bred means pigs are born in outdoor systems in straw-bedded arks, with access to a large outdoor paddock. They move indoors for growing and finishing at or shortly after weaning – usually into straw-bedded systems in large airy barns. The sows remain in the outdoor system throughout their lives.

Outdoor-reared means the pigs are born and reared in outdoor systems, usually spending around half their lives outside. During this time, they may not have access to pasture, but will have an outside pen and a straw-bedded ark.

PROJECT Set up a pig pen

The focal point of a pig pen will be the house or sty. Always place the door facing away from the prevailing wind and build it on a raised pallet to stop rising damp. The shell is then covered by some corrugated iron. Don't go spending much money on their housing, but do use large nails or screws and make it strong enough to withstand a fully grown pig enjoying a good back scratch!

YOU WILL NEED

Electric fence	Stone weight
Pig sty	Trough
Straw	Organic feed
Water bowl	

1. Before moving your pigs into their pen, surround the area with an electric fence (see page 208). **2. Keep the pig sty** topped up with plenty of fresh straw, putting a whole bale in at a time. **3. Ensure there is plenty of water** in the pen to satisfy a pig's thirst. If possible, sink the container into the ground, as well as putting a weight in it, to prevent the pigs knocking it over. **4. To prevent the feed** from being trampled into the ground in their pig-like enthusiasm to get at it, place it into a trough twice a day.

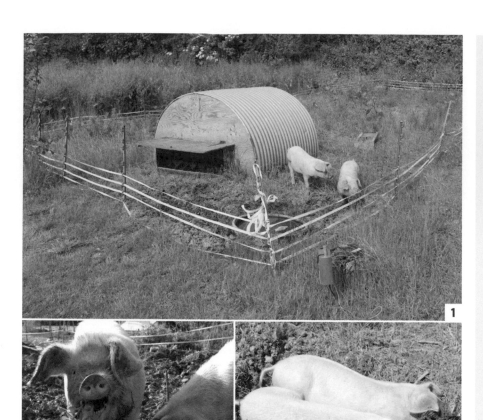

1

2

3

1. Keep your pigs in overgrown areas so they can root around for treats and clear the ground. **2. Throw fresh greens** into the pen. **3. In hot weather,** make a muddy patch to serve as a wallow.

POTENTIAL PROBLEMS

We never found that keeping pigs on a small scale presented many problems. However, here are some common ailments to watch out for:

Erysipelas can be fatal if not treated quickly. Symptoms include raised diamond patterns on the back and sides, which turn purple. Within 48 hours the pig can die, but a shot of penicillin in time can cure it.

Lice are not particularly dangerous to the pig, but they do cause great irritation. Buy lice wash from an agricultural supplier to treat them.

Meningitis will cause a pig to go off its food, lie on the floor, and hold its head to one side as it walks around. An injection of antibiotics can produce a rapid cure, so act fast.

Scour, or diarrhoea, mainly occurs in piglets. Reduce their feed and give them a course of antibiotics.

The bottom line with any health problems is: if in doubt, call a vet out.

HOUSING AND BASIC CARE

Most of the equipment needed to keep pigs is readily available from agricultural suppliers, but some equipment, such as feeders and electric fencing, can be expensive to buy new, so look out for it at farm auctions. If you are buying second-hand equipment, make sure you disinfect it first and scour it with a wire brush before use. When choosing the pig sty (see page 249), ensure it is well insulated against the cold. The more energy your pigs put into growing instead of keeping warm, the better.

PIGS AND WATER

Pigs drink copious volumes of water and their container will need topping up regularly. We positioned it near the edge of the fence so that we could easily fill it. In hot weather you should also create an area of wet mud that the pigs can use as a "wallow". Wallowing helps the pigs to cool down and the layer of dried mud on their skin offers some protection against sunburn.

FEEDING PIGS

The feed that we used was an organic pellet that contained plenty of protein and carbohydrates. This was in addition to a foraging diet of grass, wild weeds, and roots.

When the weaners are young, allow them to eat as much as they want from an automatic feeder placed next to their house. When they get to 10 weeks, start to feed them twice a day. Regardless of age we fed them 1kg (2¼lb) of food each morning and evening. For us that meant giving our two pigs a large full scoop of food twice daily and they did fine. We continued with this method of feeding until slaughter (see opposite).

You can throw the food over the ground for them to forage for, but as it is the most expensive part of the pig-rearing process, and they often end up urinating on it or trampling it into the ground, we poured it into their trough to avoid waste. Conversely, we did make a point of throwing fresh greens and fodder crops all over their enclosure for them to search for.

We liked to give our pigs access to fresh pasture with plenty of roots for them to dig up and eat. It is also a good idea to plant fodder crops on the land before they move in to save money on their feed bills.

Serving your pigs household waste is illegal, as there are now many regulations restricting the feeding of pigs with any waste that has been in your kitchen. This is to avoid the spread of disease. However, if you top and tail vegetables in the garden and position your beds near to where the pigs live, then you will find they can still eat plenty of fresh peelings and spoiled vegetables.

When you feed your pigs, call loudly each time with a distinctive noise. This will train them to come to you and makes it much easier to shepherd them if they escape. Our pig-calling noise sounded like "shoooEEee!", but the call you choose is completely up to you...

SLAUGHTERING PIGS

Once your pigs weigh 65–80kg (140–180lb) you can slaughter them at any time. We normally kept our girls for 9–10 months before slaughter, which is much longer than commercial farms, but gave us more of their digging and manuring power to help cultivate bits of land.

There are so many regulations covering the killing and eating of pigs on your own land, that we recommend sending them to a local abattoir instead. Visit the abattoir to check you are happy with the price and services they offer. This also reduces stress on the day of the slaughter, as you'll know what's in store. To move your pigs, tempt them into the trailer using pig boards, your signature pig call, and some food. Once in the abattoir, take them for their appointment and say your goodbyes. The next day you can collect a scraped carcass split into halves to butcher at home, or ask a butcher to prepare the cuts for you.

1. Curing your own meat to make bacon is a delicious way of using belly pork and back bacon. **2. Spare ribs are one of our favourite cuts.** They are ideal slow-cooked on the barbecue before sharing with friends and family.

CUTS OF PORK

We believe in nose-to-tail eating: every cut of pork can be delicious if it's cooked properly.

Shoulder is a big, juicy cut. Whole or boned, it is good slow-smoked for feasts or diced or minced for sausages.

Spare ribs are cut from the upper part of the shoulder with plenty of marbled fat. They are delicious with a barbecue rub and slow-smoked with water in a pan.

Belly is ideal for sticky burnt ends or curing and smoking your own bacon. Slow roast whole with the bones or remove them with a boning knife.

Back bacon comes from the joint stretching from the middle of the back to the belly, so you get both eye and streaky in one cut.

Leg is used for making ham, the king of cured meats. The leg can be broken down for smaller hams or rolled for a roast. Good fat marbling will keep the leg moist during brining and smoking.

KNIVES

Investing in a good set of blades for different jobs makes for a much more straightforward butchery experience.

A boning knife is sharp, narrow, and flexible. Easy to move around meat to remove bones, it can transform your DIY butchery.

A meat slicer is a long thin, slightly flexible blade that cuts long even strokes through brisket and salmon perfectly. Not essential, but it certainly makes you look the part!

A scimitar is heavier than most knives and curved at the tip to rock back and forth for easy chopping. Excellent for general butchery and raw meat prep as well as slicing.

A meat cleaver or butcher's cleaver should be strong and relatively heavy. Ideal for cutting bone without causing it to splinter.

A chef's knife is an all-round knife and a must for butchery. You will find that it becomes key in your kitchen.

A steel is used to keep your knives sharp. You'll find that regular sharpening, even whilst working, makes a real difference. Remember to keep your steel dry.

Using sharp knives makes an enormous difference when doing your own butchery at home. Remember to sharpen with a steel regularly in-between cuts.

MAKING SAUSAGES AND SALAMI

Sausage making is great fun and an easy thing to try at home, either with your own pork or meat from your local butcher. You don't even need to buy any specialist equipment (though it can be a great help).

The difference between sausages and salami all comes down to the volume of salt you use, as salami is essentially a cured sausage. This also means it has the added bonus of lasting longer!

GETTING STARTED

If you enjoy making your own sausages, salamis, and burgers, consider investing in a meat mincer and sausage filler.

Filling sausages is much easier to do with a machine than by hand. Before we bought our piece of kit, we used a rudimentary funnel and wooden spoon and achieved great sausages, but it took

so much longer and, for us, making delicious home-made sausages more than justified the expense of a new kitchen gadget.

There are hand-powered versions available for both mincers and sausage fillers, but an electric option will save you lots of energy and time, ultimately allowing you to achieve more. We opted

for a combined mincer and sausage filler machine, which does both jobs well enough. The key thing to look for is the option of grinding meat with different width plates so that you can choose between coarse-, medium-, or fine-grade mince.

RECIPE **Pork sausages**

Ever since we made our first sausages together, we've very rarely resorted to shop-bought bangers. We like to experiment with the flavours and add seasonal ingredients such as chopped leeks, apple, and spices. You can also use the same equipment to make your own vegetarian sausages using veg, beans, and pulses instead of pork.

YOU WILL NEED

500g (1lb 2oz) pork shoulder, slightly frozen, diced
500g (1lb 2oz) pork belly, diced
1 tbsp salt
1 tsp thyme

1 tsp ground mustard
1 tsp cracked black pepper
1 tsp paprika
150g (5½oz) breadcrumbs
Natural sausage casings

1. Mix your ingredients well in a big bowl. Set up your meat grinder with a 4mm (⅛in) mincing plate and slowly feed the mixture into the chute. Repeat for a smoother texture. **2. Hold one damp natural sausage casing** against the nozzle of your sausage machine, then start extruding the sausage meat slow and evenly. **3. Form the meat** into a large ring, then twist into 15cm (6in) sausages. Hang the sausages to dry in the fridge over a tray for 1–2 days, then store in a sealed container in the fridge for 1–2 weeks.

RECIPE Salami

The word salami comes from the Italian "sale" meaning salt. Roman legionnaires were often paid with salt (hence the word salary), and we certainly wouldn't mind getting paid in salt as its ability to cure or season food is phenomenal. We always use at least 2–4 per cent weight of salt in our salami, and it must be mixed in well. You want the salt to penetrate all of the pork: this is where a salami either succeeds or fails. We use table salt or a fine sea salt flake for this type of recipe because it gets right into the minced meat. To make salami that is ready to eat sooner, use less salt and enjoy after a day hanging in fridge.

YOU WILL NEED

2kg (4½lb) pork shoulder (or pork mince if you don't have a mincer)

50g (1¾oz) salt

8 garlic cloves, finely chopped

2 tbsp fennel seeds

2 tbsp smoked paprika

100ml (3½fl oz) red wine

500g (1lb 2oz) pork fat, diced and cured if possible

Natural sausage casing

Butcher's twine and S-hooks

1. If you have a sausage mincer, start by chopping your pork into 2.5cm (1in) cubes – roughly big enough to be squeezed through the mincer. Mince the meat and place in a large mixing bowl. Add the salt, garlic, herbs, and spices. Mix together well and add the red wine and diced pork fat. **2. Soak your natural sausage** casing in a bowl of water. **3. Using a sausage machine,** feed the mince mix into the casing (see opposite). Twist the sausages at 20–30cm (8–12in) intervals, then tie off securely with butcher's twine, separating the salamis. **4. Hang in a well-ventilated,** undercover area, and leave for one month to air dry. During the hanging process the salami ferments, which gives it natural health benefits. For example, the lactic acid bacteria it contains can help maintain a healthy digestive system. If any white mould appears on the salamis, clean them with a vinegar and water solution and a piece of muslin.

TYPES OF SALAMI TO TRY

German
Juniper berries and lots of black pepper make this a perfect pairing with bread and a pint of beer for a rustic lunch.

Pepperoni
With its Italian-American influence, this salami has become world-famous. A dry sausage originating from southern Italy, it is often a blend of cured pork and beef. It has spicy elements from the paprika and sweetness from capsicum peppers.

Garlic saucisson
Adding huge amounts of garlic into the mix results in an intense salami that can stand out amongst an array of other flavours in stews, stuffing, or mixed platters.

MAKING BACON AND BRINING HAM

Using salt to draw the water out of fresh food is one of the oldest and simplest method of food preservation. The processes of curing and brining inhibit the enzymes and microbes that cause food to perish, which make it possible to store meat and fish without refrigeration. The strong, salty flavours of cured and brined food remain popular, and are often enhanced with herbs, spices, and other ingredients.

SALTS FOR CURING

The techniques used to preserve food using salt may have changed over the centuries, but the chemical properties of salt, which produce a unique reaction when left in contact with raw food, remain the same and are as useful today as they were a thousand years ago. Many varieties of salt are available, some better suited to curing and brining than others.

Sea salt contains a complex combination of minerals, which gives regional salts their distinctive flavours.

Pure dried vacuum (PDV) has very fine granules and up to 99.9 per cent purity. It is perfect for large salt-box curing.

Himalayan salt is naturally pink, of high quality, and makes tasty cures.

DRY CURING

The most accessible starting point for preserving your food, dry curing can be used to preserve all sorts of meat and fish. It's also a key skill to master before you start smoking food. Meat is vulnerable during the curing process so it must be kept in a cool place. Sugar is also often added to counteract loss of colour and maintain a good flavour.

Daily dry curing, often used for bacon, involves rubbing a cure into the meat over the course of 5–6 days.

The salt-box method works just as it sounds: place your meat into a plastic or ceramic box and cover with the cure.

The total immersion method is an expensive and long curing process done prior to air-drying. It requires lots of salt, but is a fairly straightforward technique.

Air-drying

After curing, most meats need to be hung up outside to draw out any remaining moisture. To air-dry meat, scrub off any salt or wash off the brine solution and pat dry. Wrap in muslin sheets or fine cloth netting secured with string, but make sure air can still flow

---

RECIPE DRY-CURED BACON

Once you have experienced a rasher of home-cured bacon sizzling in the pan, you will never go back to brine-injected bacon that leaches grey water. Curing bacon takes time and planning, so it can be useful to have cuts of bacon at different stages: some can be curing in the fridge, while others are air-drying. You can also cold-smoke your bacon joint for 6–8 hours after air-drying it to impart a strong, smoky flavour into the meat (see pages 224–25).

YOU WILL NEED
500g (1lb 2oz) salt
500g (1lb 2oz) brown sugar
2 tbsp juniper berries, crushed
1 tbsp cracked pink peppercorns
2 tsp chilli flakes
1kg (2¼lb) pork belly (middle bacon)

1. Mix together all your ingredients, except the pork, to make the cure. Reserve 100g (3½oz) and set the rest aside in an airtight container.   **2. Rub the reserved cure** into the pork evenly, using about a quarter on the fat side.   **3. Place the pork** in a food-safe bag or container in the fridge and leave for 24 hours. Pour off any water that has been drawn out of the meat and rub with another 100g (3½oz) of the cure. Repeat this process for five days. Your bacon should firm up and darken slightly in colour. Rinse the joint in a little water, pat dry, and wrap in muslin.   **4. Hang the bacon** in a cool, well-ventilated place and leave for two days, then slice into rashers ready to grill or fry.

around the food to enable evaporation. Hang in a sheltered, shady spot with good air circulation and a fairly constant temperature of 10–18°C (50–65°F).

The longer produce is dried, the longer it will keep and the more its flavour intensifies. Check it regularly and trust your nose – if something smells wrong, then don't risk eating it.

BRINING

Soaking meat, fish, or vegetables in a strong brine solution is another effective way to preserve food. Though not the most attractive option, the brined meat stays moist and is ready to cook or take on flavour from smoking. Adding herbs and spices into the saline solution can also really penetrate the meat or fish and enhance the taste, while adding sugar makes for a less salty result.

BRINING ESSENTIALS

You only need salt and water to brine food – a solution of 6–10 per cent salt is ideal – along with a suitable container in which to submerge and seal the meat. A plastic freezer bag laid on a tray can work well, but the bigger the meat, the bigger the container required – a leg of ham may need an ice box or a washtub. The meat needs to be refrigerated below 5°C (41°F).

Sweet brines, as the name suggests, feature a quantity of sugar as well as the salt/water mixture. This not only helps reduce the salty flavours of brined food but increases the growth of *Lactobacillus* – beneficial bacteria in the curing process.

Spices and herbs add extra flavour to your brine. While all you strictly need in a brine is salt, you can't go wrong with bay leaves, black pepper, cloves, coriander seeds, garlic, juniper berries, root ginger, rosemary, star anise, or thyme.

RECIPE ROASTED HAM

There's nothing more satisfying to carve and serve than a ham you've made from scratch. The process also enables you to add real flavour. This recipe uses a sweet cure and a glaze made from orange juice and mustard, a classic combination that works with cloves.

YOU WILL NEED

500g (1lb 2oz) salt
750g (1lb 10oz) sugar
5 litres (8¾ pints) hot water
2 bay leaves
1 tsp black peppercorns
2 star anise
1 cinnamon stick

4 cloves, plus extra for studding the ham
1kg (2¼lb) gammon joint
For the glaze
75ml (2½fl oz) orange juice
2 tbsp brown sugar
1 tbsp mustard powder

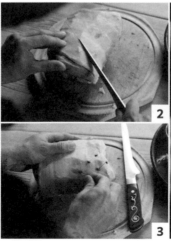

1. Dissolve the salt and sugar in the water to make the brine, then add the spices and allow to cool. Submerge the gammon in the brine and leave for 2–3 days in the fridge. Remove the gammon from the brine, pat dry, and submerge in a pan of cold water. Simmer for 1 hour. To avoid your ham tasting too salty, pour away half of the water after the first hour of cooking and top up with fresh boiling water. **2. Remove the ham** from the pan when the fat starts to separate from the meat and allow it to cool slightly. Preheat the oven to 180°C (350°F). Remove the skin from the ham and use a sharp knife to score the fat in a diamond pattern. **3. Place a clove** into each cross-section, then set the ham aside. **4. Mix the glaze ingredients** in a pan and bring to the boil. Cook for 10–15 minutes until the glaze starts to thicken, then brush onto the ham generously. **5. Roast the ham** for 25 minutes, then baste with more glaze. Cook again until caramelized.

KEEPING CATTLE

Imagine how great it would feel to supply all of your own protein and dairy products. In our opinion, keeping cattle is the epitome of self-sufficiency, but before you take on any animals, remember that it is a serious commitment and there are important regulations to follow (see page 248). Although one of the more complicated and time-consuming challenges, it can be extremely rewarding.

CHOOSING WHAT TO BUY

Keeping cattle involves much more effort and a larger financial outlay than buying a few chickens. You will need extra sheds and an acre of pasture for each cow. Visit local farmers and experienced neighbours to learn the ropes and think carefully about how many you'll want. One cow will provide all the milk you could need.

You will have to decide between dairy cows, beef cows, or more traditional dual-purpose cows. We didn't keep cattle due to lack of space, but we know many people who favour older, dual-purpose breeds and natural rearing techniques.

Our favourite breeds of cattle for smallholdings include:

Ayrshires can be white and red and are a tough breed that produces plenty of high-quality milk. They fatten early and can remain productive for up to 12 years.

Dexter is our favourite breed. Their beef is delicious. Their small size makes them easier to butcher and preserve.

Jerseys are docile, hardy, and produce excellent milk that's ideal for making butter and cheese. They live for a long time and can become part of the family.

Short Horns are excellent dual-purpose cows, suitable for a small herd. They are hardy and long-lived, rear calves very easily, and make good foster mothers.

HOW TO BUY CATTLE

The first time you visit an auction, go with an experienced friend or farmer. You can find good deals at auctions, and the best cows to choose between are a "first-calver" and an older cow.

A first-calver is a heifer that has had just one calf and therefore produces lots of milk. The farmer will probably have bred her to sell, so you know she will be a good cow and you can start milking her straightaway. The disadvantage is that you are both beginners, and she may be a bit flighty and nervous at first. Also, her teats will be smaller and harder to squeeze than those of a mature cow.

An older cow is more likely to have a placid nature, as well as elongated, well-milked teats, making her easier to milk. The drawback is that she won't be with you for so long.

HOUSING AND BASIC CARE

Cattle usually only need to come in under cover to be milked or to eat dry hay. Watch what your neighbours do at different times of the year; they will know the local climate and conditions, and will be able to advise you about extra shelter that your cattle may need.

Fussier breeds that produce high milk yields will want to go inside to eat extra food when the weather is bad or if the grass is not in prime condition. If you intensively rear cattle indoors over winter, you can have great results, but it is much more expensive to supply them with straw and extra food.

The amount of food your cattle will need depends on the breed, their size, and their ability to adapt to the environment. Equally, they will need different-sized rations at different times. Normally, cattle receive a "maintenance" ration and a little bit more, known as a "production" ration, when they are being milked. Add food according to how much milk they are giving. Hardy breeds, such as Dexters, thrive on grass and a simple supplement of hay.

If your cows are looking fatter or thinner, adjust their feed of hay and pellets accordingly. Watch how much milk they produce, how hungry they are, and if they appear healthy. If in doubt, ask local farmers, vets, or agricultural suppliers for advice.

MATING

Smaller cows can be mated after 15 months, but most bigger cows should wait until 20 months. They can only be mated when they are in heat, or "bulling". At this time, a cow will have a slightly swollen vulva, moo more noisily than normal, and mount other cattle or allow another cow to mount her. Bulling occurs at 21-day intervals, and only lasts

EXAMINING A HEIFER

As well as examining the younger heifer, ask to see her mum; this will give you an indication of what she will end up like.
Ensure she is calm and reasonably tame.
Feel the udder carefully to detect any lumps; these can be caused by mastitis, which blocks the flow of milk.
If the cow is in milk, ask if you can try milking her. Check the workings of each teat and see how calm she remains. She is likely to be a bit frisky as you will be a stranger to her, so cut her some slack before passing judgement.
Check the cow's teeth to determine her age. A mature cow at five years old has eight incisors on the lower jaw. The age of an older cow can be gauged by the wear, but this takes an experienced eye.
Ask if she is tuberculosis tested (TT) and also free from brucellosis.
Don't buy her if the farmer tells you she has a problem, such as a blind quarter, where one part of her udder has no milk.

for about 18 hours, so stay alert. Once you have chosen a cow that you want a bull to mate with, take her to a neighbour's farm or bring the bull to her. Alternatively, call your local artificial insemination authority.

CALVING

Outdoor hardy cattle will normally calve on their own, with no real trouble. You're fairly likely to go out one day and see your cow licking a newly born calf. If, however, the calf has not sucked from its mother's teat within an hour of birth, hold it to its mother and make it suck. It may be necessary to tie up the mother while you do this. If you bring your cows in, make sure it is draught-free, so the calves don't get too cold. Never turn out calves without their mothers in winter.

MILKING

Milk your cows twice a day, preferably at 12-hour intervals. Normally one cow will have to produce one calf a year if

1. Regular handling ensures that cattle are calm and easy to look after. **2. High-fibre cattle pellets** top up the basic dietary requirements of cattle producing high milk yields. **3. Keep a salt lick** in your cattle's field; it's an easy source of sodium and other minerals.

she is to produce milk, so you'll also have calves to keep an eye on.

To get milk from your cows, you will have to take calves away from their mothers early on. The mother will bellow loudly for a few days. Substitute yourself

for the young one, so your cow allows you to milk her as a calf would. Feed the calf separately with milk in a bucket. This may seem cruel, but cows aren't human; they have shorter memories and it's just one of the realities of milking your herd.

PROJECT   How to milk a cow

Before you start milking, wash the cow's udder, massaging it to stimulate the flow of milk as you clean. Then wash the rear end of the cow to prevent any nasties dripping into your nice clean milk pail. Also wash the teats and your hands before you begin, and make sure that you give the cow something tasty to eat to distract her.

YOU WILL NEED
Washing equipment
Small stool and pail

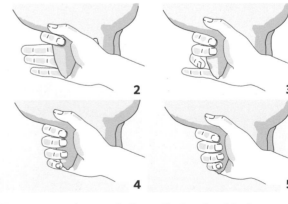

1. Position a small stool on the right-hand side of the cow and grip the pail between your knees.   **2. Grasp the two front teats,** one in each hand, and squeeze them at the top with each thumb and forefinger to trap the milk in the teats.   **3. Move your hands down,** bringing in your middle fingers.   **4. Bring in your little fingers** so your whole hand is around the teat.   **5. Squeeze the milk** from the bottom of the teats, then release and repeat in a rhythmic rolling action. Use common sense to squeeze at the right pressure.

MAKING BUTTER AND KEFIR MILK

If you have a cow or goat, then making cream, butter, and yogurt is a great way to make the most of their milk. It's also enormously satisfying to make your own dairy products, even if you live in the city with no animals and have to go out and buy the milk. We often whisk up a batch of butter to go with some home-baked bread, and always make enough to freeze some for later.

SKIMMING MILK FOR CREAM

If you milk a cow or goat and then leave a bowl of the milk to stand at room temperature, the cream will separate naturally and rise to the surface. You can then skim it off with a skimmer – a flat, saucer-shaped utensil about 20cm (8in) across, traditionally made out of wood or tin, and perforated with small holes to allow the excess milk to drain away.

The alternative way to separate cream from milk is to use a specially designed shallow trough, often made of slate, with a plughole in the base. Pour the milk in and wait for the cream to rise. Once the cream is almost solid, pull the plug and the milk runs away into a vessel below, leaving you with a trough of cream.

If you are buying milk to make cream, you need non-homogenized whole milk, such as Gold Top. Some supermarkets, farm shops, and home-delivery dairies sell it. You can't use homogenized milk, as the homogenization process stops the fat, or cream, from separating out.

SOURING MILK FOR YOGURT

Yogurt is milk that has been soured, or cultured, with lactic-acid bacteria. Simply stir around 30ml (2 tbsp) of shop-bought live yogurt into about 1.2 litres (2 pints) of whole or semi-skimmed milk that's been sterilized, then cooled to blood temperature. Cover the container and leave in a warm environment to work overnight. We recommend using a hay box (see pages 58–59). The yogurt is ready to eat when it has a thick consistency, at which point it's probably best to move it to the fridge. If you want to keep your yogurt culture alive to make more than one batch, keep it in the hay box and every time you take out some yogurt to eat, replace it with the same quantity of fresh milk.

SOURING CREAM FOR BUTTER

To turn cream into butter you first need to sour, or ripen, it by encouraging bacteria to turn some of the lactose into lactic acid. In warm weather this will happen naturally; for a quicker result and on cooler days, add a few teaspoons of already soured cream or yogurt to the cream and stir to mix.

CLOTTED CREAM

Clotted cream is made by heating unpasteurized or non-homogenized milk so that the cream becomes very thick and gains a yellow crust. Leave the fresh milk for 12 hours at room temperature, then heat it to 92°C (198°F). Cool it immediately by pouring it into a bowl. Once cool, leave it in the fridge for a further 24 hours, then skim off the delicious home-made clotted cream.

KEFIR MILK

A popular fermented dairy product, kefir is a refreshing drink and can also be used to make fruity smoothies or spicy marinades. It contains beneficial strains of bacteria, so is excellent after antibiotic treatments to restore intestinal balance. It's also easy to make yourself.

How to make kefir milk
Heat 1 litre (1¾ pints) milk to 32°C (90°F) and then stir ½ tsp kefir grains into the milk. Leave at room temperature for 12–24 hours for the milk to ripen and the kefir to develop flavour. The longer you leave it, the stronger the taste.
Store in the fridge and consume within 10 days. Drink a little at a time if you are new to the drink, then drink more as you get used to it over a week or two.
To make more kefir milk, simply pour 50ml (1¾fl oz) of your fermented kefir into 1 litre (1¾ pints) of warm milk and leave for a further 12 hours. Rinse the grains in between batches in cold water and avoid strong cleaning agents for the container as these could taint the fermentation. This can be repeated for as long as you look after your kefir "mother". If you are going away or want to slow down the process, leave your kefir grains in a little milk in the fridge until you return to activate them.

1. Add salt and herbs to fresh butter and roll up in greaseproof paper. **2. Kefir grains** are extremely easy to look after, and can produce a healthy fermented milk.

RECIPE Home-made butter

Shaking soured cream (see opposite) turns it into butter. You can simply use an electric whisk to do this – you don't need a butter churn. If the cream is at a temperature of about 20°C (68°F) then the butter will "come" (change from cream to butter) in a matter of minutes.

YOU WILL NEED

Electric whisk
Butter pats and stamp
Wooden chopping board
Greaseproof paper

1.2 litres (2 pints) double cream
15ml (3 tsp) live natural yogurt
Fine sea salt

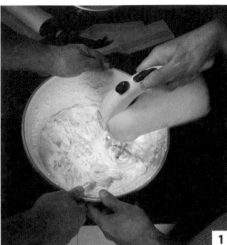

1. Pour the cream into a clean sterilized bowl. Add the live yogurt and whisk for a few minutes. The consistency starts to change to make soft peaks **2. Continue whisking** until the cream looks like scrambled eggs and turns pale yellow. After 2–3 minutes more, small globules of butter form. Add a little cold, clean water when the mixture looks like a firm mass of butter globules and carry on whisking for 1 minute on a low speed. Pour off the milky by-product (buttermilk) and keep it to use in another recipe.

3. Use grooved butter pats or wooden spatulas to transfer the butter to a wooden chopping board for washing. The pats need to be cold and wet, so dip them in ice and water first. **4. Mix and squeeze** the butter with the pats. Keep adding more water to wash the butter – this prevents it turning rancid later on. Collect the buttermilk as it washes off. Continue until the liquid runs clear.

5. Salt the butter with at least 2 per cent of its weight in salt, sprinkling a little at a time from a height for good coverage. Mix thoroughly. Use the pats to press the butter into shape. Push firmly to drive out air bubbles.
6. Stamp to decorate, and wrap in greaseproof paper.

MAKING YOUR OWN CHEESE

Making cheese originally began as a way of using up surplus milk and it's a useful technique to learn if you keep cows or goats. Even if you don't produce your own milk, making cheese is still extremely creative and gives you the chance to experiment by adding home-grown herbs to flavour your own delicious soft and hard cheeses.

WHAT IS CHEESE?

Leaving milk in a warm place or mixing additives into it increases its acidity and causes curds and whey to form. Cheese is made from the curds. It is delicious and a great source of protein, but also very high in calories. Soft cheese made from semi-skimmed milk is a less fattening option.

WHAT MILK TO USE

If you don't own milking animals, buy whole or semi-skimmed milk to make cheese (avoid skimmed milk). You can use homogenized milk.

Pasteurized milk is fine for cheese-making, as long as you use an effective starter (see right) and allow the lactic-acid bacteria to develop overnight before you begin.

MAKING SOFT CHEESE

You can make soft cheese by allowing some milk to curdle naturally on warm days in summer, or by adding a coagulant (see opposite). It is fairly tasteless, which gives you the opportunity to be adventurous with flavourings (see recipe opposite). Cream cheese is a soft cheese made with curdled cream instead of milk. It has a smoother texture and more buttery taste.

MAKING HARD CHEESE

A great way to preserve the summer glut of fresh milk into winter, hard cheese was traditionally made from the milk from more than one cow, and from an evening and a morning's milking. The process was started off with evening milk and morning milk was added the next day. But don't worry if you haven't got any cows – we have achieved some excellent results by simply using organic supermarket milk.

Using a starter

"Starters" are batches of milk that are very high in lactic-acid bacteria, which speeds up the cheese-making process. You can buy milk-based starters or try making your own. We make ours by leaving a litre (1¾ pints) of semi-skimmed milk at 27–30°C (81–86°F) for 24 hours to go sour – this is known as the "mother culture". Then skim off the top layer of the starter and add it to another pint of semi-skimmed milk. Cover this with a cloth and leave for another 24 hours at 21°C (70°F) to create the working culture.

Using a coagulant

Rennet is a key ingredient in cheese-making. It comes from the stomach of a calf, goat, or lamb, and is the enzyme that causes milk to coagulate into curds and whey. Milk that is curdled with rennet without further processing is known as junket.

To produce vegetarian cheese, buy rennet made from fermented microorganisms, plant extracts, or synthetic animal rennet – or simply use lemon juice.

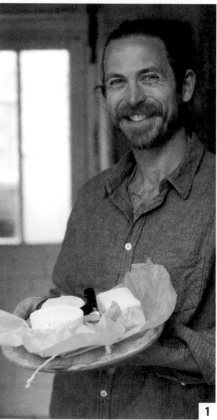

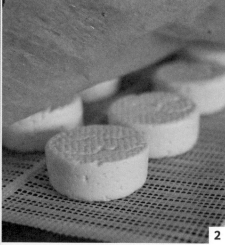

1. Home-made hard cheese will keep for months if stored correctly, but in practice it gets eaten all too quickly. Take it out of the refrigerator around an hour before serving to let it come up to room temperature. **2. Soft cheese** doesn't need rennet to help the milk curdle – lemon juice will do the job – making it an ideal vegetarian option. Serve with slices of hot buttered toast.

RECIPE Soft cheese

Soft cheese is quick and easy to make, and is ready to eat in less than 24 hours. It doesn't keep for long and needs to be eaten up quickly. Its mild taste gives you the chance to jazz it up with different flavourings, from classic crushed garlic and chopped fresh herbs to coarsely crushed peppercorns.

YOU WILL NEED

Large pan	Juice from a lemon
Slotted spoon	Chopped herbs
Muslin bag	Chopped garlic
String	Salt and pepper
1 litre (1¾ pints) milk	

1. Bring a pan of milk to a gentle simmer. Take off the heat immediately and add the lemon juice. Stir the milk, which will start to curdle.   **2. Use a slotted spoon** to put the curds into a muslin bag. Tie up the bag with string.   **3. Hang the curds** above a bowl or sink overnight, to allow the whey to drip out.

4. Unwrap the curds and you'll see they have turned into a home-made soft cheese.   **5. Spread the cheese** out on a work top and mix in some flavourings. We used chopped fresh chives, crushed garlic, salt, and pepper. The soft cheese lasts a few days when stored in the fridge.

CREAM CHEESE

Warm some cream slowly until it is curdled. Leave the curds in the whey overnight. Then drain the whey off and cut up the curd with a long-bladed knife. Add some salt and some butter if you want it to taste a bit richer, and tie it up in a muslin bag. Hang the bag up in the refrigerator or another cool place for a day to drip. The following day, tighten the bag up and leave it to hang for a month. You can leave it for up to four months to mature, but we are never that patient!

BUY LOCAL

We don't have time to make as much cheese as we would like, so we buy a selection of local cheeses. There has been a revival in small-scale dairies who take pride in producing truly delicious cheese. Look around and do some research into which regional specialities are available in your area, instead of opting for a bland block of Cheddar from some distant factory. Another option when choice is limited is to buy ordinary, uninspiring cheese and smoke it at home to add another rich, exciting flavour (see pages 224–225).

KEEPING SHEEP

Rearing sheep reduces our reliance on imported meat and wool that travel thousands of kilometres to reach us. It takes commitment and skill to look after a small flock so it's not something to undertake lightly, and you'll need to check the regulations that apply in your region before you get started (see page 248). It's well worth the effort, though, and a big step towards becoming self-sufficient.

CHOOSING SHEEP

Consider whether you want a specialized breed – one that produces good wool, for instance – or a more general-purpose breed. Geography is also important. On an exposed site with poor grazing, a primitive hardy sheep, such as Soay or Ouessant, will be best.

Dorset Horns can lamb more than once a year. They produce good quality wool and a high milk yield.

Border Leicester is an elegant English sheep. A prolific breeder, they produce good mutton and wool.

Southdown is a manageable size and matures early.

Ouessant are a hardy, primitive breed, which means they haven't been overbred.

Soays are similar to wild sheep, with good milk yields and lean meat.

How to buy sheep

Buying good-quality sheep can be daunting. Visit local livestock auctions, and ask a farmer for advice before you start. Always look at the seller's records, and check that the sheep have good teeth – they should be broad, short, and fit squarely in the mouth.

Start small, with three ewes for example, or two orphaned lambs, until you feel confident that keeping sheep will fit in with your lifestyle.

HOUSING AND BASIC CARE

The land needed for sheep varies greatly depending on the size of the breed and the quality of pasture – the grassland and mixed vegetation that grazing animals feed on. As a rough guide you could keep about 10 ewes on 0.4 hectares (1 acre), but you must have more land available so you can move them on. Moving sheep onto new pasture on a six-to-eight week rotation reduces the build up of parasites on the land and also conserves the grass.

Your flock shouldn't require much additional feed if they have good grazing. If you have restricted access to grass, give them pellets – in troughs to avoid waste – or a rack filled with good-quality hay. They also need a clean, fresh supply of water in a bowl at least 0.5m (20in) off the ground to prevent them fouling it.

Sheep shouldn't need any housing beyond the natural shelter offered by outcrops of rocks and trees, although some rare breeds are susceptible to wet and cold weather. It may pay to have a shed for ewes when they are lambing.

To secure your animals temporarily (for instance, when you need to examine their feet, give them medication, or round them up for shearing or lambing), construct small enclosures using hurdles. For more permanent boundaries, see pages 206–209).

BREEDING SHEEP

Mate or tup sheep in the autumn. It's too expensive to keep a ram for less than a dozen sheep; we suggest borrowing one. Rub some reddle (dark-coloured earth) on his chest so you can see when he has served all the ewes. Keep your ewes on poor pasture for a few weeks beforehand – called flushing the flock – then move them onto good pasture with the ram.

A sheep's gestation period is 147 days and lambing normally starts in late February or March. A good shepherd allows sheep to get on with it alone, while remaining vigilant. The first signs of a ewe giving birth will be that they choose a place away from the flock and repeatedly lie down. A water bag appears first, followed by the lamb's feet and nose. When the lamb is born, the ewe licks away the membrane and mucus from its head to stimulate it. If a ewe has been in labour for more than an hour and is obviously in discomfort, call a neighbouring farmer or vet.

When all the ewes have lambed, move them to the best pastures you have. New mothers and offspring will thrive on good nourishment.

Ouessant sheep from France are small and easy to carry – ideal for a smallholder.

PROJECT Shear a sheep and turn fleece into wool

Shearing takes lots of practice. Try learning the ropes on a neighbouring farm first. To turn raw fleece into wool, you need to wash and card it before sending it off to be spun or trying to spin it yourself. Cleaning fleece is an easy process, but you'll need to prepare it first by "skirting". This is when you pick out any dirt or grass stuck to it. Once the fleece is clean, you'll need to card it, which involves ripping the fibres apart to turn it into a fluffy piece of wool. You'll either need a drum carder, carding brushes, or dog brushes with steel teeth.

HOW TO SHEAR A SHEEP

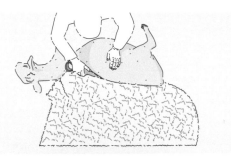

1. Clip all the wool off the stomach, down to the udders, taking care not to chop off her teats.   **2. Next shear her down one side,** starting at the throat, and around the side of her neck and head. Keep going as far as you can reach; then roll the sheep over and shear the other side. The fleece should now come right off her body except at the hindquarters.
3. Lay her flat on the ground and shear the fleece away from the back and hindquarters. Tidy her up with a trim of her hind legs and tail so that the other sheep don't laugh at her.

WASHING AND CARDING FLEECE

1. Fill a sink or bathtub with hot water and washing up liquid; this use of detergent will remove some of the lanolin and some of the fleece's waterproofing qualities but will ensure your fleece is clean. Push a few handfuls of fleece under the water, but don't swish it or the wool will felt together. Leave it for 30–45 minutes so the soap can break down the grease and dirt, then drain the water and gently lift out the wool. Refill the sink or bath with more hot soapy water and soak the wool for another 30 minutes. Repeat once more if it still looks dirty, otherwise place your wool onto a towel and pat it dry, then hang it carefully to finish drying on a clothes dryer.   **2. Next, card your wool.** Place some wool onto one carder. Hold it in your left hand, then pull the other carder gently across the lower part of the first carder until the fibres separate between the two. Repeat three or four times, gradually working the whole of the carder. Don't push the carders together forcefully, as the teeth of one could cut into the cloth of the other.   **3. Hold the carders angled in a V-shape,** with the bottom edges together. Use small upward right-left, right-left flicking actions to ease the fibres away from the carders. Replace all the fibres on the left-hand carder and repeat the carding process until you are happy with the blend.

SPINNING WOOL

To turn your carded wool into yarn, you need to spin it. It takes skill and practice to master a traditional spinning wheel, but hand-spinning with a drop spindle is easy to learn and doesn't involve any expensive kit.

A drop spindle is essentially a wooden stick with a weight on one end and a hook on the other; they're cheap to buy, but also straightforward to make if you have basic woodworking tools.
To use a drop spindle, attach the hook to the carded wool. Hold the wool in one hand, and spin the spindle with the other. The fibres will twist together to form a string of yarn.

KEEPING GOATS

Goats tend to be hardier than sheep and are great at reclaiming a neglected smallholding as they eat anything from brambles to weeds. Their meat is a very sustainable source of protein, but goats also produce excellent milk, which can be used to make delicious cheese. Don't forget to check local regulations for rearing and transporting livestock in your area (see page 248).

CHOOSING GOATS

Goats are gregarious animals, so you should keep at least two.

Saanen can get quite big, but have high milk yields on good grazing.

Anglo-Nubian have lower milk yields but of top-notch quality.

Toggenburg are fairly small, but do well on larger, free-range areas.

Golden Guernsey are small, with lower milk yields than large breeds.

Angora are renowned for their soft mohair fleeces and high milk yields.

1. Well-brought-up goats shouldn't butt their handlers. **2. Goats are masters** of escape, so make sure you keep your fencing in good condition.

How to buy goats

As a rule, goats with strong legs, a wedge-shaped rear end, and strong backs are likely to be good animals. You can get more information by joining a goat-keeper's organization.

Buy weaned kids or goatlings between 1 and 2 years old. Despite being a bit more expensive to buy, they are cheaper to feed than young kids that need milk.

HOUSING AND BASIC CARE

Goats don't like rain or cold. Most need a draught-proof shelter to sleep in at night. If you are building one from scratch, put in a concrete floor to make cleaning easy – the manure is great to add to compost. Also, try to provide your goats with an outdoor daytime shelter as a bolt hole in case of bad weather.

Securing your animals

Good fencing for goats is essential (see pages 206–209). Bear in mind a wise old saying: "Goats spend 23 hours of the day planning their escape and the last hour executing it." Anyone who keeps goats will soon become aware of the constant battle to stop them from escaping and eating the contents of the fruit and vegetable garden.

Tethering goats is another option, but it is time-consuming; as with sheep, you need to move them regularly to prevent the land from becoming contaminated with a build-up of parasites.

Feeding and drinking

Goats are best fed by grazing them on scrubland or pasture. They will thrive on heather- or gorse-covered hillsides and in deciduous woodland, but will not do so well in coniferous woods. Goats need huge amounts of fibre, which they get from browsing on brambles, thistles, and twigs. Supplement grazing with a special goat mix from an agricultural supplier. Suspended feed troughs stop goats fouling in their food and then refusing to eat it. Provide them with hay all year round – on average, a milking goat will need about a bale a week. Also give your goats a salt lick and a constant supply of fresh water.

BREEDING GOATS

Take goats to be mated – it is usually cheaper than keeping a billy goat yourself. The advice for kids is the same as for lambing (see page 262), although the gestation is a few days longer.

Castrating goats

If you plan to eat billy kids, castrate them before they're three months old, as this stops their meat from tasting too strong. The most humane method is to use an Elastrator, available from agricultural suppliers.

POTENTIAL PROBLEMS

Foot rot can be a problem for both goats and sheep on damp ground. Keep hooves trimmed. If animals get foot rot, walk them through a foot bath of formalin.
Abscesses in goats are best lanced and kept clean with antiseptic solutions. Ask a vet to help with this procedure.
Dose goats against fluke if they have access to wet or marshy land.

RECIPE Chevre

Of all the semi-soft cheeses in the world, this is probably the most famous. Its simple and iconic shape, smooth texture, and elegant taste are all perfectly balanced, resulting in an understated but flavoursome cheese.

YOU WILL NEED

Muslin

Moulds

5 litres (1 gallon) whole goats' milk

1 sachet of chevre starter

1. Heat the milk slowly to 30°C (86°F), then add the starter. Wait for 5 minutes, then stir for 2–3 minutes. Cover the pan, allow to cool, and ripen at room temperature for 12–14 hours. **2. Ladle the curds** into a muslin liner and leave to drain for 15–20 minutes. **3. Spoon the curds** into cheese moulds and leave to drain for 12 hours at room temperature. **4. Place the moulds** on a draining rack and flip them once during the 12-hour draining period. Place in a sealed container in the fridge and eat within 1 week. Sprinkle with herbs and spices before serving. Our favourites include: tarragon, honey, and paprika, chervil and rose petals, pink peppercorns and toasted fennel seeds, and smoked salt, dill, and lemon zest.

RECIPE Halloumi

When we first made our version of halloumi, we were amazed at the simplicity of the process. We had always assumed that there were secret techniques safe-guarded by Cypriot goat-herders and shrouded in mountain mystery. In fact, you don't even need starters to make it at home, just some rennet and the correct equipment. This cheese's structure is ideal for cooking and its salty, tangy taste is robust enough to pair with spices and strong flavours.

YOU WILL NEED

Large stock pan

Cheesecloth

Colander or cheese mat

Tomme mould and weights, or cheese press

Sterilized glass jars

5 litres (1 gallon) whole goats' milk

5 drops of rennet

Medium brine for storing, optional (100g/3½ oz salt mixed with 1 litre/ 1¾ pints water)

1. Heat the milk to 25°C (77°F), add the rennet, then turn off the heat and leave in a warm place to coagulate for an hour. The curds will separate from the whey. Test the curds with your finger and if they break cleanly, cut into 1cm (½in) pieces. Leave them to rest for 5–10 minutes in the pan. **2. Lift the curds** carefully out of the whey and into a colander lined with cheesecloth or onto a sterilized cheese mat. Sprinkle with salt and leave to drain for 15 minutes. Set aside the whey, and put it in the fridge. **3. Press the curds** into a tomme cheese mould and leave under a heavy weight for 3 hours or use a cheese press. **4. Heat the whey** slowly over half an hour to 87°C (189°F), but don't allow the whey to boil. Cut the curds into smaller blocks and place into the whey. Cook for 30–40 minutes or until the curds float. **5. Skim out the cheese** and air dry for 1 hour, flipping occasionally. Store in a sterilized container (glass jars are ideal) either in whey in the fridge, where it will keep for 2–5 days, or in a medium brine solution for up to 2 months.

CATCHING AND PREPARING FISH

Our fishmonger stocked fish that had been caught within a radius of 64 kilometres (40 miles), so we were spoilt for choice for fresh local fish. But every once in a while we popped over the hill for a bit of fishing. There are laws and regulations governing freshwater fishing, but shore fishing is free to all. Wherever you fish, make sure you're doing so sustainably (see mcsuk.org for advice).

GETTING STARTED

You don't need any fancy or expensive gear to try your hand at fishing. We have a selection of rods, reels, and tackle that we have picked up over the years from the classified ads or car boot sales, so we have not spent much money on our equipment. That said, it's more than good enough. There are also some key considerations to bear in mind depending on where you plan to fish.

FRESHWATER FISHING

Most freshwater fishing is for trout or salmon, which are classed as game fish, but that is only because we have forgotten that other types of fish are edible. British stock ponds used to be filled with species that are now only caught for sport and seem to have gained the reputation of being inedible. But pike, carp, and perch – to name but a few – are eaten in many parts of the world today, and are still classed as good food.

Before you cast out your rod, you'll need to buy a rod fishing licence specific to the types of fish you plan to catch. You'll also have to buy a permit to fish your chosen waters, and it will have to be in season. There are some local variations to fishing seasons, so do your research before you plan to go fishing. All you will then need is the correct tackle, some skill, and a little luck – easy really.

SEA FISHING

When sea fishing, people tend to divide into groups, as follows:

Those who wish to stay active will either use spinners or mackerel feathers and cast and retrieve to their hearts' content. When there are mackerel around this is a very successful method, and they make great fish for a barbecue or to stock the freezer. Truly fresh mackerel, simply filleted (see box, right), seasoned, and popped onto the grill, are usually tasty enough to convert even the most sceptical of fish eaters.

The less active group bait fish, either on a float rig or ledgering on the bottom. This would appear to be the simpler option, but it does involve a little organizing of the bait. We always kept supplies of sand eels, squid, and mackerel in the freezer so we could go out fishing at any time.

CHOOSING FISH AND SEAFOOD

Whether you're visiting the fishmonger or making the most of your fresh catch, you'll need to consider the following guidelines for fish and seafood.

Fresh fish

Skin should look shiny and be moist, and the fish should look alive. The scales should adhere tightly to the body. Any coating on the fish (many secrete a slime over their skin that serves as a protective

We have years of experience fishing rivers for trout and, in Cornwall, salmon. If we were to be very honest, the time spent fishing compared to the catch we achieve is far from productive, but it is very relaxing.

armour) should be transparent. Lift up the gills and they should be cherry-red, not brownish.

Flesh should be firm to the touch and when pressed it should be taut enough to spring back, without leaving a depression in the surface. The grain of the flesh should be dense, without gaps between the layers, and if the fish is pre-wrapped in plastic, make sure there is no liquid leaking, which is a sign of age.

Saltwater fish should smell briny and like the sea, while freshwater fish should smell clean with no muddy or ammonia aroma.

Eyes should be clear and protruding and the tail should be moist and flat, not dried up or curled.

Seafood

Live seafood, such as crabs, mussels, or oysters, should be kept at 4.4°C (40°F). Surround them with seaweed, damp newspaper, or sea grass to insulate them and prevent them getting too cold, which will kill them. Keep non-live seafood buried in ice, if possible.

Never eat dead shellfish. Only consume shellfish that have unbroken, tightly closed shells before cooking.

SUSTAINABLE FISH AND SEAFOOD

Whether you choose to catch your own fish or buy it from a fishmonger or supermarket, make sure you choose sustainable species. More than 90 per cent of global fish stocks are overexploited to some extent, including popular species like cod, salmon, haddock, tuna, and prawns. To help ease pressure on these stocks, why not opt for something a little different, such as sustainably caught mackerel, pollock, Dover sole, or Devon brown crab. Not only would this help ease pressure on at-risk species, but it could also encourage you to be a little more creative with your Friday fish suppers.

Of course, what counts as sustainable can change from year to year, as some fish stocks recover while others become at-risk. Keep informed by checking the latest advice from the Canal & River Trust (canalrivertrust.org) for freshwater fish, and the Marine Conservation Society (mcsuk.org) for sea species.

HOW TO PREPARE A FRESH FISH

To turn your prize catch into a delicious meal, you'll first need to fillet it. This basic method works well for smaller fish that you are planning to cure or smoke. The fish does not have to be gutted first. Always prepare fish and seafood separately from other foods, on a washable chopping board, to avoid cross-contamination.

Lay the fish on a clean chopping board with its backbone towards you and its head in your hand.

Make a deep diagonal cut behind the gill, angling the knife towards the front of the fish to reach the flesh behind the skull. Using a thin and flexible blade, cut down the spine, keeping the blade parallel to the board, to remove the first fillet. Stop at the tail to slice it away cleanly.

To remove the second fillet, turn the fish over and make a diagonal cut behind the gill. This time, use the palm of your hand to carefully guide the knife by pressing it down close to the spine, keeping the blade parallel to the board as you cut.

Alternatively, butterfly a fish by placing it belly-down on a board and cutting down the backbone from head to tail. (Don't cut all the way through or you will puncture the stomach and guts.) Then use scissors to remove the spine, rib cage, and viscera. Clean the cavity by rinsing under cold water and patting dry.

Filleting fish is a skill worth mastering so you can enjoy your own catch of the day. A sharp filleting knife makes all the difference.

Don't be intimidated by the thought of preparing fish at home. There are many videos online that give more detailed advice on how to fillet specific fish.

CURING AND SMOKING FISH

It's hard to think of anything that compares to a fish supper enjoyed after a successful afternoon's angling. To make the rest of our catch last a bit longer, we'd always rather cure and smoke it than bury it in the freezer where it'll lose its delicate texture – or worse, get forgotten about. But these traditional methods are about more than just preservation: they also create exquisite flavour.

CURING FISH

Curing draws out excess water, leaving you with a firmer, longer-lasting fish. You can use either a salt water brine or a dry salt cure (see pages 254–255). Experiment with adding herbs, spices, or flavourings like citrus zest – just about anything works, but stick to only one or two to avoid muddling the flavours.

Cured fish, such as gravlax (see opposite), can be delicious on its own, but it's also crucial to cure fish if you plan to smoke it. This is because it forms a pellicle on the fish – a sticky, salty surface layer that helps the smoke particles stick.

Dry curing fish

The simplest cure for fish, and the one we use most often, is a 50:50 mix of granulated brown sugar and coarse sea salt or rock salt (avoid fine salt as it gives an aggressive cure that leaves the fish too salty). A typical measure is 100g (3½oz) salt and 100g (3½oz) sugar per 1kg (2¼lb) of fish. Generally, a handful of cure will be enough for a couple of small fish, a mackerel, or a salmon fillet steak.

Sprinkle a layer of the cure into a plastic or ceramic container, place the fish on top, and sprinkle over more cure. Wrap the fish in baking parchment and leave in the fridge overnight.

The next day, remove from the fridge, rinse off the cure, and pat dry with kitchen paper. Leave in the fridge uncovered for at least 6-8 hours.

HOT SMOKING FISH

Unlike meat, which can be cold-smoked (see pages 225–227), fish needs a different approach. Hot smoking is far quicker than cold smoking, and produces much more intense flavours, meaning that you can add real flavour to the fish without ruining its delicate flaky texture by overcooking it. The only downside is that the fish is not dried or preserved in the same way as hot-smoked meat would be, meaning that it has a shorter shelf life and needs to be refrigerated.

When starting out, it's tempting to add lots of wood chips or extend the cooking time, but this gives you a heavy smoke that we personally dislike. Fish doesn't take much cooking in the pan and the same's the case for hot smoking. Less is also more with the smoke: the flavour of the fish should work with it rather than being overpowered by it.

Choosing your fish

Pretty much any fish can be hot smoked, including shellfish, but in our opinion the tastiest and best at absorbing flavour are oily fish, such as trout (see right), because the oil provides a permeable coating so the smoke can easily penetrate it. Other favourites are salmon, mackerel, eel, and smaller fish, such as sardines.

Equipment

To hot smoke, you need two things: heat and a lidded container that will allow the smoke to drift over the fish. A charcoal or gas barbecue, with the temperature kept low and the lid on, is perfect. You could buy a stove-top smoker, but we find DIY contraptions work just as well: try a lidded wok (see box, right), or a roasting tin or casserole, as described opposite.

Wood chips

When you hot-smoke fish, the woodsmoke itself acts as a kind of seasoning. Almost any kind of wood will work, but you might wish to use woods like alder or fruit woods. Salmon is often smoked with alder – a tradition started by the indigenous peoples of the northwestern United States. Experiment by gradually adding more wood chips until you have a flavour you like, taking care not to use too many, as it can be easy to overpower the fish and end up with a bitter taste. These are our favourite woods for smoking fish:

Oak has a classic, bold flavour. Great with mackerel and salmon, but easy to overdo.

Beech is the best all-rounder: subtle, fragrant, and not too overpowering.

Apple is mild, fruity, and wonderful.

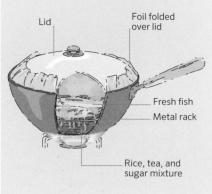

HOT SMOKING ON A HOB

Line a wok with foil and place 110g (4oz) tea leaves, 250g (9oz) rice, and 2 teaspoons of sugar in the bottom. Add a rack and lay the fish on top. Fold the foil over the lid to seal. Cook on high for 5 minutes and a further 10 minutes on a lower heat. Never leave the wok unattended and eat the cooked food straight away.

Lid

Foil folded over lid

Fresh fish

Metal rack

Rice, tea, and sugar mixture

RECIPE Cured salmon

A traditional gravlax cure is flavoured with lots of dill and black pepper, but this recipe puts a colourful twist on the Scandinavian classic with the addition of beetroot, which turns the fish a vivid purple. Not only does the finished dish look visually impressive, but the whisky and orange add a real depth of flavour.

YOU WILL NEED

75ml (2½fl oz) whisky
4 tbsp salt
2 tbsp granulated sugar
2 tbsp freshly chopped dill
Zest of 1 orange

1 beetroot, grated
750g (1lb 10oz) side of salmon, skinned and deboned

1. Mix the whisky, salt, sugar, dill, and orange zest with the grated beetroot. **2. Rub the cure** into the salmon, covering it completely. Wrap the salmon tightly in baking parchment and leave in the fridge to cure for 24–36 hours, pressed between two plates with a weight on top (evenly spaced tins from the larder are ideal). **3. Scrape off the cure** carefully and rinse the salmon under cold water. Pat dry gently. **4. Slice the side of salmon** thinly on an angle and serve with a fresh watercress salad and a slice of toasted pumpernickel bread.

RECIPE Hot smoked trout

What could be better than delicious trout that you've caught and smoked yourself? Delicious trout that you've caught and smoked yourself on a hot smoker. If you don't have a stovetop hot smoker, build your own using a large roasting tin, a wire rack, and some foil. The only downside with this DIY method is that you can get quite a strong and dense smoke that can be very intense. When it comes to wood chips, our advice is: less is more.

YOU WILL NEED

450g (1lb) trout
2 tbsp apple wood chips
Stovetop hot smoker, or large roasting tin and wire rack
Ramekin and tin foil (optional)

1. Gut the fish and remove its head, then clean it under cold running water. **2. Carefully score the skin** to allow the smoke to penetrate the fish more easily during cooking. **3. Sprinkle the wood chips** into the base of your hot smoker, ensuring even coverage. **4. Place your trout** on a wire rack and put it in the smoker. There must be space for the smoke to circulate, so you may need to raise the wire rack on an upturned ramekin. Seal the smoker with a lid or tin foil, and then place onto a gas hob. **5. Cook over a high heat** for 5–7 minutes.

KEEPING BEES

Honey is the food bees produce for themselves from nectar. We started keeping bees over 25 years ago in the garden of our suburban home because the honey we produce is far superior to anything from the supermarket, and the majority of imported honey. However, be wary of taking up beekeeping if you suffer from anaphylactic reactions to stings or have extreme allergic reactions.

ESTABLISHING YOUR COLONY

There are several ways to get started. You can order "package bees", which are essentially a swarm in a box – queen included – that arrives without brood or frames, or you can establish a colony from a nucleus or small hive of about five frames that includes a laying queen. Another option is to buy a colony that is already up and running, which is what we did.

Secondhand hives regularly come up for sale and your local beekeeping association will know what equipment is available. If you have never kept bees before, it is essential to spend time with a beekeepers' group, where you can get advice and share tips.

THE ELEMENTS OF A HIVE

There are several styles of hive, but they all have the same main parts.

The brood box is located at the base and is where the queen lives and new bees are hatched and tended. The queen lays eggs in the brood nest and, once hatched, the larvae are fed with pollen and honey stored around the nest.

The supers stand above the brood and this is where the honey is stored. Some honey is stored in the brood, but a beekeeper takes the harvest of surplus honey from the supers. Make sure that there is a 6mm (¼in) gap, also known as "beespace", between each frame in the super so the bees can move through the hive. Spacers are needed between some types of frame to maintain good "beespace". Smaller spacers are used in the brood box and larger spacers in the gaps of the supers.

The queen excluder is a grill between the brood and supers and does just what it says: it prevents the queen from going up to the supers and laying eggs. It has holes wide enough for the worker bees to travel up and down but, due to her larger abdomen, the queen can't pass through the narrow gaps.

Frames sit inside the brood and supers, and are the building blocks that enable your bees to start the honey-storing process. They normally have wooden sides enclosing a sheet of wax in hexagonal cell shapes to give the bees a head start.

SITING YOUR HIVE

The position of your hive is of prime importance. Point it away from the prevailing wind and, if possible, facing east or southeast to catch the early morning sun.

If you live in an urban area, put a barrier made out of a willow hurdle in front of the entrance to your hives. This will help to force the bees' flight path upwards into the air when they exit the

1. Bees leave and enter the hive through small "beespaces" in the sides of the super. **2. An open hive** exposes the frames that stand upright within the super as it rests on the brood below. **3. Wear a bee suit** with everything tucked in, as bees tend to crawl upwards. Wear boots and gloves too. Hives are best positioned at least 1m (3ft) apart.

PROJECT **Gather honey**

Harvest your honey in late August or September, but seek advice from a local beekeeper before you enter a hive. A smoker and centrifugal extractor are essential pieces of equipment. The smoker simulates a forest fire and encourages the bees to eat honey, making them nice and docile. The centrifugal extractor can be expensive to buy, but often you can borrow one from a beekeepers' association.

YOU WILL NEED
Smoker
Sharp knife
Centrifugal honey extractor
Glass jars

1. Fill your smoker with a fuel that smokes for a long time. We used dried, rotted wood and rolled-up corrugated cardboard.
2. Blow a few wisps into the front of the hive and leave it for a few minutes while word spreads through the colony that it's time to get eating.   **3. Take out the heavy frames** and start scraping out the honey as quickly as possible – before the bees find out what you're up to!   **4. Use a sharp knife** heated in boiling water to cut just under the surface of the capping on each cell. This will allow the honey to ooze out.   **5. Secure four frames** that have had the caps cut off into a centrifugal honey extractor. Gently spin the handle until all the honey is flung onto the edges of the tub and drips down the sides to a sump with a tap.   **6. Filter the honey** through a sieve to remove waxy bits and store in sealed glass jars in a dark place.

hive, rather than straight across your neighbours' gardens. Provide a source of water near the hive so the bees don't become a nuisance around someone else's pond.

MAINTAINING THE COLONY
Regularly check on the health and progress of your bees. The best time to inspect your hives is on warm afternoons when most of the colony is out gathering nectar and pollen. Make a routine inspection every week from April to mid-October, checking to see that:
The queen is still laying.
The brood look healthy and disease-free.
The bees have enough space for their honey – be prepared to add an extra super.

If you don't visit your hive regularly, you may find your bees have produced queen cells (by feeding selected larvae royal jelly) and they may swarm, which means you will lose a large portion of

your honey-producing worker bees. In addition, your bees will have glued everything together with layers of propolis. Moving them then jolts the other frames and you are likely to disturb your bees.

In years when the weather has been bad or the bees have suffered from disease, we don't harvest honey as it is the best food to give your colony. When you do harvest, or when your bees are short of honey, you will need to feed them so they survive the winter. Fill a small plastic container with cold syrup made from 1kg (2¼lb) sugar heated in 1 litre (1¾ pints) water. Cover the container with fine gauze and then pop on a lid with a small hole in it. Turn it upside down and place on the crown board on top of the uppermost super. The vacuum formed by the syrup in the container stops it dribbling onto the hive.

POTENTIAL PROBLEMS
Stay up-to-date with information issued by beekeeping groups and DEFRA.
Varroa mite is a tiny parasite that rides on a worker bee's body, enters a brood cell, and feeds on the larvae. Check for infestation by inspecting brood cells and counting dead mites that have dropped out the bottom of a hive. Control with pyrethroid-based chemicals on slow-release strips.
European foulbrood is a disease that creates malformed larvae, but can be treated with a technique called "shook swarm"; literally shaking bees into a new brood chamber.
Swarming is more likely in hives that contain an older queen. A colony that is about to swarm can be distracted by moving the hive 2m (6½ft) and turning it through 180°. Place an empty hive in the original position and the bees should fill up the new hive. Feed both hives until the colonies re-establish themselves.

USING HONEY AND PROPOLIS

Its natural sweetness combined with its delicate flavours from local blossom make home-made honey a real treat. Raw honey that hasn't been pasteurized also has fantastic health benefits: eating local honey, for example, has been shown to help people who suffer from hayfever, and propolis, which you can harvest from beehives, has antiseptic qualities.

LOCAL FLAVOUR

The flavour, colour, aroma, and crystalline structure of honey are impacted by the hive's environment, particularly the flowers the bees feed on. If your hives are located near an apple orchard, for example, your honey will have heady floral notes with a hint of apple and will be light-to-medium amber in colour. Alternatively, if your local farm grows rapeseed, the colour is likely to be much whiter with an intense flavour. Some of the most popular monofloral honeys include manuka, acacia, rapeseed, borage, and apple blossom.

HONEY INFUSIONS

Honey is a wonderful carrier of flavour, and we've always enjoyed infusing honey to use in cooking and in homemade remedies when we need to give our immune systems a boost.

For flavoursome smoothies or dressings, infuse sprigs of fresh thyme or turmeric in a pot of honey. Or if you have a cold, try adding a few raw garlic cloves to a pot of honey and leaving them to infuse to make a home-made medicine. The raw garlic becomes palatable and works with the healing properties of the honey.

To maintain its beneficial properties, it's best to keep honey raw. So if you are adding it to a drink, don't spoon it into boiling water; instead allow the drink to cool slightly before stirring it in.

RECIPE Spiced propolis infusion

We've been making this drink to soothe sore throats since we started keeping bees. Propolis is a fantastic bee product. It is used in the hive as a sort of glue, but it also has great antiseptic properties. To collect it, install a propolis net above a super or brood box (see pages 270–271). The bees don't like the holes and will start to fill them with propolis. To harvest it, remove the net and place in the freezer for a few hours. Scrunch it up and collect the valuable pieces. You will need at least a 60 per cent proof alcohol, such as Stroh rum, to dissolve the propolis and make the tincture. For smaller amounts, use 2 tbsp alcohol to each 10g (¼oz) of propolis.

YOU WILL NEED
Glass jar
Coffee filter
100g (3½oz) propolis
400ml (14fl oz) 60+ per cent alcohol
1 lemon
Cloves
Fresh ginger
1–2 tsp honey

1. Harvest the propolis as described above.   **2. For the tincture,** put the propolis and alcohol in a jar and shake every day for 6–8 days until the propolis has dissolved. Strain the liquid through a coffee filter to remove bits of bee that may be mixed in with it.   **3. To make the drink,** squeeze the juice from half the lemon and slice the rest, studding each slice with cloves. Cut some slices of ginger.   **4. Pour 1 tsp of propolis** into a mug and add boiling water with the honey, lemon juice, and slices of ginger and lemon.

RECIPE Honey mead

Mead is one of the oldest drinks known to humankind, and was especially popular in northern Europe where grapes didn't grow well. It was drunk after Norse weddings on a holiday known as a "honeymoon". Every year we make our own mead using honey from our beehives and it's as good a reason as any to keep bees (see pages 270–271).

YOU WILL NEED
Funnel
Demijohn with airlock seal
2kg (4½lb) honey
Juice of 2 lemons
Juice of 2 oranges
1 heaped tsp yeast

1. Mix the honey with 4.5 litres (1 gallon) of water in a large pan and warm it until the honey has dissolved. **2. Add the fruit juices** to the honey mixture and leave to cool. **3. Add the yeast** and mix well. **4. Strain the golden liquid** into a demijohn and leave the mead to ferment with an airlock seal attached to it (see page 197). When fermentation stops and the mead stops bubbling away through the airlock, you can siphon off the mead into sterilized bottles (see pages 196–197), and "lay down" for 6 months to mature.

INDEX

PUBLISHER ACKNOWLEDGMENTS

First edition: DK would like to thank James Strawbridge for the artwork
concepts; Stephanie Jackson for commissioning; Kat Mead, Adèle Hayward, and
Helen Spencer for the set-up; Zia Allaway, Pip Morgan, and Diana Vowles for
editorial assistance; Sue Bosanko for indexing; and Lucy Claxton at DK images.

Second edition: DK would like to thank James Strawbridge for art direction and
food styling, Tia Tamblyn for food styling assistance, John Hersey and Simon
Burt for additional photography, Oreolu Grillo, Lucy Philpott, and Millie Andrew
for editorial assistance, Barbara Zúñiga and Sophie State for design assistance,
and Vanessa Bird for indexing.

PICTURE CREDITS

The publisher would like to thank the following for their kind permission to
reproduce their photographs:
(key: b-bottom; c-centre; l-left; r-right; t-top)
41 Photolibrary: Johnny Bouchier (tr). **44 Alamy Images:** Paul Glendell (tl).
45 iStockphoto.com: Smitt (cr). **46 Photoshot:** David Wimsett (tr). **Science
Photo Library:** Alex Bartel (tl). **Still Pictures:** Martin Bond (b). **47 Alamy Images:**
The Garden Picture Library (t). **48 Science Photo Library:** David Hay Jones (bl).
49 Corbis: Dietrich Rose (b). **Still Pictures:** Martin Bond (t). **50 Alamy Images:**
Jeff Morgan 05 (t); Steven Poe (b). **51 Alamy Images:** camera lucida environment
(b); Steven Poe (t). **54 Dreamstime.com:** Hywit Dimyadi/Photosoup (cb); Sergiy
Bykhunenko/Sbworld4 (bc). **68 Dreamstime.com:** Derektenhue (bc).
78 Science Photo Library: Martin Bond (l). **Dreamstime.com:** Helen Hotson

(bc). **80 123RF.com:** Ievgenii Biletskyi (tr). **86 Dreamstime.com:** Freerlaw (cl).
90 Science Photo Library: Simon Fraser (tl). **Alamy Images:** Wolfgang Polzer (bl)
98 Photolibrary: Bananastock (bl). **99 Alamy Images:** Doug Houghton (tr).
113 GAP Photos: Jerry Harpur (tr). **132 Getty Images:** Dianna Jazwinski (cr).
133 Getty Images: Frederic Pacorel (br). **135 Getty Images:** Lawrence Lawry (tr);
Andrew Parkinson (tl). **211 Garden World Images:** MAP / Alain Guerrier (tr).
212 Dreamstime.com: Lynn Watson/Luckydog1 (bl). **214 Alamy Stock Photo:**
Raymond Wood (bl). **215 Getty Images:** Thomas Kitchin & Victoria Hurst (tr).
242 Dreamstime.com: Ivan Kurmyshov (cr). **248 Alamy Stock Photo:** Edd
Westmacott (crb). **Shuttershock:** Citrusaid (clb). **258 123RF.com:** Luis Mario
Hernández Aldana (br).
Simon Burt: **157** (br); **162** (bl); **164**; **166** (cb); **224** (cl, bl, r); **225** (tr, cr); **226** (bl,
br); **229** (tl); **254** (l, cl, cr).
John Hersey: **26** (br); **30** (t); **150**; **152**; **156** (tr); **167** (c); **180**.

The following images are from the personal archive of Dick and James
Strawbridge and are used with permission:
12 (ct); **13** (c); **15** (tl, cl); **17** (r); **19** (cl, tl); **26** (bl); **27** (tr, cr); **29** (bl, br); **31** (tl);
37 (bl); **58**; **70** (tl); **75** (l); **93** (tl); **111**; **129** (bl, cr); **130**; **133** (cr); **168** (l); **172** (l, r),
174 (br), **204–205**; **239** (b); **244** (bl, br); **250** (bl).

All other images © Dorling Kindersley. For further information, see www.
dkimages.com

ABOUT THE AUTHORS

DICK STRAWBRIDGE, a television presenter and retired army engineer, has always been fascinated by how things work. His desire for a greener life led him to put his practical skills to use in converting a derelict farm in Cornwall into a modern, ecologically friendly home. This journey to a self-sufficient lifestyle was documented in the BBC series *It's Not Easy Being Green*, which ran for three seasons.

Life has moved on, and he can currently be seen taking on an even bigger challenge in *Escape to the Chateau* (Channel 4). For more information about Dick's adventures in France, visit www.thechateau.tv.

JAMES STRAWBRIDGE, Dick's son, has inherited his father's enthusiasm for inventing and DIY, and co-presented *It's Not Easy Being Green*. Following the programme, James and his father presented other popular series including *The Hungry Sailors* (ITV 2012, 2013) and *Saturday Farm* (ITV 2013). James has also co-written a series of books on artisan kitchen crafts, preserving food, and traditional skills: *Made at Home* (Octopus, May 2012). He currently lives in Cornwall and works as a development chef. www.jamesstrawbridge.com

AUTHOR ACKNOWLEDGMENTS

OUR THANKS

The knowledge we have collected in this book comes from years of experimenting and playing outside. We have to thank everyone who has given their time to teach us, and those who have allowed us to watch and learn.

DICK'S PERSONAL THANKS

I'm not sure that I should be thanking my co-author, but James has continued to be my conscience, personal organizer, and artistic director. While I have been off gallivanting, he has kept the project on schedule and it has been a privilege working with him. It's not often a father gets a chance to spend such quality time with his son. James and I both believe life is for living, and somehow I've been lucky enough to find myself living in a chateau, in rural France, with Angela, my wonderful wife, and our young children, Arthur and Dorothy, whose boundless curiosity reminds me how important it is to share our knowledge. This book is very

special to us as every page contains reminders of the experiences, hard work, fun and pleasures of our lives so far... I am a very lucky man.

JAMES'S PERSONAL THANKS

To my wife Holly, thank you for tending, watering, and weeding our garden, and nurturing me. My children Indy, Pippin, and Arrietty – I love you loads and hopefully this book will help you in some way with your own lives long after I'm gone. I hope it serves as a good reminder of what made Daddy happy.

I'd like to thank my family for inspiring my journey towards a more sustainable lifestyle: my dad and co-author for his practical, just-do-it attitude, which has filled me with the self-confidence to try anything; my mum for teaching me to appreciate nature and especially her inspiring passion for pollinating insects; and my sister for providing a human context, putting fair trade and ethical

consumerism back into the mix rather than just getting lost in nature.

A huge thank you to all my friends in Cornwall. Andrew and Sue at Charlie Harris butchers in Tywardreath for advice with our sausages, Cornish Sea Salt for their seasoning, Katie Wood at Polmarkyn Dairy, Ty from ProQ Smokers in Bodmin, Natalie and Josh at I O Shen knives, and Rose Greene for her fermentation expertise. I'm sure there are others who have helped along the way and I hope you know I'm extremely grateful to everyone I have the pleasure to work with.

Finally, thank you to Simon Burt for the excellent photographs that bring this book up to date and breathe fresh life into its subjects, and to Tia, Richard, and Julie at Botelet for providing the stunning location in Cornwall where we shot lots of the recent recipes. Your hospitality and friendship was invaluable. MC, Amy, Stephanie, and all the DK team: thank you for all your hard work and positivity!